McGraw-Hill Ryerson

PRISM MATH

CANADIAN EDITION

purple

Teacher's Edition

McGraw-Hill Ryerson

Toronto Montréal Boston Burr Ridge, IL Dubuque, IA Madison, WI New York San Francisco
St. Louis Bangkok Bogotá Caracas Kuala Lumpur Lisbon London Madrid
Mexico City Milan New Delhi Santiago Seoul Singapore Sydney Taipei

McGraw-Hill
A Division of The McGraw·Hill Companies

Prism Math – Purple Teacher's Edition
Copyright © 2005, McGraw-Hill Ryerson Limited, a Subsidiary of The McGraw-Hill Companies. All rights reserved. No part of this publication may be reproduced or transmitted in any form or by any means, or stored in a data base or retrieval system, without the prior written permission of McGraw-Hill Ryerson Limited, or, in the case of photocopying or other reprographic copying, a licence from The Canadian Copyright Licensing Agency (Access Copyright). For an Access Copyright licence, visit www.accesscopyright.ca or call toll free to 1-800-893-5777.

ISBN-13: 978-0-07-096048-0
ISBN-10: 0-07-096048-8

http://www.mcgrawhill.ca

5 6 7 8 9 10 MP 0 9 8 7

Printed and bound in Canada

Care has been taken to trace ownership of copyright material contained in this text. The publishers will gladly accept any information that will enable them to rectify any reference or credit in subsequent printings.

Library and Archives Canada Cataloguing in Publication

Prism Math – Purple. – Teacher's ed.

Teacher's guide to one of a series of non-grade specific workbooks for use in grades 1–12. The level of difficulty increases throughout the series in the following order: gold, brown, red, orange, yellow, green, blue and purple.
ISBN: 0-07-096048-8

1. Mathematics–Problems, exercises, etc. 2. Mathematics–Study and teaching (Elementary) 3. Mathematics–Study and teaching (Secondary)

QA36.5.P758 2005 Suppl.   510'.76   C2005-900446-0

PUBLISHER: Diane Wyman
MANAGER, EDITORIAL SERVICES: Linda Allison
SUPERVISING EDITOR: Kristi Moreau
COPY EDITORS: Julia Cochrane; Write On!
JUNIOR DEVELOPMENTAL EDITOR: Megan Elston
EDITORIAL ASSISTANT: Erin Hartley
PRODUCTION SUPERVISOR: Yolanda Pigden
PRODUCTION COORDINATOR: Janie Deneau
COVER DESIGN: Dianna Little
ELECTRONIC PAGE MAKE-UP: SR Nova Pvt. Ltd., Bangalore, India

COPIES OF THIS BOOK MAY BE OBTAINED BY CONTACTING:

McGraw-Hill Ryerson Ltd.

WEB SITE:
http://www.mcgrawhill.ca

E-MAIL:
orders@mcgrawhill.ca

TOLL-FREE FAX:
1-800-463-5885

TOLL-FREE CALL:
1-800-565-5758

OR BY MAILING YOUR ORDER TO:
McGraw-Hill Ryerson
Order Department
300 Water Street
Whitby, ON L1N 9B6

Please quote the ISBN and title when placing your order.

Student text ISBN:
0-07-096047-X

Contents

PURPLE BOOK PRETESTS 1–4
PROBLEM-SOLVING STRATEGIES
   Multi-Step 5
   Draw a Picture 6
   Look for a Pattern 7
   Guess and Check 8
   Identify Missing Information 9
   Make a Table 10
   Make a List 11
   Solve a Simpler Problem 12
   Work Backward 13
   Use Estimation 14
   Use a Formula 15
   Use Logical Reasoning 16

1 CHAPTER 1 Whole Numbers/Decimals
CHAPTER 1 PRETEST 17
  1 Addition (whole numbers) 19
  2 Subtraction (whole numbers) 21
  3 Multiplication (whole numbers) 23
  4 Division (whole numbers) 25
  5 Problem Solving 27
  6 Addition and Subtraction (decimals) ..29
  7 Multiplication (decimals) 31
  8 Division (decimals) 33
CHAPTER 1 PRACTICE TEST 35

2 CHAPTER 2 Fractions
CHAPTER 2 PRETEST 36
  1 Fractions and Mixed Numerals 37
  2 Renaming Numbers 38
  3 Adding and Subtracting Fractions ..39
  4 Adding and Subtracting Mixed Numerals 41
  5 Multiplication 43
  6 Division 45
CHAPTER 2 PRACTICE TEST 47

3 CHAPTER 3 Pre-Algebra Equations
CHAPTER 3 PRETEST 48
  1 Number Phrases PRE-ALGEBRA 49
  2 Writing Equations PRE-ALGEBRA 50
  3 Solving Equations (division) PRE-ALGEBRA 51
  4 Solving Equations (multiplication) PRE-ALGEBRA .53
  5 Solving Equations (subtraction) PRE-ALGEBRA .55
  6 Solving Equations (addition) PRE-ALGEBRA 57
  7 Solving Equations Review PRE-ALGEBRA 59
CHAPTER 3 PRACTICE TEST 61

4 CHAPTER 4 Using Pre-Algebra
CHAPTER 4 PRETEST 62
  1 Combining Terms PRE-ALGEBRA 63
  2 Solving Equations PRE-ALGEBRA 64
  3 Problem Solving PRE-ALGEBRA 65
  4 Problem Solving PRE-ALGEBRA 66
  5 Problem Solving PRE-ALGEBRA 67
  6 Problem Solving PRE-ALGEBRA 69
CHAPTER 4 PRACTICE TEST 71

5 CHAPTER 5 Ratio, Rate, Proportion, and Percent
CHAPTER 5 PRETEST 72
  1 Ratio 73
  2 Proportions (recognizing) 74
  3 Proportions PRE-ALGEBRA 75
  4 Proportions PRE-ALGEBRA 76
  5 Scale Drawings 77
  6 Problem Solving PRE-ALGEBRA 79
  7 Percent PRE-ALGEBRA 81
  8 Fractions and Percent PRE-ALGEBRA 82
  9 Decimals and Percent 83
10 Fractions, Decimals, and Percent 84
11 Comparing and Ordering 85
12 Percent of a Number PRE-ALGEBRA 87
13 Percent of a Number PRE-ALGEBRA 89
14 Percent of a Number PRE-ALGEBRA 91
15 Percent PRE-ALGEBRA 93
CHAPTER 5 PRACTICE TEST 95

6 CHAPTER 6 Simple/Compound Interest
CHAPTER 6 PRETEST 96
  1 Simple Interest PRE-ALGEBRA 97
  2 Simple Interest PRE-ALGEBRA 99
  3 Compound Interest 101
  4 Compound Interest 103
CHAPTER 6 PRACTICE TEST 105

7 CHAPTER 7 Metric Measurement
CHAPTER 7 PRETEST 106
  1 Metric Prefixes 107
  2 Length 108
  3 Units of Length 109
  4 Capacity 110
  5 Units of Capacity 111
  6 Units of Mass 112
  7 Problem Solving 113
  8 Temperature 114
CHAPTER 7 PRACTICE TEST 115

8 CHAPTER 8 More Metric Measurement and Estimation
CHAPTER 8 PRETEST 116
  1 Units of Length 117
  2 Units of Capacity, Time, Mass 119
  3 Adding Time 121
  4 Subtracting Time 123

T3

- 5 Multiplying Measures 125
- 6 Measurement 127
- 7 Rounding Numbers 128
- 8 Estimating Sums and Differences 129
- 9 Estimating Products 130
- CHAPTER 8 PRACTICE TEST 131

9 CHAPTER 9 Geometry

- CHAPTER 9 PRETEST 132
- 1 Lines, Line Segments, and Rays 133
- 2 Circles 134
- 3 Angles 135
- 4 Angle Measurement 136
- 5 Angle Measurement 137
- 6 Congruent Angles 138
- 7 Parallel and Perpendicular Lines 139
- 8 Triangles 140
- CHAPTER 9 PRACTICE TEST 141

10 CHAPTER 10 Similar Triangles

- CHAPTER 10 PRETEST 142
- 1 Similar Triangles 143
- 2 Similar Triangles 144
- 3 Similar Triangles PRE-ALGEBRA 145
- 4 Squares and Square Roots 147
- 5 Squares and Square Roots (table) 148
- 6 Squares and Square Roots (table) 149
- 7 The Pythagorean Theorem PRE-ALGEBRA 151
- 8 Using the Pythagorean Theorem PRE-ALGEBRA 153
- 9 Similar Right Triangles PRE-ALGEBRA 155
- CHAPTER 10 PRACTICE TEST 157

11 CHAPTER 11 Perimeter, Area, and Volume

- CHAPTER 11 PRETEST 158
- 1 Perimeter PRE-ALGEBRA 159
- 2 Circumference PRE-ALGEBRA 160
- 3 Area of a Rectangle PRE-ALGEBRA 161
- 4 Area of a Triangle PRE-ALGEBRA 162
- 5 Area of a Circle PRE-ALGEBRA 163
- 6 Area of a Parallelogram PRE-ALGEBRA 165
- 7 Surface Area of a Rectangular Prism 166
- 8 Surface Area of a Triangular Prism 167
- 9 Surface Area of a Cylinder 168
- 10 Volume of a Rectangular Prism PRE-ALGEBRA 169
- 11 Volume of a Triangular Prism PRE-ALGEBRA 170
- 12 Volume of a Cylinder PRE-ALGEBRA 171
- 13 Volume of a Cone PRE-ALGEBRA 173
- 14 Volume of a Pyramid PRE-ALGEBRA 174
- 15 Perimeter, Area, and Volume PRE-ALGEBRA 175
- CHAPTER 11 PRACTICE TEST 177

12 CHAPTER 12 Graphs

- CHAPTER 12 PRETEST 178
- 1 Multiple Bar Graphs 179
- 2 Misleading Bar Graphs 180
- 3 Multiple Line Graphs 181
- 4 Misleading Line Graphs 182
- 5 Scatter Plots 183
- 6 Scatter Plots 184
- 7 Circles 185
- 8 Circles 186
- 9 Circle Graphs 187
- 10 Circle Graphs 188
- CHAPTER 12 PRACTICE TEST 189

13 CHAPTER 13 Probability

- CHAPTER 13 PRETEST 190
- 1 Probability 191
- 2 0 and 1 Probabilities 193
- 3 Sample Spaces 195
- 4 Probability Experiments 197
- 5 Probability and Percent PRE-ALGEBRA 199
- 6 Predicting with Probability 201
- 7 More Probability Experiments 203
- 8 Problem Solving 205
- CHAPTER 13 PRACTICE TEST 206

MID-TEST Chapters 1-6 207
FINAL TEST Chapters 1-13 209–212
CUMULATIVE REVIEWS 213–238

ALGEBRA READINESS

- Variables, Expressions, Equations 239
- Properties of Numbers 240
- The Distributive Property 241
- Evaluating Expressions 242
- Solving Equations 243
- Solving Equations 244
- Solving Two-Step Equations 245
- Solving Equations 246
- Solving Inequalities 247
- Inequalities on a Number Line 248
- Integers 249
- Absolute Value 250
- Adding and Subtracting Integers 251
- Multiplying and Dividing Integers 252
- Powers and Exponents 253
- Negative Exponents 254
- Multiplying and Dividing Powers 255
- Scientific Notation 256
- Ordered Pairs 257
- Graphing in Four Quadrants 258
- Making Function Tables 259
- Graphing Linear Functions 260
- Slope 261
- Slope-Intercept Form 262

PRISM MATH Assessment 263–264
CHAPTER TESTS 265–277
Answers to CHAPTER TESTS 278–281

PRISM MATH

✔ **Straightforward** ✔ **Solid** ✔ **Comprehensive**

Mathematics Instruction and Practice that leads to student success in mathematics.

Eight levels provide thorough development of basic mathematics skills and strategies.

The straightforward design and colour level designations on the student books allow for flexible use at **grade level** or for **remediation** from first grade through adult education.

gold	Level 1
brown	Level 2
red	Level 3
orange	Level 4
yellow	Level 5
green	Level 6
blue	Level 7
purple	Level 8

PRISM MATH
GRADES 1-ADULT

Level 1, Gold Student Edition	0-07-096029-1
Level 1, Gold Teacher's Edition	0-07-096022-4
Level 2, Brown Student Edition	0-07-096023-2
Level 2, Brown Teacher's Edition	0-07-096024-0
Level 3, Red Student Edition	0-07-096025-9
Level 3, Red Teacher's Edition	0-07-096026-7
Level 4, Orange Student Edition	0-07-096027-5
Level 4, Orange Teacher's Edition	0-07-096028-3
Level 5, Yellow Student Edition	0-07-096049-6
Level 5, Yellow Teacher's Edition	0-07-096030-5
Level 6, Green Student Edition	0-07-096021-6
Level 6, Green Teacher's Edition	0-07-096032-1
Level 7, Blue Student Edition	0-07-096033-X
Level 7, Blue Teacher's Edition	0-07-096034-8
Level 8, Purple Student Edition	0-07-096047-X
Level 8, Purple Teacher's Edition	0-07-096048-8

Using PRISM MATH

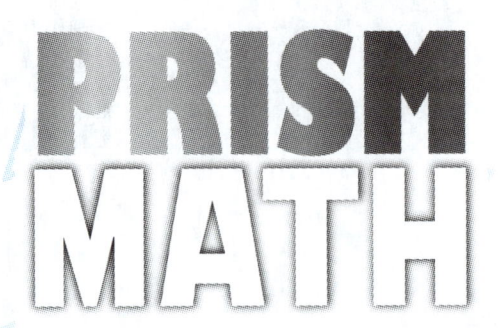

A straightforward structure for a variety of classroom situations.

SUPPLEMENTAL

▶ Many schools use **Prism Math** for extra practice of core mathematics to supplement other math instruction. **Prism Math** makes sure the basics are taught and practiced to ensure student achievement.

INTERVENTION

▶ Teachers have successfully used **Prism Math** to intensively reteach and practice crucial mathematics. **Prism Math** specifically addresses the needs of struggling students.

AFTER SCHOOL

▶ **Prism Math** is a perfect after school math reinforcement and practice program with its specific and concentrated lessons.

SUMMER SCHOOL

▶ **Prism Math's** focused and explicit lessons make it an ideal summer school program.

HOME SCHOOL

▶ **Prism Math** is frequently used as a comprehensive math curriculum in home school environments.

Student Edition

Simple, direct, and explicit lessons

▶ Lesson title identifies the skill.

▶ **Solved problems** model procedures and solutions.

▶ Every lesson has plenty of practice opportunities with **computation** and **problem-solving applications.**

▶ Includes instruction in **Problem-Solving Strategies** at the beginning of every book.

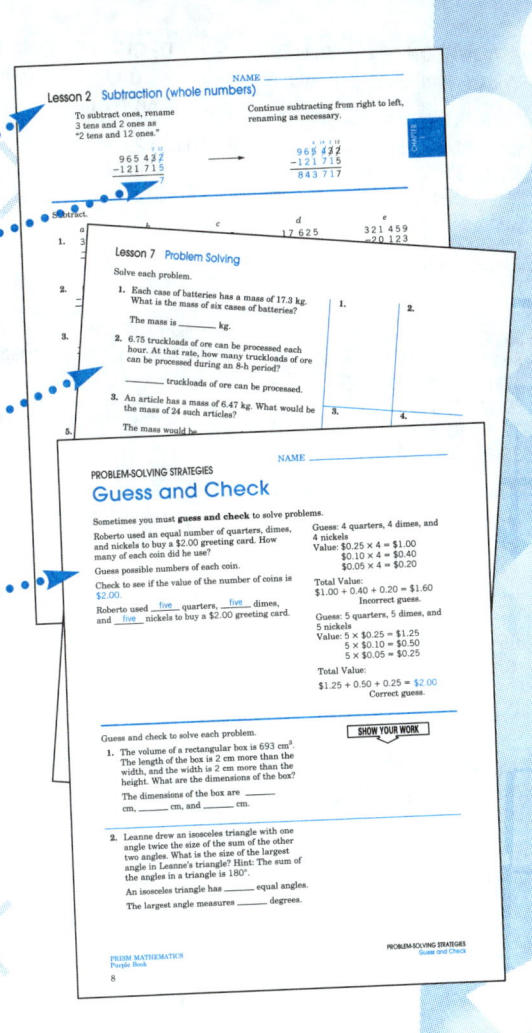

Teacher's Edition

Prerequisite skills are clearly identified.

◀ All answers are conveniently provided.

◀ References prompt when to use the **Mid-Test, Final-Test,** and **Cumulative Review.**

◀ **Lesson Follow-up and Error Analysis** gives suggestions for responding to different levels of student performance.

✓ Also includes **Blackline Masters** of **Chapter Tests** for effective assessment.

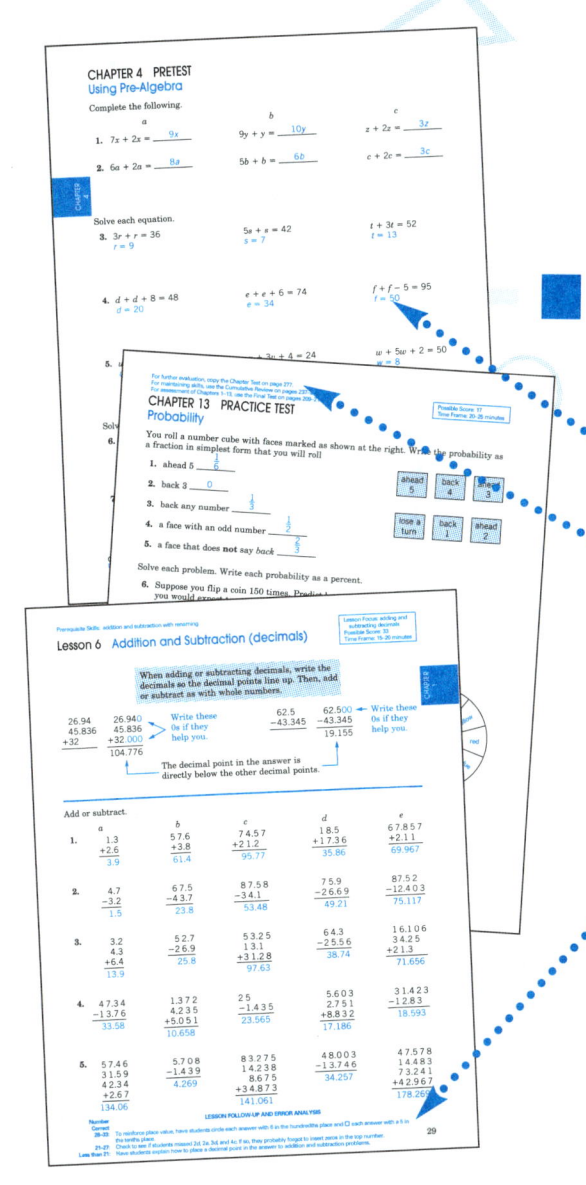

T7

	ADDITION (Whole Numbers)	ALGEBRA (Readiness)	DECIMALS and MONEY	DIVISION (Whole Numbers)	FRACTIONS	GEOMETRY	MEASUREMENT	MULTIPLICATION (Whole Numbers)
LEVEL 1 Gold Book	pages 21-22, 25, 27, 31, 33, 35, 37-47, 49-50, 53-54, 57-59, 105-112, 117-122, 125-127, 129-142		pages 26-28, 55-58, 61-65, 67-69			pages 97-102	pages 77-94	
LEVEL 2 Brown Book	pages 8-14, 29-30, 33, 35-36, 49-52, 57-58, 61-64, 69-70, 73-76, 79-82, 85-86, 89-92, 107-109		pages 17-19, 24, 67-88, 91	readiness pages 161-164	pages 45-48	pages 49-52	pages 55-70	readiness pages 155-160
LEVEL 3 Red Book	pages 23, 25-26, 29-32, 35-36, 49-52, 57-58, 61-64, 69-70, 73-76, 79-82, 85-86, 89-92, 107-108		pages 102, 106, 107-108	pages 141-148, 151-158	pages 187-192	pages 195-202	pages 61-70, 73-82, 85-94, 135-138	pages 121-128, 131-138
LEVEL 4 Orange Book	pages 23, 25-28, 33-38, 43-44, 47-54		pages 97-102	pages 105-112, 115-126, 129-132, 135-136, 139-142, 145-146	pages 149-156	pages 159-168	pages 171-184, 187-202	pages 61-70, 73-82, 85-94, 135-138
LEVEL 5 Yellow Book	pages 24, 27-34, 89	pages 245-252	pages 87-98	pages 52-62, 65-74, 76-84, 95-96	pages 139-148, 151-162, 165-180, 183-196	pages 199-207	pages 113-124, 127-136	pages 39-50
LEVEL 6 Green Book	pages 25, 27-28, 31-34	pages 159-162, 237-250	pages 97-116, 119-126, 129-140	pages 38, 43-50	pages 53-68, 71-82, 85-94	pages 191-198	pages 143-152, 155-164	pages 37, 39-42, 47-50
LEVEL 7 Blue Book	pages 19-20, 23-26	pages 91-94, 97-100, 106, 117-126, 129-136, 171-184, 189-194, 245-262	pages 67-86, 109-112	pages 31-40	pages 43-64, 106-107	pages 156-168, 171-186	pages 138-144, 147-154, 171-186, 189-194	pages 27, 39-40
LEVEL 8 Purple Book	pages 19-20, 27-28	pages 49-60, 63-70, 75-76, 79-82, 87-94, 97-100, 145-146, 151-156, 159-165, 169-176, 199-200, 239-262	pages 29-34, 84	pages 25-28	pages 37-46, 82, 84	pages 133-140, 159-175	pages 107-144, 117-130, 159-175	pages 23-24, 27-28

PRISM MATH SCOPE AND SEQUENCE

NUMBER and NUMERATION	PATTERNS and RELATIONS	PROBABILITY	PROBLEM SOLVING and STRATEGIES	RATIO, PROPORTION and PERCENT	STATISTICS and GRAPHING	SUBTRACTION (Whole Numbers)
pages 7-18, 61-72	pages 16, 73-74, 100-101		pages 1-6, 27-28, 38, 40, 42, 50, 52, 54, 57-58, 108, 110, 112, 114, 116, 118, 120, 126, 128, 130, 134, 138, 140		pages 81-88	pages 23-25, 27-30, 32, 34, 36-44, 46, 48, 51-54, 56-59, 105-108, 113-116, 119-125, 127-140, 143-144
pages 7, 17-26, 115-122	pages 23-25, 51		pages 1-6, 14, 30, 32, 36, 38, 40, 42, 60, 80, 82, 84, 86, 88, 92, 94, 102, 104, 108, 110, 112, 124, 126, 128, 130, 132, 134		pages 71-74	pages 8-14, 31-32, 34, 37-42, 77-78, 85-88, 92-94, 97-98, 105-112, 129-134
pages 39-46, 105	pages 43, 99-100	pages 211-212	pages 13-20, 26, 28, 30, 32, 34, 36, 40, 42, 46, 50, 52, 54, 56, 58, 62, 64, 66, 68, 70, 74, 76, 78, 80, 82, 84, 86, 90, 92, 94, 96, 98, 100, 108, 114, 116, 118, 122, 124, 126, 128, 132, 136, 138, 144, 146, 148, 152, 154, 156, 158, 162, 164, 166, 168, 172, 174, 176, 178, 180, 182-184, 200, 202, 212		pages 205-210	pages 24, 27, 29, 33-36, 53-58, 65-70, 73-74, 77-80, 83-86, 93-96
pages 55-58, 143-146		pages 213-214	pages 13-20, 26, 28, 30, 34, 36, 38, 40, 42, 48, 50, 52, 54, 56, 58, 62, 64, 66, 68, 70, 74, 76, 78, 80, 82, 86, 88, 90, 92, 94, 100, 102, 106, 108, 110, 112, 116, 118, 120, 122, 124, 126, 132, 136, 138, 140, 142, 144, 146, 154, 156, 160, 162, 168, 174, 176, 182, 190, 192, 194, 196, 198, 201-202, 206, 207, 210, 212, 214		pages 205-212	pages 24-26, 29-30, 39-44, 47-50, 53-54
pages 35-36, 49-50, 61-62, 141-142		pages 109-110	pages 13-22, 28, 30, 32, 34, 40, 42, 44, 46, 48, 50, 66, 68, 72, 80, 82, 88, 90, 92, 94, 104, 106, 108, 110, 112, 114, 116, 124, 126, 130, 132, 134, 136, 154, 156, 158, 160, 162, 174, 176, 178, 180, 186, 188, 190, 192, 194, 196		pages 101-108	pages 25-32, 35-36, 91-92
pages 33-34, 49-50, 58-61, 113-116		pages 183-186	pages 13-22, 28, 30, 32, 34, 40, 42, 44, 46, 48, 50, 64, 66, 68, 72, 80, 82, 88, 90, 92, 94, 104, 106, 108, 110, 112, 114, 116, 124, 126, 130, 132, 136, 138, 140, 146, 148, 156, 158, 160, 162, 170, 172, 174, 182, 184, 186, 188	pages 167-174	pages 177-181, 187-188	pages 26, 29-34
pages 29-30, 44		pages 205-210	pages 5-16, 20, 22, 24, 26, 28, 32, 34, 36, 38, 40, 50, 54, 56, 60, 62, 64, 70, 72, 74, 78, 80, 82, 84, 86, 92, 94, 96-98, 100, 114, 118, 120, 122, 124, 126, 130, 132, 134, 136, 144, 150, 154, 172, 174, 176, 178, 180, 182, 184, 190, 192, 194, 204, 206, 208, 210-211	pages 89-100, 103-104, 117-125, 129-135	pages 197-211	pages 21-26
pages 128-130		pages 191-205	pages 5-16, 20, 22, 24, 26-28, 30, 32, 34, 40, 42, 44, 46, 52, 54, 56, 58, 60, 65-70, 78-80, 86, 88, 90, 92, 94, 98, 100, 102, 104, 113, 118, 120, 122, 124, 126, 146, 154, 156, 164, 172, 176, 192, 194, 196, 198, 200, 202, 204-205	pages 73-80	pages 179-188	pages 21-22, 27-28

PRISM MATH SCOPE AND SEQUENCE

Error Analysis

As students develop math skills and understanding, they invariably will make mistakes. These errors provide important insight into student thinking and level of skill. All errors are not equal. Sometimes a student knows how to solve the problem but makes a simple mistake in copying the problem. Other times a student error can reveal that the student does not understand the fundamental skill necessary to solve a problem. In that case, error analysis can be used to identify the source of the problem and reteach the prerequisite skill.

Below are some common errors that students make. Identifying the source of the error and then reteaching the skill will eliminate frustration and build confidence.

Choosing the Wrong Operation

8 + 4 = 4

Simple mistakes are made when students add instead of subtract or multiply instead of divide.

Misaligning Numbers in Computation

$$\begin{array}{r}895\\-24\\\hline 655\end{array} \qquad \begin{array}{r}34\\\times 7\\\hline 28\\21\\\hline 49\end{array}$$

Many mistakes occur when students do not line up their numbers clearly, even when they understand the mathematics.

Regrouping

Regrouping is a very difficult concept for many students who do not have solid place value skills. When they are simply following a procedure rather than understanding why it works, errors occur.

- Subtracting the lower digit from the higher number regardless of whether it is the subtrahend.

$$\begin{array}{r}495\\-67\\\hline 432\end{array}$$

- Subtracting from left to right before regrouping.

$$\begin{array}{r}495\\-67\\\hline 438\end{array}$$

PURPLE BOOK PRETESTS
Readiness Check

Complete.

	a	b	c	d	e
1.	25 647 8 016 +95 648 ——— 129 311	65 137 −9 849 ——— 55 288	65 000 −13 296 ——— 51 704	847 ×60 ——— 50 820	9 365 ×478 ——— 4 476 470
2.	342 8)2 736	84 r3 63)5 295	$16.85 2.16 +4.78 ——— $23.79	0.3694 +0.87 ——— 1.2394	$14.03 −7.57 ——— $6.46
3.	0.9 −0.375 ——— 0.525	0.456 ×1.7 ——— 0.7752	25.37 ×6.02 ——— 152.7274	0.262 6)1.572	330 0.05)16.5
4.	7.72 0.45)3.474	$\frac{4}{7} \times \frac{7}{10}$ $\frac{2}{5}$	$1\frac{2}{3} \times 2\frac{1}{2}$ $4\frac{1}{6}$	$\frac{5}{6} \div \frac{5}{9}$ $1\frac{1}{2}$	$5\frac{3}{5} \div 4\frac{1}{5}$ $1\frac{1}{3}$
5.	$\frac{7}{12}$ $+\frac{2}{3}$ ——— $1\frac{1}{4}$	$3\frac{4}{5}$ $+2\frac{2}{3}$ ——— $6\frac{7}{15}$	$\frac{7}{12}$ $-\frac{1}{3}$ ——— $\frac{1}{4}$	5 $-1\frac{5}{6}$ ——— $3\frac{1}{6}$	$2\frac{5}{18}$ $-1\frac{5}{6}$ ——— $\frac{4}{9}$

Solve each of the following.

	a	b	c	d
6.	$\frac{n}{4} = \frac{6}{8}$ n = 3	$\frac{5}{n} = \frac{15}{21}$ n = 7	$\frac{3}{7} = \frac{27}{n}$ n = 63	$\frac{9}{27} = \frac{n}{81}$ n = 27
7.	$\frac{12}{17} = \frac{120}{n}$ n = 170	$\frac{7}{10} = \frac{n}{100}$ n = 70	$\frac{9}{n} = \frac{75}{100}$ n = 12	$\frac{n}{25} = \frac{64}{100}$ n = 16

PRISM MATHEMATICS
Purple Book

PURPLE BOOK PRETESTS
Readiness Check (continued)

Complete. Write each fraction in simplest terms.

a

	percent	fraction	decimal
8.	16%	$\frac{4}{25}$	0.16
9.	75%	$\frac{3}{4}$	0.75
10.	9%	$\frac{9}{100}$	0.09

b

percent	fraction	decimal
8%	$\frac{2}{25}$	0.08
5%	$\frac{1}{20}$	0.05
220%	$2\frac{1}{5}$	2.2

Complete the following.

a   b   c

11. __21__ is 25% of 84.   7 is __5__% of 140.   8.5 is 17% of __50__.

12. 64.2 is __42.8__% of 150.   3.2 is 8% of __40__.   __144__ is 90% of 160.

	principal	interest rate	time	interest	total amount
13.	$800	7%	1 year	$56	$856
14.	$1000	$9\frac{1}{2}$%	2 years	$190	$1190

Complete each of the following. Use 3.14 for π.

a   b 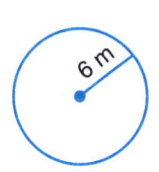   c

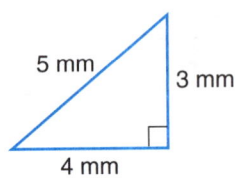

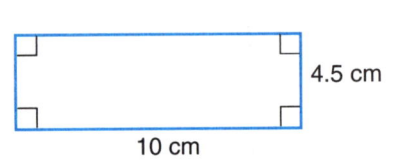

	Perimeter/Circumference	12 mm	37.68 m	29 cm
15.				
16.	Area	6 mm²	113.04 m²	45 cm²

Find the volume of each.

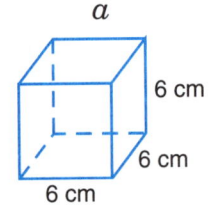

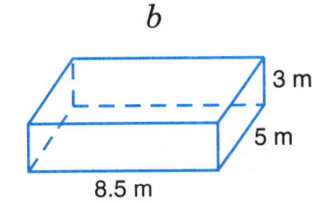

a   b   c

17.

__216__ cm³   __127.5__ m³   __60__ mm³

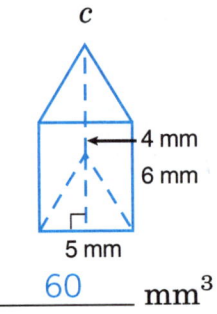

PURPLE BOOK PRETESTS
Mixed Facts Pretest

Add, subtract, multiply, or divide. Watch the signs.

	a	b	c	d
1.	58 + 76 = 134	83 − 27 = 56	68 × 9 = 612	3696 ÷ 8 = 462
2.	503 − 86 = 417	9457 ÷ 7 = 1351	857 + 98 = 955	80 × 97 = 7760
3.	8974 ÷ 40 = 224 r14	678 + 954 = 1632	608 × 79 = 48 032	520 − 298 = 222
4.	698 × 605 = 422 290	5948 + 673 = 6621	4012 − 689 = 3323	4760 ÷ 70 = 68
5.	865 × 79 = 68 335	940 ÷ 39 = 24 r4	875 + 903 + 678 = 2456	8001 − 4237 = 3764
6.	35 174 − 6 498 = 28 676	9617 + 5289 = 14 906	3078 × 95 = 292 410	6643 ÷ 91 = 73

PRISM MATHEMATICS
Purple Book

PURPLE BOOK PRETESTS
Mixed Facts Pretest (continued)

	a	b	c	d
7.	1793 8065 +7689 ――― 17 547	48 205 −13 978 ――― 34 227	973 ×100 ――― 97 300	13 r2 21)275
8.	598 ×507 ――― 303 186	23 986 6 070 +34 689 ――― 64 745	76 000 −19 356 ――― 56 644	31 r8 82)2550
9.	73 91)6643	679 ×896 ――― 608 384	63 742 86 009 +73 658 ――― 223 409	826 300 −97 513 ――― 728 787
10.	7095 ×846 ――― 6 002 370	3606 27)97 362	842 000 −267 194 ――― 574 806	706 052 +39 868 ――― 745 920
11.	986 745 +776 988 ――― 1 763 733	531 006 −89 438 ――― 441 568	2019 r23 43)86 840	4706 ×800 ――― 3 764 800

PRISM MATHEMATICS
Purple Book

4

PURPLE BOOK PRETESTS
Mixed Facts Pretest

PROBLEM-SOLVING STRATEGIES

Lesson Focus: Problem-Solving Strategy—Multi-Step
Possible Score: 6
Time Frame: 15 minutes

Prerequisite Skills: perimeter; average; addition, subtraction, multiplication facts, and division facts

Multi-Step

Sometimes it takes **multiple steps** to solve problems.

A wooden fence is to be built around a 30-m by 50-m garden. If the wood for the fence costs $36.95 per metre, how much will the wood for the entire fence cost?

The total distance around the garden is ___160___ m.

It will cost ___$5912___ to fence the garden.

Find the total number of metres of fencing needed.

30 + 30 + 50 + 50 = 160

Next, find the total cost of fencing that distance.

160 × $36.95 = $5912

Solve each problem.

SHOW YOUR WORK

1. On her first five science tests, Maria scored the following: 92, 86, 78, 94, and 95. What must she score on the sixth test so that her average for all six marks is 1 point higher than her average is right now?

   Maria has a total of ___445___ points on her first five tests.

   One point higher than her average on the first five tests is ___90___.

   Maria needs a total of ___540___ points on all six tests.

   Maria must score a(n) ___95___ on her sixth test.

   Find the average of her first five tests.
   92 + 86 + 78 + 94 + 95 = 445
   445 ÷ 5 = 89
   Add 1 to that average.
   89 + 1 = 90
   To find the total points for an average of 90, multiply by 6.
   90 × 6 = 540 points
   Subtract the points.
   540 − 445 = 95

2. A carpet cleaning company has eight homeowners that want their carpets cleaned. It takes one worker 12 h per home to clean a house full of carpet. If a team of three workers is assigned to each house, how many hours will it take to clean all eight homes?

   Three workers can clean one house in ___4___ h.

   The carpet cleaning company can have the carpet in all eight homes cleaned in ___32___ h.

   Find the times it takes 3 workers to clean 1 house.
   12 ÷ 3 = 4 hours.
   Find the total number of hours to clean the carpets in all 8 houses.
   8 × 4 = 32 hours.

PRISM MATHEMATICS
Purple Book

PROBLEM-SOLVING STRATEGIES
Multi-Step

5

PROBLEM-SOLVING STRATEGIES

Prerequisite Skills: volume; perimeter; addition, subtraction, and multiplication facts

Draw a Picture

Lesson Focus: Problem-Solving Strategy—Draw a Picture
Possible Score: 4
Time Frame: 10 minutes

Sometimes you can **draw a picture** to solve problems.

A rectangular piece of cardboard is 8 cm wide and 16 cm long. Squares measuring 2 cm on each side are cut from each corner of the cardboard. Then, the sides of the cardboard are folded to make a box. What is the volume of the box?

After cutting a 2-cm square from each corner and folding the cardboard into a box, the new width is __4__ cm, and the new length is __12__ cm.
The height of the box is __2__ cm.
The volume of the box is __96__ cm³.

Draw a picture of the rectangular cardboard with squares cut from the corners.

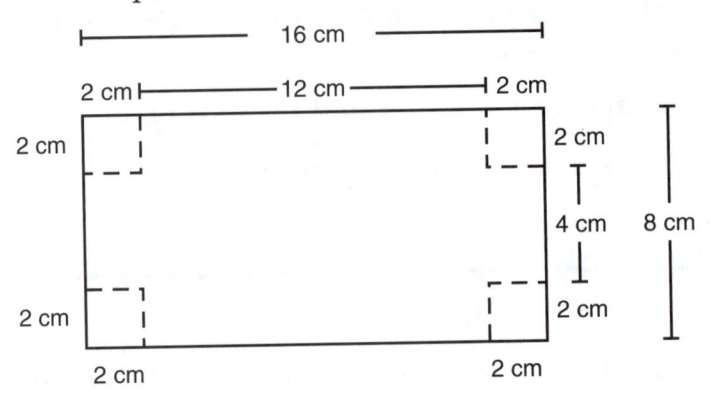

Then find the volume.

$V = \text{length} \times \text{width} \times \text{height}$
$V = 12 \times 4 \times 2$
$ = 96$

Draw a picture to solve each problem.

SHOW YOUR WORK

1. A baseball diamond can be described as a square that measures about 27 m on each side. What is the perimeter of a square whose sides are drawn inside the baseball diamond at a distance of 1 m from the baselines?

    Each side of the inner square is __84__ m long.

    The perimeter of the inner square is __336__ m.

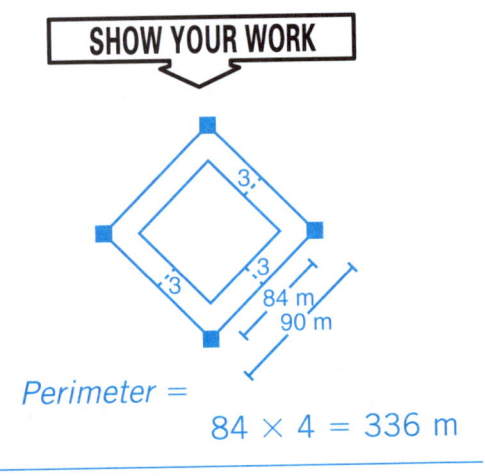

    Perimeter =
    $84 \times 4 = 336$ m

2. Mr. Story designed a flag that is 250 cm by 375 cm. After the flag was completed, he decided to attach 15-cm-long fringe around the edges of the flag. What are the dimensions of the flag including the fringe?

    With the fringe attached, the flag has a width of __280__ cm and a length of __405__ cm.

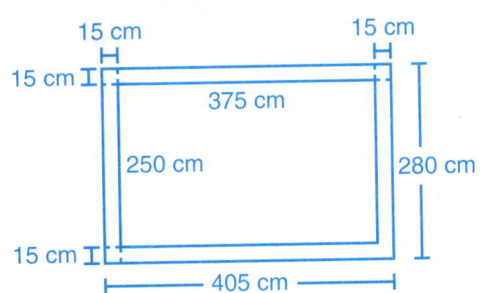

PRISM MATHEMATICS
Purple Book

6

PROBLEM-SOLVING STRATEGIES
Prerequisite Skills: addition facts

Look for a Pattern

Lesson Focus: Problem-Solving Strategy—Look for a Pattern
Possible Score: 4
Time Frame: 10 minutes

Sometimes you must **look for a pattern** to solve problems.

Jordan designed the following pattern.

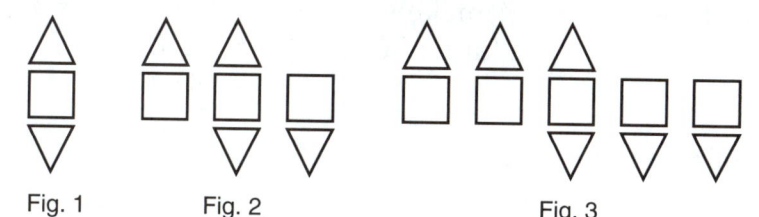

Fig. 1   Fig. 2   Fig. 3

What will the next figure in his pattern look like?

The number of triangles on the bottom of each figure increases by ___1___.

The number of triangles on the top of each figure increases by ___1___.

The number of squares in each figure increases by ___2___.

The next figure in Jordan's pattern will look like this:

Look for a pattern.

Number of triangles on bottom:
1 2 3 4
 +1 +1 +1

Number of triangles on top:
1 2 3 4
 +1 +1 +1

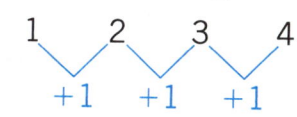

Number of squares:
1 3 5 7
 +2 +2 +2

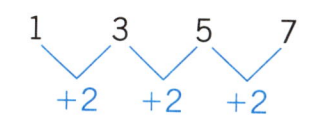

Solve each problem.

1. Brianne designs jewellery. Below are three of the five steps she followed in making a silver barrette from links.

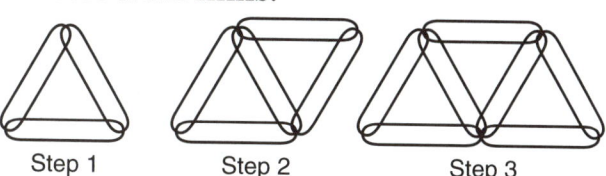

   Step 1   Step 2   Step 3

   How many links did Brianne use to make the barrette?

   Step 1 uses ___3___ silver links.

   Step 2 uses ___5___ silver links.

   Step 3 uses ___7___ silver links.

   After finishing step 5, Brianne used ___11___ silver links.

SHOW YOUR WORK

Look for a pattern.
Step: 1   2   3   4   5
Number
of links: 3   5   7   9   11
Pattern: +2 +2 +2 +2

PRISM MATHEMATICS
Purple Book

PROBLEM-SOLVING STRATEGIES
Look for a Pattern

7

PROBLEM-SOLVING STRATEGIES

Prerequisite Skills: addition and multiplication facts, volume; money; triangles

> Lesson Focus: Problem-Solving Strategy—Guess and Check
> Possible Score: 5
> Time Frame: 10 minutes

Guess and Check

Sometimes you must **guess and check** to solve problems.

Roberto used an equal number of quarters, dimes, and nickels to buy a $2.00 greeting card. How many of each coin did he use?

Guess possible numbers of each coin.

Check to see if the value of the number of coins is $2.00.

Roberto used __five__ quarters, __five__ dimes, and __five__ nickels to buy a $2.00 greeting card.

Guess: 4 quarters, 4 dimes, and 4 nickels
Value: $0.25 × 4 = $1.00
$0.10 × 4 = $0.40
$0.05 × 4 = $0.20

Total Value:
$1.00 + 0.40 + 0.20 = $1.60
Incorrect guess.

Guess: 5 quarters, 5 dimes, and 5 nickels
Value: 5 × $0.25 = $1.25
5 × $0.10 = $0.50
5 × $0.05 = $0.25

Total Value:
$1.25 + 0.50 + 0.25 = $2.00
Correct guess.

Guess and check to solve each problem.

SHOW YOUR WORK

1. The volume of a rectangular box is 693 cm³. The length of the box is 2 cm more than the width, and the width is 2 cm more than the height. What are the dimensions of the box?

   The dimensions of the box are __11__ cm, __9__ cm, and __7__ cm.

   Guess: 11, 9, 7
   V = length × width × height
   Check:
   V = 11 × 9 × 7 = 693
   Correct guess.

2. Leanne drew an isosceles triangle with one angle twice the size of the sum of the other two angles. What is the size of the largest angle in Leanne's triangle? Hint: The sum of the angles in a triangle is 180°.

   An isosceles triangle has __2__ equal angles.
   The largest angle measures __120__ degrees.

   The triangle is isosceles, so two of the angles are equal.
   Guess: 30°
   Use your guess to find the third angle.
   2(30 + 30) = 2(60) = 120
   Check:
   30 + 30 + 120 = 180
   Correct guess.

PRISM MATHEMATICS
Purple Book

PROBLEM-SOLVING STRATEGIES
Guess and Check

8

PROBLEM-SOLVING STRATEGIES

Prerequisite Skills: scale drawings; percents; distance formula; time

Lesson Focus: Problem-Solving Strategy—
Identify Missing Information
Possible Score: 4
Time Frame: 10 minutes

Identify Missing Information

Sometimes there is **not enough information** to solve the problem.

On a map, Shelly used a ruler to determine the distance from her town to her grandmother's town. She measured the distance as 5 cm. How many metres is Shelly's town from her grandmother's town?

　　Not enough information

Missing information: how many kilometres 1 cm represents on the map

Use a proportion to determine the number of actual kilometres equivalent to 5 cm on the map.

$$\frac{1 \text{ cm}}{? \text{ km}} = \frac{5 \text{ cm}}{? \text{ km}}$$

Information on how many kilometres on the map each centimetre represents is missing.

Identify the missing information in each problem.

1. Last softball season, the Jets won nine of the games they played. What percent of games played did they win?

    　　Not enough information

    Missing information: the number of games the Jets played in all

    SHOW YOUR WORK

    Divide the number of games won by the number of games played. Then, multiply that answer by 100 to find the percent.

    $$\frac{9}{\text{games played}} \times 100 = n\%$$

    The number of games played is not given.

2. Carl leaves home at 7:30 A.M. to travel to the lake where he has a summer job. If he drives at an average rate of 70 km/h, when will he arrive at the lake?

    　　Not enough information

    Missing information: the distance from Carl's home to the lake

    Divide Carl's driving distance by his speed to find the time it takes him to drive to the lake. Then, add this time to 7:30.
    distance × 75 = time
    7:30 + time = arrival time
    The distance Carl drives is not given.

PRISM MATHEMATICS
Purple Book

PROBLEM-SOLVING STRATEGIES
Identify Missing Information

PROBLEM-SOLVING STRATEGIES

Prerequisite Skills: calendar

Make a Table

Lesson Focus: Problem-Solving Strategy—Make a Table
Possible Score: 3
Time Frame: 10 minutes

Sometimes you can **make a table** to solve problems.

Archie mows Mr. Chun's lawn every third day. Every Monday and Thursday he has trumpet lessons. Every fourth day Archie helps his grandfather repair clocks. If he does all three activities on Monday, June 1, when is the next date that he will do all three activities?

Archie does all three activities again on June ___25___.

Make a table to determine what date Archie does all three activities.

Mon	Tues	Wed	Thurs	Fri	Sat	Sun
1	2	3	4	5	6	7
8	9	10	11	12	13	14
15	16	17	18	19	20	21
22	23	24	25	26	27	28
29	30					

○ = mow　　／ = repair clocks
＼ = trumpet lessons

Make a table to solve each problem.

SHOW YOUR WORK

1. Marshal puts $2 into his savings account every 3 days. Karen puts $3 into her savings account every 4 days. They both began their savings account on Monday, September 1. At the end of 30 days, how much money will they each have in savings?

   After 30 days, Marshal will have ___$20___ in his savings account.

   After 30 days, Karen will have ___$24___ in her savings account.

Mon	Tues	Wed	Thur	Fri	Sat	Sun
1 M–$2 K–$3	2	3	4 M–$4	5 K–$6	6	7 M–$6
8	9 K–$9	10 M–$8	11	12	13 M–$10 K–$12	14
15	16 M–$12	17 K–$15	18	19 M–$14	20	21 K–$18
22 M–$16	23	24	25 M–$18 K–$21	26	27	28 M–$20
29 K–$24	30	M = Marshal K = Karen				

2. At a store's grand opening, every third person entering the store wins a $5 gift certificate. Every fifth person wins a free ice-cream cone. Out of 30 people, how many people will get both a $5 gift certificate and an ice-cream cone?

   Out of 30 people, ___2___ people will get both a $5 gift certificate and an ice-cream cone.

1	2	3	4	5	6
7	8	9	10	11	12
13	14	15	16	17	18
19	20	21	22	23	24
25	26	27	28	29	30

／ = win $5 certificate
＼ = win an ice cream cone

PRISM MATHEMATICS
Purple Book

PROBLEM-SOLVING STRATEGIES

Prerequisite Skills: organizing data; addition facts; money

Lesson Focus: Problem-Solving Strategy—Make a List
Possible Score: 3
Time Frame: 15 minutes

Make a List

Sometimes you can **make a list** to solve problems.

There is a certain town called Quatreville where every family has exactly four children. In how many different orders of birth can boys and girls be born into the families in Quatreville?

There are ___16___ possible orders of birth in Quatreville.

Make a list of all possible combinations. Let B stand for boy and G stand for girl. Start by listing the oldest child first.

4 boys	3 boys	2 boys	1 boy	0 boys
BBBB	BBBG	BBGG	BGGG	GGGG
	BBGB	BGBG	GBGG	
	BGBB	BGGB	GGBG	
	GBBB	GBGB	GGGB	
		GBBG		
		GGBB		

Count the combinations in each column.

  1   4   6   4   1

Make a list to solve each problem.

SHOW YOUR WORK

1. A vending machine at Riley's school sells yogourt cups that cost $0.55. The machine accepts only loonies or correct change. It only gives nickels, dimes, and quarters for change. How many different combinations of nickels, dimes, and quarters could it give?

   The machine must give ___45¢___ in change when a loonie is used.

   There are ___8___ different combinations that the machine can give change in quarters, dimes, and nickels.

   Make a list of all the possible combinations $0.45 can be made using quarters, dimes, and nickels.

quarters	dimes	nickels
1	2	0
1	1	2
1	0	4
0	1	7
0	2	5
0	3	3
0	4	1
0	0	9

   Count the combinations.

2. Two number cubes, one numbered 1 to 6, one numbered 7 to 12, are tossed. How many different ways are there to get a sum of 14?

   There are ___5___ different ways to make a sum of 14 when rolling the two number cubes.

   Make a list of all the possible ways to get a sum of 14 using cubes.
   2 + 12   5 + 9
   3 + 11   6 + 8
   4 + 10

PRISM MATHEMATICS
Purple Book

PROBLEM-SOLVING STRATEGIES
Make a List

PROBLEM-SOLVING STRATEGIES

Prerequisite Skills: perimeter; area; patterns; multiplication and subtraction facts

Lesson Focus: Problem-Solving Strategy—
Solve a Simpler Problem
Possible Score: 13
Time Frame: 10 minutes

Solve a Simpler Problem

Sometimes you can **solve a simpler problem** to solve problems.

Motega is helping the scouts with a project. He needs to cut a 200-cm length of wire into 1-cm pieces. How many cuts will he have to make?

Motega would need to make __1__ cut in a 2-cm length of wire.

Motega would need to make __5__ cuts in a 6-cm length of wire.

Motega would need to make __9__ cuts in a 10-cm length of wire.

Motega would need to make __199__ cuts in a 200-cm length of wire.

Solve a simpler problem. Determine how many cuts Motega will need to cut in a smaller length of wire.

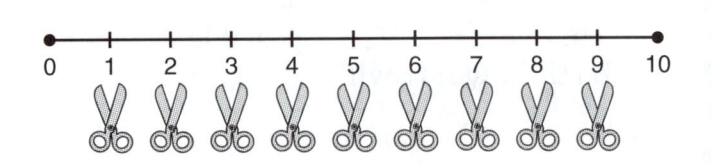

Solve a simpler problem to solve each problem.

SHOW YOUR WORK

1. For its fundraiser, the soccer team sold raffle tickets. The tickets were numbered in order. The first ticket Jorge sold was numbered 299 and the last ticket he sold was numbered 355. How many tickets did Jorge sell?

   Jorge sold __57__ tickets.

   Determine how many tickets Jorge sold if the first ticket was numbered 9 and the last was numbered 15.
   9 10 11 12 13 14 15
   (7 tickets)
   15 − 9 = 6, which is less than 7
   355 − 299 = 56, which is one less than 57

2. Alma and her family are going to plant 28 fir trees around a square lot. They want to plant one tree at each corner and then space the remaining trees 3 m apart around the perimeter of the lot. What is the area of the lot?

   If 4 trees are planted, the lot has
   __3__ × __3__ = __9__ m².

   If 8 trees are planted, the lot has
   __6__ × __6__ = __36__ m².

   If 12 trees are planted, the lot has
   __9__ × __9__ = __81__ m².

   If 28 trees are planted, the lot has
   __21__ × __21__ = __441__ m².

   Use smaller lots with fewer trees. Analyze the pattern.

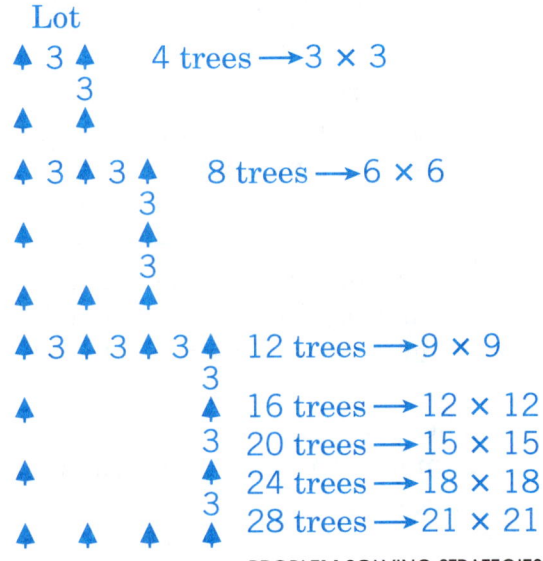

   4 trees → 3 × 3
   8 trees → 6 × 6
   12 trees → 9 × 9
   16 trees → 12 × 12
   20 trees → 15 × 15
   24 trees → 18 × 18
   28 trees → 21 × 21

PRISM MATHEMATICS
Purple Book

PROBLEM-SOLVING STRATEGIES
Solve a Simpler Problem

Lesson Focus: Problem-Solving Strategy—Work Backward
Possible Score: 5
Time Frame: 15 minutes

PROBLEM-SOLVING STRATEGIES
Prerequisite Skills: proportions; measurement; addition, subtraction, and division facts

Work Backward

Sometimes you can **work backward** to solve problems.

A company that made a very good profit this year gave half to a local charity. They gave an overseas charity half of the remaining money and kept $3.4 billion. How much profit did the company have this year?

The company had a profit of $\underline{\$13.6\ billion}$ this year.

Work backward.

Before they gave half to an overseas charity:
$3.4 billion + $3.4 billion
= $6.8 billion

Before they gave half to a local charity:
$6.8 billion + $6.8 billion
= $13.6 billion

Work backward to solve each problem.

SHOW YOUR WORK

1. Michael cashed his paycheque. He spent half his pay on new clothes. He spent a third of what he had left on a gift for his mother. He put the remaining $25 in his savings account. How much was Michael's paycheque worth?

   Michael's paycheque was worth $__75__.

   The gift is $\frac{1}{3}$ of remaining money after buying clothes, so the savings amount is $\frac{2}{3}$.
   $\frac{25}{n} = \frac{2}{3}$; n = $37.50
   $\frac{1}{2}$ of pay = $37.50
   $37.50 × 2 = $75

2. After deductions, LaToya's paycheque for 40 h of work was worth $372. She paid $68 in provincial taxes, $18 in federal taxes, and $42 in city taxes. How much does LaToya get paid per hour?

   LaToya earned $__500__ before deductions.
   LaToya is paid $__12.50__ per hour.

   Total taxes:
   $42 + 18 + 68 + 372 = $500
   $500 ÷ 40 = $12.50

3. Building A is 3.5 m taller than Building B.
   Building B is 2.5 m taller than Building D.
   Building D is 15 m taller than Building C.
   Building C is 37 m tall. How tall is Building A?
   Building A is __58__ m tall.

      37 m   Building C
   + 15 m
      52 m   Building D
   + 2.5 m
     54.5 m   Building B
   + 3.5 m
   Building A 58 m

PRISM MATHEMATICS
Purple Book

PROBLEM-SOLVING STRATEGIES
Work Backward

Lesson Focus: Problem-Solving Strategy—Use Estimation
Possible Score: 3
Time Frame: 10 minutes

PROBLEM-SOLVING STRATEGIES
Prerequisite Skills: addition, subtraction, multiplication, and division facts; rounding; comparing decimals

Use Estimation

Sometimes you can **use estimation** to solve problems.

When Tamoko visited Mexico, $1 in Canadian money could be exchanged for 6.875 pesos. If Tamoko exchanged $95 in Canadian money, about how many pesos should she receive?

Estimate. Round 6.875 pesos to 7 and 95 to 100, then multiply.

7 × 100 = 700 pesos

Tamoko should receive about __700__ pesos.

Use estimation to solve each problem.

SHOW YOUR WORK

1. A 400-g box of Corn Frosties costs $3.79. Another brand, Sugared Corn Flakes, costs $4.99 for a 600-g box. Which cereal costs less per gram? About how much less per gram does it cost?

   __Sugared Corn Flakes__ costs less per gram.

   It costs about __$0.002__ less per gram.

   Round $3.79 to $4.
   $4 ÷ 400 = $0.01 (Rounded to nearest penny)
   Round $4.99 to $5.
   $5 ÷ 600 = $0.008
   Sugared Corn Flakes costs less per gram.
   $0.01 − $0.008 = $0.002

2. When Leona went to Japan, $1 in Canadian money could be exchanged for 91.845 yen. If Leona exchanged $45 in Canadian money, about how many yen should she have received?

   Leona should have received about __4500__ yen.

   Round 91.845 yen to 90 and $45 to $50, then multiply.
   90 × 50 = 4500 yen

   Or round 91.845 yen to 100 (the nearest hundred) and $45 to $50, then multiply.
   100 × 50 = 5000 yen

3. A 675-g jar of Best spaghetti sauce costs $3.59. Another brand, Better spaghetti sauce, costs $4.89 for a 950-g jar. Which brand of spaghetti sauce costs less per gram? About how much less per gram does it cost?

   __Better__ spaghetti sauce costs less per gram.

   It costs about __$0.001__ less per gram.

   Round 675 to 700 and $3.59 to $4.
   $4 ÷ 700 = $0.006
   Round 950 to 1000 and $4.89 to $5.
   $5 ÷ 1000 = 0.005 (Rounded to nearest penny)
   Better spaghetti sauce costs less per ounce.
   $0.006 − $0.005 = $0.001

PRISM MATHEMATICS
Purple Book

PROBLEM-SOLVING STRATEGIES
Use Estimation

PROBLEM-SOLVING STRATEGIES

Lesson Focus: Problem-Solving Strategy—Use a Formula
Possible Score: 4
Time Frame: 10 minutes

Prerequisite Skills: exponents; volume of a cylinder; multiplication facts; formulas; solving equations

Use a Formula

You can **use a formula** to help solve problems.

A formula for the number of apples in a box is $N = S^3$, where S is the number of apples along each side of the box. Find N when there are 11 apples along one side of the box.

There are __1331__ apples in this box.

Use the formula given to find the number of apples in the box.

$N = S^3$ or $S \times S \times S$
$N = 11 \times 11 \times 11$
$ = 1331$

Use a formula to solve each problem.

SHOW YOUR WORK

1. Kyle uses the formula $C = 15x + 25y$ to find the cost C, in cents, of x oranges and y apples. How much would Kyle pay for 9 oranges and 7 apples?

    Kyle would pay __$3.10__ for 9 oranges and 7 apples.

    Use the formula given to find the cost of 9 oranges and 7 apples.
    $C = 15x + 25y$
    $C = (15 \times 9) + (25 \times 7)$
    $ = 135 + 175$
    $ = 310$ cents or $3.10

2. A box has a volume of 240 cm³. The height of the box is 3 cm and the width of the box is 5 cm. What is the length of the box?

    The formula for the volume of a box is __$V = l \times w \times h$__.

    The length of the box is __16__ cm.

    Use the formula for the volume of a rectangular solid. Find l, the length.
    $V = l \times w \times h$
    $240 = l \times 5 \times 3$
    $\dfrac{240}{15} = \dfrac{15l}{15}$
    $16 = l$

3. Using $\pi \doteq 3.14$, find the volume of a cylindrical tank having a diameter of 14 m and a height of 10 m.

    The cylindrical tank has a volume of about __1540__ m³.

    Use the formula for volume of a cylinder. The radius equals half of the diameter, or 7 m.
    $V = Bh$
    $V = \pi r^2 h$
    $V \approx \dfrac{22}{7} \times 7^2 \times 10$
    $ \approx 22 \times 7 \times 10$
    $ \approx 1540$ m³

PRISM MATHEMATICS
Purple Book

PROBLEM-SOLVING STRATEGIES
Use a Formula

PROBLEM-SOLVING STRATEGIES

Prerequisite Skills: factors; even and odd numbers; prime numbers; use a table

Lesson Focus: Problem-Solving Strategy—
    Use Logical Reasoning
Possible Score: 5
Time Frame: 10 minutes

Use Logical Reasoning

You can **use logical reasoning** to help solve problems.

Ann, Seung, Yolanda, and Denny each have a pet. They each have a different pet: a beagle, a cat, an angelfish, and a gerbil. Yolanda lives next door to the person with the gerbil. Denny and Seung have pets that live in small habitats. Ann cannot have a dog because of allergies. Seung is afraid of animals that bite. Which animal does each person own?

Ann has a(n) __cat__.
Seung has a(n) __angelfish__.
Yolanda has a(n) __beagle__.
Denny has a(n) __gerbil__.

Use a table to keep track of the facts. Then, indicate your conclusions.

	Beagle	Cat	Angel-fish	Gerbil
Ann	no	yes	no	no
Seung	no	no	yes	no
Yolanda	yes	no	no	no
Denny	no	no	no	yes

Use logical reasoning to help solve each problem.

SHOW YOUR WORK

1. Jarod, Scott, and Rick each ate either a hamburger, cheeseburger, or a cheese pizza for lunch. Rick did not have any meat for lunch. Jarod does not like cheese. What did each person eat for lunch?

   Jarod ate the __hamburger__.
   Scott ate the __cheeseburger__.
   Rick ate the __cheese pizza__.

	HB	CB	CP
Jarod	yes	no	no
Scott	no	yes	no
Enrique	no	no	yes

---

2. Four letters, A, B, C, and D, are each written with a number, 1, 3, 5, or 8, though not necessarily in that order. The letter A is written with a prime number. The letter B is written with a number less than 5. Neither A nor D is written with a number that is an odd factor of 40. What numbers are the letters B and D written with?

   Letter B is written with the number __1__.
   Letter D is written with the number __8__.

	1	3	5	8
A	no	yes	no	no
B	yes	no	no	no
C	no	no	yes	no
D	no	no	no	yes

PRISM MATHEMATICS
Purple Book

CHAPTER 1 PRETEST
Whole Numbers/Decimals

Add or subtract.

	a	b	c	d	e
1.	4815 +7263 ――― 12 078	15.8 +23.9 ――― 39.7	165 975 45 336 +64 821 ――― 276 132	48.62 3.94 +5.75 ――― 58.31	64.559 61.947 +83.827 ――― 210.333
2.	1286 +576 ――― 1862	45.88 +8.90 ――― 54.78	126 783 +267 854 ――― 394 637	14.892 −7.698 ――― 7.194	1000 −256 ――― 744
3.	4296 −853 ――― 3443	8.3 −2.7 ――― 5.6	526 794 −64 288 ――― 462 506	15.21 −5.76 ――― 9.45	561.341 −279.648 ――― 281.693

Multiply or divide.

	a	b	c	d
4.	264 ×13 ――― 3432	957 ×816 ――― 780 912	16.23 ×4.8 ――― 77.904	1.867 ×23.1 ――― 43.1277
5.	1658 ×623 ――― 1 032 934	21.79 ×25.4 ――― 553.466	321 49)15 729	13 r26 35)481
6.	22 r2 113)2488	165 2.5)412.5	610 0.088)53.68	2.46 3.81)9.3726

PRISM MATHEMATICS
Purple Book

CHAPTER 1 PRETEST

17

CHAPTER 1 PRETEST Problem Solving

Solve each problem.

> **Park Ticket Prices**
> $1.50 per ride
> $13.00 day pass

1. There were 4129 ride tickets sold today at the park. How much money was collected for the ride tickets?

   $ __6193.50__ was collected.

2. There were 8765 day passes sold at the park. How much money was collected for the day passes sold?

   $ __113 945.00__ was collected.

3. How much money was collected today for both ride tickets and day passes?

   $ __120 138.50__ was collected.

4. How many more day passes were sold today than ride tickets?

   __4636__ more day passes were sold than ride tickets.

5. A group spent $4108.00 on day passes. If each person purchased one day pass, how many people were in their group?

   There were __316__ people in their group.

CHAPTER 1
Whole Numbers/Decimals

Prerequisite Skills: addition with renaming

Lesson Focus: adding up to five addends with 6 digits
Possible Score: 32
Time Frame: 25–30 minutes

Lesson 1 Addition (whole numbers)

Add the ones. Rename 27 as "2 tens and 7 ones."

Continue adding from right to left.

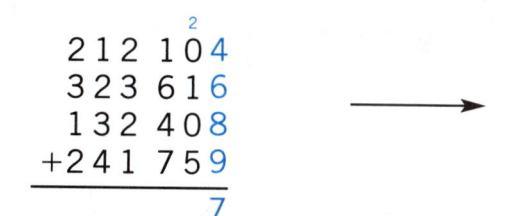

```
   ²
 212 104           ¹ ¹ ²
 323 616         212 104
 132 408         323 616
+241 759         132 408
────────        +241 759
       7        ────────
                 909 887
```

Add.

	a	b	c	d	e
1.	23 +14 ── 37	224 +73 ─── 297	312 +5324 ──── 5636	43 214 +4 325 ────── 47 539	41 321 +612 314 ─────── 653 635
2.	16 47 +13 ── 76	217 316 +142 ─── 675	3317 2154 +1212 ──── 6683	21 016 14 527 +51 202 ────── 86 745	260 316 217 327 +411 342 ─────── 888 985
3.	31 70 14 +52 ── 167	273 162 253 +210 ─── 898	1131 2262 3473 +1051 ──── 7917	41 370 2 151 33 225 +11 118 ────── 87 864	121 065 302 432 304 144 +41 213 ─────── 768 854
4.	36 75 84 31 +17 ── 243	633 710 821 502 +221 ─── 2887	1123 2651 1762 2873 +1411 ──── 9820	11 616 12 573 21 412 40 331 +13 214 ────── 99 146	232 362 351 171 64 221 71 141 +182 314 ─────── 901 209
5.	34 76 58 67 +73 ── 308	542 624 852 715 +316 ─── 3049	7067 8458 5312 2521 +1214 ──── 24 572	31 145 14 214 3 142 76 125 +3 214 ────── 127 840	212 304 321 456 214 672 523 214 +314 235 ─────── 1 585 881

Number Correct — **LESSON FOLLOW-UP AND ERROR ANALYSIS**

27–32: To reinforce numeration, have students circle every even-numbered answer.
22–26: Check to see if students made errors renaming. If so, ask them to explain how to record renaming.
Less than 22: Before they redo the page, have students practice basic addition facts using flash cards.

Lesson 1   Problem Solving

Solve each problem.

1. On Thursday, 1335 books were borrowed from the library. On Friday, 1852 books were borrowed. How many books were borrowed in all?

   _____3187_____ books were borrowed.

2. A carpenter ordered pieces of plywood. One box contained 158 pieces, another 232 pieces, and the third 116 pieces. How many pieces were received in all?

   The carpenter received _____506_____ pieces in all.

3. Three new homes were sold last week. The prices were $215 422, $199 554, and $218 432. What were the total sales for the week?

   The total sales were $_____633 408_____.

4. Joan drove 187 km in one day. She drove 207 km the next day. In all, how far did she drive?

   Joan drove _____394_____ km.

5. Sophia drove 415 km on the first day of her vacation. On the second day, she drove 520 km. How many kilometres did she drive in all?

   Sophia drove _____935_____ km in all.

6. Matthew is going to the school dance on Friday night. He bought a shirt for $31 and a pair of pants for $42. How much did he spend in all?

   Matthew spent $_____73_____.

7. During a three-year period Mrs. Newman drove her car the following distances: 8456 km, 9754 km, and 7652 km. How many kilometres did she drive her car during the three years?

   She drove _____25 862_____ km.

1.

2.

3.

4.

5.

6.

7.

CHAPTER 1
Whole Numbers/Decimals

Lesson 1
Addition (whole numbers)

20

Prerequisite Skills: subtraction with renaming

Lesson Focus: subtracting 1- to 6-digit numbers
Possible Score: 42
Time Frame: 25–30 minutes

Lesson 2 Subtraction (whole numbers)

To subtract ones, rename 3 tens and 2 ones as "2 tens and 12 ones."

Continue subtracting from right to left, renaming as necessary.

$$\begin{array}{r} 965\ 43\overset{2\ 12}{\cancel{3}\cancel{2}} \\ -121\ 715 \\ \hline 7 \end{array} \longrightarrow \begin{array}{r} 9\overset{4\ 14\ 2\ 12}{6\cancel{5}\ \cancel{4}\cancel{3}\cancel{2}} \\ -121\ 715 \\ \hline 843\ 717 \end{array}$$

Subtract.

	a	b	c	d	e
1.	37 −6 ――― 31	327 −16 ――― 311	4325 −214 ――― 4111	17 625 −3 214 ――― 14 411	321 459 −20 123 ――― 301 336
2.	59 −14 ――― 45	847 −231 ――― 616	6875 −1534 ――― 5341	87 654 −14 123 ――― 73 531	582 785 −131 524 ――― 451 261
3.	70 −49 ――― 21	968 −159 ――― 809	8752 −4127 ――― 4625	78 547 −31 218 ――― 47 329	495 627 −314 518 ――― 181 109
4.	97 −78 ――― 19	523 −72 ――― 451	5963 −2172 ――― 3791	25 753 −14 182 ――― 11 571	457 245 −112 158 ――― 345 087
5.	83 −45 ――― 38	675 −289 ――― 386	5028 −4917 ――― 111	86 743 −21 892 ――― 64 851	675 247 −321 482 ――― 353 765
6.	45 −27 ――― 18	607 −299 ――― 308	8207 −3149 ――― 5058	74 003 −21 456 ――― 52 547	900 435 −417 624 ――― 482 811
7.	81 −27 ――― 54	700 −287 ――― 413	6732 −865 ――― 5867	67 524 −29 689 ――― 37 835	351 257 −165 268 ――― 185 989

Number Correct **LESSON FOLLOW-UP AND ERROR ANALYSIS**

29–35: To reinforce place value, have students circle every odd-numbered answer.
22–28: Ask students to explain how to rename in subtraction problems.
Less than 22: Before students redo the page, have them practice basic subtraction facts using flash cards.

Lesson 2 Problem Solving

Solve each problem.

1. The population of Westerville is 54 552 and the population of Pickerington is 48 964. How many more people live in Westerville than live in Pickerington?

   __5588__ more people live in Westerville.

2. Violet Elementary School parents and teachers raised $2507 at the Spring Fair. At the Winter Carnival they raised $3465. How much more money did they raise at the Winter Carnival?

   They raised $__958__ more at the Winter Carnival.

3. Last year, the cost of a movie ticket at the Palace Theatre was $8. This year, the cost is $9. How much more does a ticket cost this year?

   A ticket costs $__1__ more this year.

4. In Newville, 2243 families receive the evening paper and 1875 receive the morning paper. How many more families receive the evening paper?

   __368__ more families receive the evening paper.

5. Roy drove 2645 km last week. This week he drove 2847 km. How many kilometres more did he drive this week?

   Roy drove __202__ km more this week.

6. Nicholas paid $45 for a book and a CD. The cost of the CD was $15. How much was the book?

   The book cost $__30__.

7. There were 316 people at last week's school dance. There were 284 people at this week's dance. How many more people were at last week's dance?

   There were __32__ more people last week.

CHAPTER 1
Whole Numbers/Decimals

Lesson 2
Subtraction (whole numbers)

Prerequisite Skills: multiplication with renaming

Lesson Focus: multiplying by 1-, 2-, or 3-digit numbers
Possible Score: 27
Time Frame: 25–30 minutes

Lesson 3 Multiplication (whole numbers)

```
    4 873
  ×   296
   29 238  ———— 6 × 4873
  438 570  ———— 90 × 4873
  974 600  ———— 200 × 4873
1 442 408      Add.
```

Multiply.

	a	b	c	d	e
1.	63 ×4 = 252	432 ×2 = 864	679 ×7 = 4753	2312 ×3 = 6936	7598 ×8 = 60 784
2.	68 ×20 = 1360	700 ×34 = 23 800	212 ×43 = 9116	1720 ×64 = 110 080	2806 ×97 = 272 182
3.	341 ×200 = 68 200	213 ×320 = 68 160	403 ×212 = 85 436	1414 ×312 = 441 168	5875 ×678 = 3 983 250
4.	700 ×426 = 298 200	646 ×925 = 597 550	925 ×436 = 403 300	9251 ×809 = 7 484 059	7487 ×869 = 6 506 203

Number Correct

LESSON FOLLOW-UP AND ERROR ANALYSIS

23–27: To reinforce place value, have students circle each answer with a zero in the tens place.

18–22: Check to see if students multiplied and renamed but forgot to add the renamed number or if they added first and then multiplied.

Less than 18: Ask students to explain how to multiply two numbers with renaming before they redo the page.

23

Lesson 3 Problem Solving

Solve each problem.

1. There were 19 bands in the parade. Each band had 33 members. How many band members were in the parade in all?

   There were ___627___ band members in all.

2. The farmer shipped 476 bags of potatoes to the market. Each bag had a mass of 26 kg. What was the mass of the potatoes in all?

   The mass of the potatoes was ___12 376___ kg.

3. The manager of the local coffee shop ordered 19 boxes of coffee stirrers. Each box contained 165 stirrers. How many coffee stirrers did the manager order?

   The manager ordered ___3135___ stirrers.

4. Mrs. Pinkerman drives 47 km each day. How many kilometres will she drive in five days?

   She will drive ___235___ km.

5. One apple costs 20¢. How much do 25 apples cost?

   25 apples cost $___5___.

6. Ms. Combs bought six boxes of note cards on sale for $7 per box. How much money did she spend?

   She spent $___42___.

7. The bus fare from Toronto to Hamilton is $19 for one person. There are seven people in the Davis family. How much will it cost them to make the trip?

   It will cost them $___133___.

CHAPTER 1
Whole Numbers/Decimals

Lesson 3
Multiplication (whole numbers)

Prerequisite Skills: division facts

Lesson Focus: dividing by 1-, 2-, or 3-digit numbers
Possible Score: 22
Time Frame: 20–25 minutes

Lesson 4 Division (whole numbers)

Think

312)66 831   312 × 1000 = 312 000   Quotient is between 100 and 1000. So its
              312 × 100 = 312 00   first digit will be in the *hundreds* place.

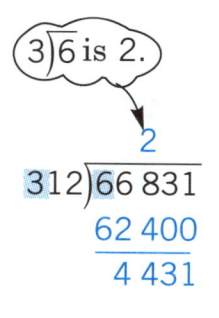

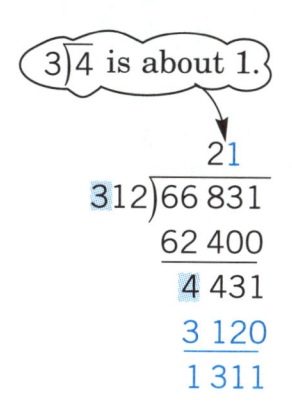

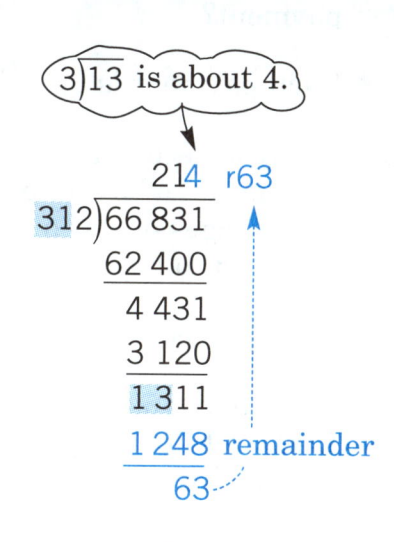

3)6 is 2.

```
      2
312)66 831
    62 400
     4 431
```

3)4 is about 1.

```
     21
312)66 831
    62 400
     4 431
     3 120
     1 311
```

3)13 is about 4.

```
    214  r63
312)66 831
    62 400
     4 431
     3 120
     1 311
     1 248  remainder
        63
```

Divide.

	a	b	c	d	e
1.	76 r3 9)6 8 7	8 r31 42)3 6 7	12 58)6 9 6	7 r114 123)9 7 5	5 r249 421)2 3 5 4
2.	203 r4 7)1 4 2 5	25 r32 57)1 4 5 7	120 r65 69)8 3 4 5	14 r1 521)7 2 9 5	19 r498 624)1 2 3 5 4
3.	6254 6)3 7 5 2 4	151 r43 83)1 2 5 7 6	2285 r31 37)8 4 5 7 6	102 r16 784)7 9 9 8 4	257 r139 379)9 7 5 4 2

LESSON FOLLOW-UP AND ERROR ANALYSIS

Number Correct
- **18–22:** Have students choose two problems from the page and write two story problems that they would solve.
- **15–17:** Check to see if students recorded answers in the wrong place-value positions. If so, have them use lined paper to redo the problems.
- **Less than 15:** Before they redo the page, ask students to explain how they estimated the first digit in each answer.

Lesson 4 Problem Solving

Solve each problem.

1. Vinny won $720 in a contest. The money will be paid in 12 equal payments. How much is each payment?

   Each payment is $ __60__ .

2. Emily is decorating for her mother's birthday. She needs to buy 42 balloons. The balloons come in packages of seven. How many packages will she need to buy?

   She will need to buy __6__ packages.

3. A group of 152 people want to ride river rafts. Each raft holds eight people. How many rafts are needed so that each person gets one ride?

   __19__ rafts are needed.

4. Nicholas and Matthew bought their mother a necklace. The necklace cost $84. They shared the cost equally. How much did each pay?

   Each paid $ __42__ .

5. The school lunch committee will need 180 servings of spaghetti for the school's annual lunch meeting. Each box of spaghetti contains nine servings. How many boxes will be needed?

   __20__ boxes of spaghetti will be needed.

6. Kathy is buying juice for the party. She needs 132 servings. If each bottle of juice holds 12 servings, how many bottles must she buy?

   She needs __11__ bottles of juice.

7. Christopher and Raymond put 84 candles on their grandmother's birthday cake. There were seven candles in each box. How many boxes of candles did they need?

   They needed __12__ boxes of candles.

1.

2.

3.

4.

5.

6.

7.

Prerequisite Skills: whole number operations

Lesson Focus: applying addition, subtraction, multiplication, and division
Possible Score: 16
Time Frame: 15–20 minutes

Lesson 5 Problem Solving

Solve each problem.

1. Luke delivered 225 flyers on Fox Street, 134 on 87th Street, 218 on Hansen Street, and 229 on 89th Street. How many flyers did he deliver?

   He delivered __806__ flyers.

2. Wayne Gretzky scored 894 goals during his career. Gordie Howe scored 801. How many more goals did Gretzky score than Howe?

   Gretzky scored __93__ more goals.

3. If Ms. Jones drives 350 km each day, how many kilometres will she drive in four days?

   She will drive __1400__ km.

4. A bricklayer laid 656 bricks in 8 h. Suppose the same number of bricks were laid each hour. How many bricks were laid each hour?

   __82__ bricks were laid each hour.

5. The load limit for a small bridge is 3000 kg. Mr. Sims' car has a mass of 1875 kg. How much less than the load limit is the mass of the car?

   The car has a mass of __1125__ kg less.

6. One week the bakery used 144 sacks of flour. Suppose each sack of flour has a mass of 55 kg. What was the total mass of the flour used?

   __7920__ kg of flour were used.

7. There are 17 370 items to be packed into boxes of 24 items each. How many boxes will be filled? How many items will be left over?

   __723__ boxes will be filled.

   __18__ items will be left over.

CHAPTER 1

Number Correct

LESSON FOLLOW-UP AND ERROR ANALYSIS

13–16: Ask students to choose one problem and write how they solved it.
10–12: Before they redo the page, have students underline what is asked for in each problem to help them focus on what operation is needed.
Less than 10: Help students identify key words that indicate which operation they should use before they redo incorrect answers.

27

Lesson 5 Problem Solving

Solve each problem.

1. The local Plumbers Union has 456 members. The local Carpenters Union has 875. How many more members does the Carpenters Union have than the Plumbers Union?

   There are ___419___ more members in the Carpenters Union.

2. Seven cars can be loaded on a transport truck. Each car has a mass of 1875 kg. What is the total mass of the cars that can be loaded on the truck?

   The total mass is ___13 125___ kg.

3. Mr. Cosgrove drove his car 10 462 km last year and 11 125 km this year. How many kilometres did he drive the car during these two years?

   He drove the car ___21 587___ km.

4. There were 1172 women at a banquet. They were seated 8 to a table. How many tables were filled? How many were at the partially filled table?

   ___146___ tables were filled.

   ___4___ women were at the partially filled table.

5. There are 2125 employees at the McKee Plant. Each works 35 h a week. What is the total number of hours worked by these employees in one week?

   The total number of hours is ___74 375___.

6. The population of Tomstown is 34 496 and the population of Janesburg is 28 574. How much greater is the population of Tomstown than the population of Janesburg?

   The population is ___5922___ greater.

7. A satellite is orbiting Earth at a speed of 28 580 km/h. At this rate, how many kilometres will the satellite travel in 1 min?

   It will travel ___476.3___ km.

CHAPTER 1
Whole Numbers/Decimals

Prerequisite Skills: addition and subtraction with renaming

Lesson Focus: adding and subtracting decimals
Possible Score: 33
Time Frame: 15–20 minutes

Lesson 6 Addition and Subtraction (decimals)

When adding or subtracting decimals, write the decimals so the decimal points line up. Then, add or subtract as with whole numbers.

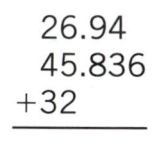

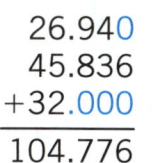

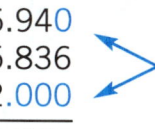

```
  26.94        26.940      Write these         62.5        62.500   ← Write these
  45.836       45.836      0s if they         −43.345     −43.345      0s if they
 +32          +32.000      help you.           _____    _____      help you.
              _____                           19.155
              104.776
```

The decimal point in the answer is directly below the other decimal points.

Add or subtract.

	a	b	c	d	e
1.	1.3 +2.6 ____ 3.9	57.6 +3.8 ____ 61.4	74.57 +21.2 _____ 95.77	18.5 +17.36 _____ 35.86	67.857 +2.11 _____ 69.967
2.	4.7 −3.2 ____ 1.5	67.5 −43.7 ____ 23.8	87.58 −34.1 _____ 53.48	75.9 −26.69 _____ 49.21	87.52 −12.403 _____ 75.117
3.	3.2 4.3 +6.4 ____ 13.9	52.7 −26.9 ____ 25.8	53.25 13.1 +31.28 _____ 97.63	64.3 −25.56 _____ 38.74	16.106 34.25 +21.3 _____ 71.656
4.	47.34 −13.76 _____ 33.58	1.372 4.235 +5.051 _____ 10.658	25 −1.435 _____ 23.565	5.603 2.751 +8.832 _____ 17.186	31.423 −12.83 _____ 18.593
5.	57.46 31.59 42.34 +2.67 _____ 134.06	5.708 −1.439 _____ 4.269	83.275 14.238 8.675 +34.873 _____ 141.061	48.003 −13.746 _____ 34.257	47.578 14.483 73.241 +42.967 _____ 178.269

Number Correct

LESSON FOLLOW-UP AND ERROR ANALYSIS

28–33: To reinforce place value, have students circle each answer with 6 in the hundredths place and □ each answer with a 5 in the tenths place.

21–27: Check to see if students missed 2d, 2e, 3d, and 4c. If so, they probably forgot to insert zeros in the top number.

Less than 21: Have students explain how to place a decimal point in the answer to addition and subtraction problems.

Lesson 6 Problem Solving

Solve each problem.

1. The water level of the lake rose 0.85 m during March, 1.30 m during April, and 0.52 m during May. How much did the water level rise during these three months?

   The water level rose ___2.67___ m.

2. In problem **1**, how much more did the water level rise during April than during May?

   It rose ___0.78___ m more during April.

3. Mr. Tadlock purchased a suit on sale for $97.95 and an overcoat for $87.50. What was the total cost of these articles?

   The total cost was $___185.45___.

4. In problem **3**, how much more did the suit cost than the overcoat?

   The suit cost $___10.45___ more.

5. Last season a certain baseball player had a batting average of .285. This season his batting average is .313. How much has the player's batting average improved?

   The player's average has improved by ___.028___.

6. The thicknesses of three machine parts are 0.514 cm, 0.317 cm, and 0.178 cm. What is the combined thickness of the parts?

   The combined thickness is ___1.009___ cm.

7. Ms. Dutcher's lot is 60.57 m long. Mr. Poole's lot is 54.73 m long. How much longer is Ms. Dutcher's lot than Mr. Poole's lot?

   Ms. Dutcher's lot is ___5.84___ m longer.

8. Ms. Jolls purchased a dress for $62.95, a pair of shoes for $19, and a purse for $11.49. What was the total amount of these purchases?

   The total amount was $___93.44___.

Prerequisite Skills: multiplication with 2- and 3-digit numbers

Lesson Focus: multiplying decimals
Possible Score: 33
Time Frame: 15–20 minutes

Lesson 7 Multiplication (decimals)

number of digits to the right of the decimal point

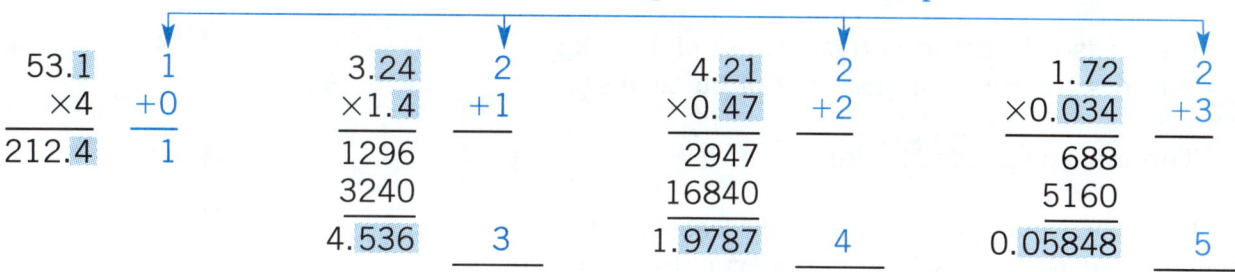

```
  53.1    1       3.24     2        4.21     2        1.72      2
  × 4    +0      ×1.4     +1       ×0.47    +2       ×0.034    +3
  ─────  ──      ─────    ──       ──────   ──       ───────   ──
  212.4   1      1296              2947              688
                 3240              16840             5160
                 ─────             ──────            ───────
                 4.536    3        1.9787   4        0.05848   5
```

Multiply.

	a	b	c	d	e
1.	32.5 ×5 ───── 162.5	4.57 ×8 ───── 36.56	672 ×0.4 ───── 268.8	1678 ×0.07 ───── 117.46	8.765 ×9 ───── 78.885
2.	31.4 ×0.09 ───── 2.826	2.23 ×0.7 ───── 1.561	417 ×0.009 ───── 3.753	0.418 ×0.6 ───── 0.2508	167.8 ×0.008 ───── 1.3424
3.	3.14 ×0.17 ───── 0.5338	67.4 ×6.7 ───── 451.58	5.09 ×0.058 ───── 0.295 22	0.724 ×0.46 ───── 0.333 04	148.6 ×2.9 ───── 430.94
4.	65.7 ×0.648 ───── 42.5736	0.584 ×35.6 ───── 20.7904	69.2 ×4.63 ───── 320.396	7.54 ×60.7 ───── 457.678	2.408 ×5.69 ───── 13.701 52
5.	3.75 ×1.24 ───── 4.6500	3.14 ×0.526 ───── 1.651 64	0.957 ×6.18 ───── 5.914 26	3.27 ×4.38 ───── 14.3226	2.123 ×4.25 ───── 9.022 75

Number Correct

LESSON FOLLOW-UP AND ERROR ANALYSIS

28–33: To reinforce place value, have students write a row of answers in words.
21–27: Before students redo incorrect answers, have them explain how decimal points are placed in the answer.
Less than 21: Ask students to explain how to multiply decimals before they redo the pages.

31

Lesson 7 Problem Solving

Solve each problem.

1. Each case of batteries has a mass of 17.3 kg. What is the mass of six cases of batteries?

   The mass is __103.8__ kg.

2. 6.75 truckloads of ore can be processed each hour. At that rate, how many truckloads of ore can be processed during an 8-h period?

   __54__ truckloads of ore can be processed.

3. An article has a mass of 6.47 kg. What would be the mass of 24 such articles?

   The mass would be __155.28__ kg.

4. Mr. Swank's car averages 7.8 L of gasoline per 100 km. How many litres of gasoline would he need to drive 1300 km?

   He would need __101.4__ L of gasoline.

5. An industrial machine uses 4.75 L of fuel each hour. At that rate, how many litres of fuel will be used in 6.5 h?

   __30.875__ L of fuel will be used.

6. What would be the cost of a 6.2-kg roast at $5.40 per kilogram?

   The cost would be $__33.48__.

7. Each sheet of paper is 0.043 cm thick. What is the combined thickness of 25 sheets?

   It is __1.075__ cm.

8. Brittany runs 1.5 km each day. How far will she run in five days?

   She will run __7.5__ km.

Prerequisite Skills: division with 2- and 3-digit divisors

Lesson Focus: dividing decimals
Possible Score: 20
Time Frame: 20–25 minutes

Lesson 8   Division (decimals)

$0.73 \overline{)21.9}$ → To get a whole number divisor, multiply both 0.73 and 21.9 by __100__. →

$$\begin{array}{r} 30 \\ 73{\overline{)2190}} \\ 2190 \\ \hline 0 \\ 0 \\ \hline 0 \end{array}$$

shorter way

$$\begin{array}{r} 30 \\ 0.73{\overline{)21.90}} \\ 2190 \\ \hline 0 \\ 0 \\ \hline 0 \end{array}$$

$0.059 \overline{)0.1357}$ → To get a whole number divisor, multiply both 0.059 and 0.1357 by __1000__. →

$$\begin{array}{r} 2.3 \\ 59{\overline{)135.7}} \\ 1180 \\ \hline 177 \\ 177 \\ \hline 0 \end{array}$$

shorter way

$$\begin{array}{r} 2.3 \\ 0.059{\overline{).135.7}} \\ 1180 \\ \hline 177 \\ 177 \\ \hline 0 \end{array}$$

Divide.

	a	*b*	*c*	*d*
1.	$0.61{\overline{)3.05}}$ = 5	$9.1{\overline{)4.55}}$ = 0.5	$0.071{\overline{)0.639}}$ = 9	$1.37{\overline{)0.959}}$ = 0.7
2.	$0.37{\overline{)0.999}}$ = 2.7	$0.95{\overline{)76}}$ = 80	$0.026{\overline{)1.378}}$ = 53	$16.7{\overline{)2.004}}$ = 0.12
3.	$0.03{\overline{)0.798}}$ = 26.6	$0.08{\overline{)2.008}}$ = 251	$0.47{\overline{)9.729}}$ = 20.7	$25.3{\overline{)0.92851}}$ = 0.0367

Number Correct — **LESSON FOLLOW-UP AND ERROR ANALYSIS**
17–20: To reinforce numeration, have students identify the greatest and least answers on page 33 and write them in words.
13–16: Ask students to explain when it is necessary to insert zeros in the dividend before they redo the pages.
Less than 13: Have students explain how to change the divisor to a whole number before dividing. (move decimal point)

Lesson 8 Problem Solving

Solve each problem.

1. A rope 40.8 m long is to be cut into four pieces of the same length. How long will each piece be?

   Each piece will be __10.2__ m long.

2. Each can of oil costs $0.92. How many cans of oil can be purchased with $23?

   __25__ cans of oil can be purchased.

3. A case of cans has a mass of 9.6 kg. Each can has a mass of 0.6 kg. How many cans are there?

   There are __16__ cans in the case.

4. Each sheet of paper is 0.016 cm thick. How many sheets will it take to make a stack of paper 18 cm high?

   It will take __1125__ sheets.

5. Amy spent $9.60 for meat. A kilogram of meat sells for $2.40. How many kilograms did she buy?

   She bought __4__ kg.

6. A machine uses 0.75 L of fuel each hour. At that rate, how long will it take to use 22.5 L of fuel?

   It will take __30__ h.

7. Each corn flake has a mass of about 0.08 g. How many flakes will it take to have a mass of 20.4 grams?

   It will take __255__ corn flakes.

8. It takes a wheel 0.6 s to make a revolution. What part of a revolution will it make in 0.018 s?

   The wheel will make __0.03__ of a revolution.

1.

2.

3.

4.

5.

6.

7.

8.

CHAPTER 1
Whole Numbers/Decimals

Lesson 8
Division (decimals)

34

For further evaluation, copy the Chapter Test on page 265.
For maintaining skills, use the Cumulative Review on pages 213–214.

Possible Score: 24
Time Frame: 15–20 minutes

CHAPTER 1 PRACTICE TEST
Whole Numbers/Decimals

Add or subtract.

	a	b	c	d
1.	7684 584 +285 **8553**	79.86 12.49 +6.91 **99.26**	86 714 56 811 +25 648 **169 173**	23.846 2.734 +45.8 **72.38**
2.	89.01 −45.99 **43.02**	76 321 −55 344 **20 977**	4.612 −1.877 **2.735**	70 001 −27 689 **42 312**
3.	2312 +5678 **7990**	12.3 +4.8 **17.1**	16.01 −8.64 **7.37**	257.314 −151.276 **106.038**

Multiply or divide.

4.	648 ×21 **13 608**	2465 ×546 **1 345 890**	46.2 ×54.1 **2499.42**	1.081 ×0.013 **0.014053**
5.	478 ×478 **228 484**	43.864 ×0.261 **11.448504**	**231 r4** 67)15 481	**4 r89** 99)485
6.	**468** 23)10 764	**7.9** 0.8)6.32	**6.9** 1.43)9.867	**0.12** 2.75)0.33

Number
Correct   LESSON FOLLOW-UP AND ERROR ANALYSIS
18–24:   Have students circle the most difficult problem on the page and explain the difficulty.
13–17:   Make sure students moved the decimal point the appropriate number of spaces, or made other errors.
Less than 13:   Ask students to explain how to figure out the number of decimal places in each multiplication and division problem.

CHAPTER 2 PRETEST
Fractions

Write each fraction in simplest form.

	a	b	c	d	e
1.	$\frac{10}{15} = \frac{2}{3}$	$\frac{3}{6} = \frac{1}{2}$	$\frac{18}{24} = \frac{3}{4}$	$\frac{40}{50} = \frac{4}{5}$	$\frac{2}{8} = \frac{1}{4}$

Rename.

2. $\frac{1}{2} = \frac{4}{8}$ $\quad$ $\frac{1}{4} = \frac{5}{20}$ $\quad$ $7 = \frac{21}{3}$ $\quad$ $4\frac{1}{5} = \frac{21}{5}$ $\quad$ $2\frac{7}{8} = \frac{23}{8}$

Write each sum, difference, product, or quotient in simplest form.

	a	b	c	d
3.	$\frac{4}{5} + \frac{3}{5} = 1\frac{2}{5}$	$1\frac{5}{12} + 4\frac{7}{12} = 6$	$3\frac{1}{2} + 8\frac{11}{15} = 12\frac{7}{30}$	$2\frac{2}{3} + 4\frac{3}{4} + 3\frac{1}{6} = 10\frac{7}{12}$
4.	$\frac{5}{8} - \frac{3}{8} = \frac{1}{4}$	$\frac{4}{6} - \frac{3}{6} = \frac{1}{6}$	$4\frac{3}{4} - 2\frac{1}{3} = 2\frac{5}{12}$	$7\frac{1}{8} - 3\frac{5}{6} = 3\frac{7}{24}$
5.	$\frac{1}{5} \times \frac{5}{7} = \frac{1}{7}$	$\frac{1}{3} \times \frac{6}{7} = \frac{2}{7}$	$1\frac{3}{5} \times \frac{2}{3} = 1\frac{1}{15}$	$2\frac{2}{5} \times 4\frac{3}{8} = 10\frac{1}{2}$
6.	$\frac{4}{5} \div \frac{5}{4} = \frac{16}{25}$	$3\frac{1}{3} \div 4\frac{3}{8} = \frac{16}{21}$	$6 \div \frac{3}{4} = 8$	$6\frac{1}{4} \div 8\frac{1}{3} = \frac{3}{4}$

Prerequisite Skills: greatest common factor

Lesson 1 Fractions and Mixed Numerals

Lesson Focus: simplifying fractions and mixed numerals
Possible Score: 15
Time Frame: 5–10 minutes

Study how to change a fraction or mixed numeral to simplest form.

Factors of 6
1, 2, 3, 6

Factors of 15
1, 3, 5, 15

$\dfrac{6}{15} = \dfrac{6 \div 3}{15 \div 3}$ Divide 6 and 15 by their greatest common factor.
$= \dfrac{2}{5}$

Factors of 8
1, 2, 4, 8

Factors of 10
1, 2, 5, 10

$7\dfrac{8}{10} = 7 + \dfrac{8}{10}$

$= 7 + \dfrac{8 \div 2}{10 \div 2}$ Divide 8 and 10 by their greatest common factor.

$= 7 + \dfrac{4}{5}$ or $7\dfrac{4}{5}$

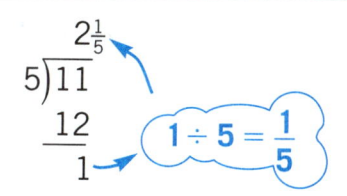

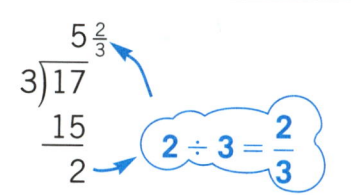

$5\overline{)11} \; 2\tfrac{1}{5}$
$\underline{12}$
1 $1 \div 5 = \dfrac{1}{5}$

$3\overline{)17} \; 5\tfrac{2}{3}$
$\underline{15}$
2 $2 \div 3 = \dfrac{2}{3}$

CHAPTER 2

Write each of the following in simplest form.

	a	b	c
1.	$\dfrac{9}{12} = \dfrac{3}{4}$	$2\dfrac{2}{4} = 2\dfrac{1}{2}$	$\dfrac{13}{5} = 2\dfrac{3}{5}$
2.	$\dfrac{2}{8} = \dfrac{1}{4}$	$3\dfrac{6}{10} = 3\dfrac{3}{5}$	$\dfrac{19}{3} = 6\dfrac{1}{3}$
3.	$\dfrac{10}{16} = \dfrac{5}{8}$	$7\dfrac{8}{12} = 7\dfrac{2}{3}$	$\dfrac{17}{2} = 8\dfrac{1}{2}$
4.	$\dfrac{18}{36} = \dfrac{1}{2}$	$5\dfrac{15}{20} = 5\dfrac{3}{4}$	$\dfrac{12}{8} = 1\dfrac{1}{2}$
5.	$\dfrac{15}{45} = \dfrac{1}{3}$	$2\dfrac{12}{28} = 2\dfrac{3}{7}$	$\dfrac{16}{10} = 1\dfrac{3}{5}$

Number Correct   LESSON FOLLOW-UP AND ERROR ANALYSIS
12–15: Have students make up five fractions and five mixed numerals and write them in simplest form.
9–11: Ask students to explain when a fraction is in simplest form. (It has no common factors except 1.)
Less than 9: Have students explain how to find common factors and greatest common factor before they redo incorrect answers.

Lesson 2 Renaming Numbers

$\dfrac{3}{4} = \dfrac{\square}{8}$ Multiply both the numerator and the denominator by the same number.

$= \dfrac{3 \times 2}{4 \times 2}$

$= \dfrac{6}{8}$ Choose 2 so the new denominator is 8.

$4 = \dfrac{\square}{8}$ Name the whole number as a fraction whose denominator is 1.

$\dfrac{4}{1} = \dfrac{4 \times 8}{1 \times 8}$

$= \dfrac{32}{8}$ Choose 8 so the new denominator is 8.

$9\dfrac{2}{3} = \dfrac{\square}{3}$

$9\dfrac{2}{3} = \dfrac{(3 \times 9) + 2}{3}$ Multiply the whole number by the denominator and add the numerator.

Use the same denominator.

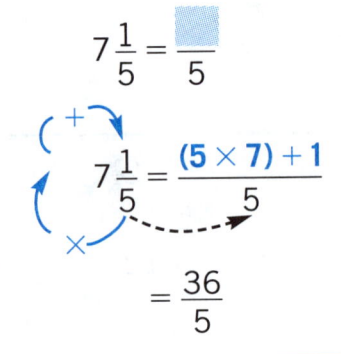

$= \dfrac{29}{3}$

$7\dfrac{1}{5} = \dfrac{\square}{5}$

$7\dfrac{1}{5} = \dfrac{(5 \times 7) + 1}{5}$

$= \dfrac{36}{5}$

Rename.

	a	b	c	d
1.	$\dfrac{1}{2} = \dfrac{3}{6}$	$\dfrac{2}{5} = \dfrac{6}{15}$	$\dfrac{3}{8} = \dfrac{6}{16}$	$\dfrac{5}{6} = \dfrac{10}{12}$
2.	$6 = \dfrac{24}{4}$	$3 = \dfrac{30}{10}$	$7 = \dfrac{21}{3}$	$5 = \dfrac{10}{2}$
3.	$4\dfrac{1}{2} = \dfrac{9}{2}$	$6\dfrac{3}{4} = \dfrac{27}{4}$	$2\dfrac{7}{10} = \dfrac{27}{10}$	$3\dfrac{5}{6} = \dfrac{23}{6}$
4.	$2 = \dfrac{24}{12}$	$\dfrac{9}{10} = \dfrac{45}{50}$	$1\dfrac{7}{8} = \dfrac{15}{8}$	$\dfrac{4}{5} = \dfrac{8}{10}$

Prerequisite Skills: greatest common factor, least common denominator

Lesson Focus: adding and subtracting fractions
Possible Score: 23
Time Frame: 25–30 minutes

Lesson 3 Adding and Subtracting Fractions

To add or subtract fractions having different denominators, rename either or both fractions so they have the same denominator. Then add or subtract.

$$\frac{4}{5} = \frac{8}{10}$$
$$+\frac{1}{2} = +\frac{5}{10}$$
$$\overline{\frac{13}{10}} = 1\frac{3}{10}$$

$$\frac{10}{12} = \frac{10}{12}$$
$$-\frac{1}{3} = -\frac{4}{12}$$
$$\overline{\frac{6}{12}} = \frac{1}{2}$$

CHAPTER 2

Write each sum or difference in simplest form.

	a	b	c	d
1.	$\frac{2}{5}$ $+\frac{2}{5}$ $\overline{\frac{4}{5}}$	$\frac{5}{8}$ $+\frac{1}{8}$ $\overline{\frac{3}{4}}$	$\frac{8}{9}$ $-\frac{7}{9}$ $\overline{\frac{1}{9}}$	$\frac{7}{8}$ $-\frac{3}{8}$ $\overline{\frac{1}{2}}$
2.	$\frac{11}{12}$ $-\frac{3}{4}$ $\overline{\frac{1}{6}}$	$\frac{13}{15}$ $-\frac{2}{3}$ $\overline{\frac{1}{5}}$	$\frac{1}{2}$ $+\frac{1}{4}$ $\overline{\frac{3}{4}}$	$\frac{1}{4}$ $+\frac{1}{5}$ $\overline{\frac{9}{20}}$
3.	$\frac{10}{12}$ $-\frac{1}{3}$ $\overline{\frac{1}{2}}$	$\frac{1}{3}$ $-\frac{1}{5}$ $\overline{\frac{2}{15}}$	$\frac{7}{10}$ $+\frac{1}{2}$ $\overline{1\frac{1}{5}}$	$\frac{4}{5}$ $+\frac{2}{4}$ $\overline{1\frac{3}{10}}$
4.	$\frac{4}{5}$ $-\frac{2}{4}$ $\overline{\frac{3}{10}}$	$\frac{1}{12}$ $+\frac{1}{3}$ $\overline{\frac{5}{12}}$	$\frac{3}{4}$ $-\frac{1}{2}$ $\overline{\frac{1}{4}}$	$\frac{5}{8}$ $-\frac{1}{6}$ $\overline{\frac{11}{24}}$

Number Correct

LESSON FOLLOW-UP AND ERROR ANALYSIS

20–23: Have students choose one problem and write a story problem it would solve.
15–19: Check to see if students tried to add both the numerators and the denominators rather than find a common denominator and rename.
Less than 15: Before they redo the pages, have students explain how to find a common denominator.

39

Lesson 3 Problem Solving

Solve. Write each answer in simplest form.

1. Kelly purchased $\frac{3}{8}$ of a round of gouda cheese and Noah purchased $\frac{3}{8}$ of a round of gouda cheese. How much gouda did they purchase?

   They purchased __$\frac{3}{4}$__ of a round of gouda.

2. Sara had $\frac{5}{6}$ of a box of raisins before she ate $\frac{1}{6}$ of a box. How much does she have left?

   She has __$\frac{2}{3}$__ of a box of raisins left.

3. One fourth of the house was painted yesterday and one half was painted today. How much of the house was painted on those two days?

   __$\frac{3}{4}$__ of the house was painted.

4. Brandon practised his clarinet each day. He spent $\frac{1}{8}$ of his time practising scales and $\frac{5}{8}$ of his time practising new pieces. What fraction of his time did he spend on scales and new pieces?

   __$\frac{3}{4}$__ of his time was spent on scales and new pieces.

5. Diane bought $\frac{3}{4}$ of a round of gouda cheese and $\frac{1}{2}$ of a round of mozzarella cheese. How much more gouda cheese than mozzarella cheese did she buy?

   She bought __$\frac{1}{4}$__ of a round more gouda cheese.

6. Eric spent $\frac{1}{4}$ h eating breakfast and $\frac{1}{3}$ h showering and getting dressed. How long did he spend getting ready altogether?

   Eric spent __$\frac{7}{12}$__ h getting ready.

7. Dan power-walked for $\frac{3}{4}$ h and jogged for $\frac{1}{3}$ h. How long did he exercise in all?

   Dan exercised for __$1\frac{1}{12}$__ h.

CHAPTER 2
Fractions

Lesson 3
Adding and Subtracting Fractions

Prerequisite Skills: greatest common factor, least common denominator

Lesson Focus: adding and subtracting mixed numerals
Possible Score: 24
Time Frame: 25–30 minutes

Lesson 4 Adding and Subtracting Mixed Numerals

To add or subtract mixed numerals having different denominators, rename either or both fractions so they have the same denominator. Then add or subtract.

$$2\frac{1}{2} = 2\frac{4}{8}$$
$$+3\frac{5}{8} = +3\frac{5}{8}$$
$$5\frac{9}{8} = 6\frac{1}{8}$$

$$4\frac{2}{3} = 4\frac{8}{12}$$
$$-1\frac{2}{4} = -1\frac{6}{12}$$
$$3\frac{2}{12} = 3\frac{1}{6}$$

Write each sum or difference in simplest form.

	a	b	c	d
1.	$1\frac{7}{10}$ $-\frac{4}{10}$ $1\frac{3}{10}$	$1\frac{4}{7}$ $+\frac{2}{7}$ $1\frac{6}{7}$	$2\frac{4}{7}$ $-1\frac{4}{7}$ 1	$2\frac{5}{6}$ $+3\frac{2}{6}$ $6\frac{1}{6}$
2.	$2\frac{1}{3}$ $4\frac{1}{6}$ $+5\frac{1}{4}$ $11\frac{3}{4}$	$\frac{2}{3}$ $\frac{1}{4}$ $+\frac{1}{2}$ $1\frac{5}{12}$	$2\frac{1}{2}$ $4\frac{1}{5}$ $+2\frac{3}{4}$ $9\frac{9}{20}$	$8\frac{3}{4}$ $-2\frac{11}{12}$ $5\frac{5}{6}$
3.	$3\frac{3}{4}$ $+1\frac{1}{2}$ $5\frac{1}{4}$	$8\frac{7}{12}$ $-2\frac{1}{6}$ $6\frac{5}{12}$	$4\frac{1}{6}$ $+1\frac{1}{2}$ $5\frac{2}{3}$	$9\frac{3}{4}$ $-4\frac{1}{6}$ $5\frac{7}{12}$
4.	$4\frac{3}{8}$ $+1\frac{1}{2}$ $5\frac{7}{8}$	$8\frac{1}{2}$ $-1\frac{1}{6}$ $7\frac{1}{3}$	$3\frac{2}{5}$ $+1\frac{7}{10}$ $5\frac{1}{10}$	$9\frac{1}{2}$ $-2\frac{1}{5}$ $7\frac{3}{10}$

Number Correct

LESSON FOLLOW-UP AND ERROR ANALYSIS

20–24: Have students choose one problem and write a story problem it would solve.
15–19: Check to see if students tried to add both the numerators and the denominators rather than find a common denominator and rename.
Less than 15: Before they redo the pages, have students explain how to find a common denominator.

Lesson 4 Problem Solving

Solve. Write each answer in simplest form.

1. The distance from Chad's house to Ryan's house is $14\frac{1}{2}$ blocks. The distance from Chad's house to Jeff's house is $7\frac{3}{4}$ blocks. How much farther is it from Chad's to Ryan's than from Chad's to Jeff's?

   It is ____$6\frac{3}{4}$____ blocks farther.

2. What is the combined distance from problem 1?

   The combined distance is ____$22\frac{1}{4}$____ blocks.

3. Yesterday Anne spent $5\frac{1}{2}$ h in school, $1\frac{3}{4}$ h playing, and $1\frac{1}{4}$ h doing homework. How much time did she spend on these activities?

   She spent ____$8\frac{1}{2}$____ h.

4. It is $6\frac{1}{4}$ blocks to the beach and $4\frac{1}{2}$ blocks to the ballpark. How much closer is it to the ballpark than to the beach?

   It is ____$1\frac{3}{4}$____ blocks closer to the ballpark.

5. Lauren can run $14\frac{7}{12}$ laps around the track in 1 h. Adam can run $12\frac{3}{4}$ laps in 1 h. How much farther can Lauren run?

   Lauren can run ____$1\frac{5}{6}$____ laps farther.

6. What is the combined distance in problem 5?

   The combined distance is ____$27\frac{1}{3}$____ laps.

7. Joan likes to volunteer. Last week she spent $4\frac{2}{3}$ h at a library and $3\frac{1}{4}$ h helping at a school. How much time did she spend last week volunteering?

   Joan spent ____$7\frac{11}{12}$____ h volunteering.

8. Carol picked $3\frac{1}{2}$ buckets of strawberries. Each bucket holds $2\frac{1}{2}$ L. How many litres of strawberries did she pick?

   Carol picked ____$8\frac{3}{4}$____ L of strawberries.

CHAPTER 2
Fractions

Lesson 4
Adding and Subtracting Mixed Numerals

Lesson 5 Multiplication

$$\frac{5}{6} \times \frac{1}{3} = \frac{5 \times 1}{6 \times 3}$$ Multiply numerators. Multiply denominators.

$$= \frac{5}{18}$$

$$5 \times \frac{3}{4} \times \frac{1}{2} = \frac{5 \times 3 \times 1}{1 \times 4 \times 2}$$

$$= \frac{15}{8} = 1\frac{7}{8}$$

$$4\frac{1}{2} \times 5\frac{2}{3} \times 1\frac{3}{5} = \frac{\overset{3}{\cancel{9}}}{2} \times \frac{17}{\underset{1}{\cancel{3}}} \times \frac{8}{5}$$ Divide a numerator and a denominator by a common factor.

$$= \frac{\overset{3}{\cancel{9}}}{\underset{1}{\cancel{2}}} \times \frac{17}{\underset{1}{\cancel{3}}} \times \frac{\overset{4}{\cancel{8}}}{5}$$

$$= \frac{3 \times 17 \times 4}{1 \times 1 \times 5}$$

$$= \frac{204}{5} \text{ or } 40\frac{4}{5}$$

Write each product in simplest form.

       a   b   c   d

1. $\frac{3}{5} \times \frac{1}{4} = \frac{3}{20}$ $\frac{2}{3} \times \frac{4}{5} = \frac{8}{15}$ $\frac{1}{3} \times \frac{1}{5} \times \frac{1}{2} = \frac{1}{30}$ $\frac{3}{4} \times \frac{1}{2} \times \frac{3}{5} = \frac{9}{40}$

2. $\frac{3}{10} \times \frac{2}{5} = \frac{3}{25}$ $\frac{2}{3} \times \frac{7}{8} = \frac{7}{12}$ $\frac{5}{8} \times \frac{3}{10} \times \frac{5}{6} = \frac{5}{32}$ $\frac{7}{12} \times \frac{6}{7} \times \frac{2}{3} = \frac{1}{3}$

3. $\frac{3}{4} \times \frac{8}{9} = \frac{2}{3}$ $\frac{3}{5} \times \frac{5}{12} = \frac{1}{4}$ $\frac{11}{12} \times \frac{3}{4} \times \frac{8}{11} = \frac{1}{2}$ $\frac{7}{16} \times \frac{4}{5} \times \frac{5}{8} = \frac{7}{32}$

4. $1\frac{1}{2} \times \frac{5}{7} = 1\frac{1}{14}$ $2\frac{1}{3} \times \frac{5}{12} = \frac{35}{36}$ $1\frac{7}{8} \times \frac{2}{3} \times \frac{3}{4} = \frac{15}{16}$ $\frac{1}{2} \times 3\frac{1}{3} \times \frac{5}{6} = 1\frac{7}{18}$

5. $\frac{3}{5} \times 3\frac{1}{3} = 2$ $\frac{5}{6} \times 3\frac{1}{2} = 2\frac{11}{12}$ $1\frac{1}{2} \times 2\frac{1}{3} \times \frac{1}{4} = \frac{7}{8}$ $\frac{2}{3} \times 2\frac{1}{2} \times 1\frac{3}{4} = 2\frac{11}{12}$

6. $2\frac{2}{3} \times 1\frac{3}{4} = 4\frac{2}{3}$ $5\frac{1}{2} \times 3\frac{1}{6} = 17\frac{5}{12}$ $1\frac{2}{3} \times 3\frac{1}{2} \times 2\frac{1}{4} = 13\frac{1}{8}$ $6\frac{1}{2} \times 2\frac{1}{3} \times 1\frac{3}{5} = 24\frac{4}{15}$

Lesson 5 Problem Solving

Solve each problem. Write each answer in simplest form.

1. The tank on Mr. Kent's lawn mower will hold $\frac{3}{4}$ of a can of gasoline. Suppose the tank is $\frac{1}{2}$ full. How much gasoline is in the tank?

    $\underline{\frac{3}{8}}$ of a can of gasoline is in the tank.

2. Band practice lasted $1\frac{1}{4}$ h. Two thirds of the time was spent marching. How much time was spent marching?

    $\underline{\frac{5}{6}}$ h was spent marching.

3. An industrial machine can make $2\frac{1}{2}$ engines an hour. How many engines will be made in $1\frac{3}{4}$ h?

    $\underline{4\frac{3}{8}}$ engines will be made.

4. In Joshua's class, $\frac{1}{4}$ of the students have blond hair. $2\frac{1}{2}$ times that fraction have brown hair. What fraction of the class has brown hair?

    $\underline{\frac{5}{8}}$ of the class has brown hair.

5. In problem 4, what fraction of the class has neither brown nor blond hair?

    $\underline{\frac{1}{8}}$ of the class has neither brown nor blond hair.

6. Steve can run $5\frac{1}{2}$ laps of the track in 12 min. His younger sister can run $\frac{3}{4}$ as far in the same time. How far can Steve's sister run in 12 min?

    She can run $\underline{4\frac{1}{8}}$ laps in 12 min.

7. In problem 6, Steve ran six 12-min runs during gym class last month. How many laps did he run in all?

    He ran $\underline{33}$ laps in all.

8. The soccer team practises $2\frac{1}{2}$ h on each of five weekday afternoons. How many hours does the team practise each week?

    The team practises $\underline{12\frac{1}{2}}$ h each week.

Prerequisite Skills: multiplication of fractions

Lesson Focus: dividing fractions and mixed numerals
Possible Score: 32
Time Frame: 20–25 minutes

Lesson 6   Division

reciprocals
$$\frac{4}{7} \times \frac{7}{4} = 1$$

reciprocals
$$5 \times \frac{1}{5} = 1$$

reciprocals
$$2\frac{3}{4} \times \frac{4}{11} = 1$$

If two numbers are reciprocals, their product is ___1___.

CHAPTER 2

Multiply by the reciprocal.

To divide any number, multiply by its reciprocal.

$$\frac{3}{8} \div \frac{4}{5} = \frac{3}{8} \times \frac{5}{4}$$
$$= \frac{15}{32}$$

$$6\frac{1}{2} \div \frac{3}{4} = \frac{13}{2} \times \underline{\frac{4}{3}}$$
$$= \underline{8\frac{2}{3}}$$

$$\frac{2}{3} \div 1\frac{1}{2} = \frac{2}{3} \times \underline{\frac{2}{3}}$$
$$= \underline{\frac{4}{9}}$$

Write each quotient in simplest form.

	a	b	c	d
1.	$\frac{1}{2} \div \frac{3}{4} = \frac{2}{3}$	$\frac{7}{8} \div \frac{2}{3} = 1\frac{5}{16}$	$\frac{4}{5} \div \frac{4}{7} = 1\frac{2}{5}$	$\frac{5}{8} \div \frac{7}{10} = \frac{25}{28}$
2.	$\frac{4}{5} \div 4 = \frac{1}{5}$	$8 \div \frac{2}{3} = 12$	$\frac{9}{10} \div 3 = \frac{3}{10}$	$9 \div \frac{3}{5} = 15$
3.	$1\frac{1}{2} \div \frac{2}{3} = 2\frac{1}{4}$	$3\frac{1}{3} \div \frac{5}{6} = 4$	$2\frac{1}{2} \div \frac{7}{10} = 3\frac{4}{7}$	$4\frac{1}{3} \div \frac{7}{8} = 4\frac{20}{21}$
4.	$\frac{7}{8} \div 2\frac{1}{2} = \frac{7}{20}$	$\frac{7}{8} \div 1\frac{3}{4} = \frac{1}{2}$	$\frac{5}{6} \div 2\frac{2}{3} = \frac{5}{16}$	$\frac{3}{4} \div 1\frac{4}{5} = \frac{5}{12}$
5.	$2 \div 1\frac{7}{8} = 1\frac{1}{15}$	$4\frac{1}{2} \div 3 = 1\frac{1}{2}$	$6 \div 1\frac{1}{8} = 5\frac{1}{3}$	$3\frac{1}{3} \div 5 = \frac{2}{3}$
6.	$1\frac{1}{2} \div 2\frac{2}{3} = \frac{9}{16}$	$3\frac{1}{4} \div 1\frac{7}{8} = 1\frac{11}{15}$	$4\frac{1}{2} \div 1\frac{1}{2} = 3$	$5\frac{1}{4} \div 1\frac{1}{8} = 4\frac{2}{3}$

Number Correct

LESSON FOLLOW-UP AND ERROR ANALYSIS

27–32: Have students choose one problem and write a story problem it would solve.
20–26: Check to see if students wrote the reciprocal of the first fraction rather than the reciprocal of the divisor (the second).
Less than 20: If the divisor is a mixed numeral, check to see if students forgot to write its reciprocal after changing it to a fraction.

45

Lesson 6 Problem Solving

Solve. Write each answer in simplest form.

1. Football practice lasted $2\frac{1}{2}$ h. An equal amount of time was spent on blocking, tackling, passing, and kicking. How much time was spent on each?

   __$\frac{5}{8}$__ h was spent on each.

2. Three-fourths of a can of gasoline was poured into four containers. Each container held the same amount. How much gasoline was poured into each container?

   __$\frac{3}{16}$__ of a can was poured into each container.

3. Suppose a motorboat uses 1 L of fuel in $1\frac{1}{4}$ h. How many litres of fuel will the boat use in 10 h?

   The boat will use __$12\frac{1}{2}$__ L of fuel in 10 h.

4. Due to a heavy rain, the water level on a lake was rising 1 cm every $\frac{2}{3}$ h. How much will the water level rise in $1\frac{1}{3}$ h?

   The water level will rise __2__ cm in $\frac{3}{4}$ h.

5. It takes Alyssa $\frac{3}{4}$ h to walk to school and back. How long does it take her to walk one way?

   It takes her __$\frac{1}{2}$__ h.

6. It takes $\frac{1}{4}$ h for 1 L of a chemical to be filtered. How many litres can be filtered in $2\frac{1}{2}$ h?

   __10__ L can be filtered.

7. In problem 6, how many litres can be filtered in $3\frac{3}{4}$ h?

   __15__ L can be filtered.

8. A flight leaves the airport every $1\frac{1}{4}$ min. How many flights will leave each hour?

   __48__ flights will leave each hour.

For further evaluation, copy the Chapter Test on page 266.
For maintaining skills, use the Cumulative Review on pages 215–216.

Possible Score: 24
Time Frame: 20–25 minutes

CHAPTER 2   PRACTICE TEST
Fractions

Write each fraction in simplest form.

	a	b	c	d
1.	$\frac{6}{15} = \frac{2}{5}$	$\frac{14}{85} = \frac{7}{9}$	$\frac{3}{12} = \frac{1}{4}$	$\frac{3}{30} = \frac{1}{10}$

Rename.

2. $\frac{3}{4} = \frac{15}{20}$ $\frac{7}{8} = \frac{21}{24}$ $5 = \frac{15}{3}$ $3\frac{1}{8} = \frac{25}{8}$

Write each sum, difference, product, or quotient in simplest form.

3. $\frac{1}{7} + \frac{2}{7} = \frac{3}{7}$ $2\frac{4}{5} + 3\frac{9}{10} = 6\frac{7}{10}$ $4\frac{2}{3} + 3\frac{1}{2} = 8\frac{1}{6}$ $3\frac{5}{6} + 5\frac{7}{8} + 1\frac{3}{4} = 11\frac{11}{24}$

4. $\frac{5}{9} - \frac{2}{9} = \frac{1}{3}$ $2\frac{1}{2} - \frac{1}{4} = 2\frac{1}{4}$ $7\frac{2}{3} - 1\frac{2}{5} = 6\frac{4}{15}$ $4\frac{1}{5} - 2\frac{7}{20} = 1\frac{17}{20}$

5. $\frac{2}{3} \times \frac{3}{4} = \frac{1}{2}$ $\frac{5}{7} \times \frac{3}{10} = \frac{3}{14}$ $6\frac{2}{3} \times \frac{1}{5} = 1\frac{1}{3}$ $3\frac{5}{9} \times 3\frac{3}{8} = 12$

6. $\frac{3}{8} \div \frac{4}{5} = \frac{15}{32}$ $4 \div \frac{1}{2} = 8$ $2\frac{2}{3} \div 5\frac{5}{7} = \frac{7}{15}$ $8\frac{5}{9} \div 1\frac{1}{10} = 7\frac{7}{9}$

Number Correct

LESSON FOLLOW-UP AND ERROR ANALYSIS

21–24: Have students reverse the division problems and solve for a new answer.
17–20: Have students explain how to change to a common denominator when adding and subtracting fractions.
Less than 17: Ask students to explain how to put fractions into simplest form.

CHAPTER 3   PRETEST
Pre-Algebra Equations

Solve each equation.

	a	b	c
1.	$2x = 12$ $x = 6$	$5y = 25$ $y = 5$	$6z = 96$ $z = 16$
2.	$\dfrac{d}{3} = 5$ $d = 15$	$\dfrac{e}{6} = 7$ $e = 42$	$\dfrac{f}{4} = 13$ $f = 52$
3.	$r + 7 = 12$ $r = 5$	$s + 3 = 25$ $s = 22$	$t + 12 = 20$ $t = 8$
4.	$g - 4 = 8$ $g = 12$	$h - 5 = 12$ $h = 17$	$j - 15 = 15$ $j = 30$
5.	$72 = 4m$ $m = 18$	$8n = 28 + 28$ $n = 7$	$9 = \dfrac{p}{5}$ $p = 45$
6.	$18 = a + 6$ $a = 12$	$b + 4 = 12 + 3$ $b = 11$	$13 = c - 4$ $c = 17$
7.	$u - 12 = 23 + 7$ $u = 42$	$v + 8 = 8$ $v = 0$	$w - 8 = 8$ $w = 16$

CHAPTER 3
Pre-Algebra Equations

48

CHAPTER 3 PRETEST

Prerequisite Skills: reading and understanding operation symbols

Lesson Focus: writing number phrases and simplifying expressions
Possible Score: 29
Time Frame: 10–15 minutes

Lesson 1 Number Phrases PRE-ALGEBRA

Letters like $a, b, n, x,$ and so on can be used to stand for numbers.

word phrase	number phrase	
Some number a added to 7	$7 + a$	If $a = 5$, then $7 + a = 7 + $ __5__ or __12__.
Some number b decreased by 4	$b - 4$	If $b = 6$, then $b - 4 = $ __6__ $- 4$ or __2__.
The product of 3 and some number n	$3 \times n$ or $3n$	If $n = 2$, then $3n = 3 \times$ __2__ or __6__.
15 divided by some number x	$\frac{15}{x}$ or $15 \div x$	If $x = 3$, then $15 \div x = 15 \div$ __3__ or __5__.

Write a number phrases for each of the following.

              a   b

1. Some number c subtracted from 11   __$11 - c$__   Five more than the number b __$b + 5$__

2. A certain number d increased by 12   __$d + 12$__   Some number t divided by 2   __$\frac{t}{2}$ or $t \div 2$__

3. The product of some number n and 8   __$8 \times n$ or $8n$__   Four less than some number x __$x - 4$__

4. Eight divided by some number j __$\frac{8}{j}$ or $8 \div j$__   The product of $\frac{1}{2}$ and y __$\frac{1}{2} \times y$ or $\frac{1}{2}y$__

Complete the following.

5. If $r = 3$, then $12 - r = $ __12__ $-$ __3__ or __9__.

6. If $s = 9$, then $7 + s = $ __7__ $+$ __9__ or __16__.

7. If $t = 3$, then $48 \div t = $ __48__ $\div$ __3__ or __16__.

8. If $u = 72$, then $\frac{1}{4}u = $ __$\frac{1}{4}$__ $\times$ __72__ or __18__.

9. If $v = 12$, then $4v = $ __4__ $\times$ __12__ or __48__.

10. If $w = 6$, then $w - 6 = $ __6__ $-$ __6__ or __0__.

11. If $x = 24$, then $\frac{x}{3} = $ __24__ $\div$ __3__ or __8__.

Number Correct

LESSON FOLLOW-UP AND ERROR ANALYSIS

25–29: Have students give a value to each variable in problems 1–4 and simplify the expression.
19–24: Ask students to explain how to simplify each expression in problems 5–11 before they redo the page.
Less than 19: Have students explain how to substitute the given value for the letter and then perform the operation indicated.

Prerequisite Skills: writing number phrases

Lesson Focus: writing and solving equations
Possible Score: 22
Time Frame: 10–15 minutes

Lesson 2   Writing Equations   PRE-ALGEBRA

An **equation** like $x + 2 = 9$ states that both $x + 2$ and 9 name the same number.

sentence	equation
The sum of some number and 2 is 9.	$x + 2 = 9$
Twelve divided by some number is 6.	$12 \div x = 6$ or $\frac{12}{x} = 6$
Seven decreased by some number is 5.	$\underline{7} - \underline{x} = \underline{5}$

$x = \underline{7}$ because $\underline{7} + 2 = 9$.

$x = \underline{2}$ because $12 \div \underline{2} = 6$.

$x = \underline{2}$ because $\underline{7} - \underline{2} = \underline{5}$.

Write an equation for each of the following.

            a   *b*

1. Some number a increased by 6 is 20. $\underline{a + 6 = 20}$ A number p decreased by 7 is 15. $\underline{p - 7 = 15}$

2. Twenty divided by some number y is 4. $\underline{20 \div y = 4 \text{ or } \frac{20}{y} = 4}$ One half of a number t is equal to 14. $\underline{\frac{1}{2}t = 14}$

3. The sum of a certain number b and 7 is 14. $\underline{b + 7 = 14}$ Twelve more than some number v is 18. $\underline{v + 12 = 18}$

4. The product of 2 and some number n is 12. $\underline{2n = 12}$ Some number d divided by 3 is equal to 14. $\underline{d \div 3 = 14 \text{ or } \frac{d}{3} = 14}$

Complete the following.

5. $x + 8 = 12$ $x = \underline{4}$ because $\underline{4} + 8 = 12$.

6. $9r = 45$ $r = \underline{5}$ because $9 \times \underline{5} = 45$.

7. $6 = \frac{1}{2}d$ $d = \underline{12}$ because $6 = \frac{1}{2} \times \underline{12}$.

8. $b - 6 = 8$ $b = \underline{14}$ because $\underline{14} - 6 = 8$.

9. $w \div 3 = 2$ $w = \underline{6}$ because $\underline{6} \div 3 = 2$.

10. $e + 16 = 18$ $e = \underline{2}$ because $\underline{2} + 16 = 18$.

11. $35 = 27 + c$ $c = \underline{8}$ because $35 = 27 + \underline{8}$.

Number Correct

50

LESSON FOLLOW-UP AND ERROR ANALYSIS

19–22: Have students express the equations in problems 5–11 in words.
14–18: Discuss that any letter can be used to represent the unknown quantity in an equation.
Less than 14: Ask students to explain what an equation is. (A number sentence in which a letter stands for the number they must find.)

Prerequisite Skills: division facts

Lesson Focus: solving equations using division
Possible Score: 28
Time Frame: 15–20 minutes

Lesson 3 Solving Equations (division)   PRE-ALGEBRA

To solve an equation, you can divide both sides of it by the same non-zero number.

$$4m = 52$$
$$\frac{4m}{4} = \frac{52}{4}$$
$$\frac{\cancel{4}m}{\cancel{4}} = \frac{\cancel{52}^{13}}{\cancel{4}}$$
$$m = 13$$

Check
$$4m = 52$$
$$4 \times 13 = 52$$
$$52 = 52$$

$$13y = 100 - 9$$
$$\frac{13y}{13} = \frac{91}{13}$$
$$\frac{\cancel{13}y}{\cancel{13}} = \frac{\cancel{91}^{7}}{\cancel{13}}$$
$$y = \underline{\quad 7 \quad}$$

To change $4m$ to m, both sides were divided by __4__.

To change $13y$ to y, both sides were divided by __13__.

Solve each equation.

         a   b   c

1. $3w = 12$ $w = 4$ $3b = 51$ $b = 17$ $8m = 100 - 4$ $m = 12$

2. $72 = 2a$ $a = 36$ $54 = 3c$ $c = 18$ $96 - 20 = 4r$ $r = 19$

3. $6e = 84$ $e = 14$ $25s = 75$ $s = 3$ $4d = 75 - 7$ $d = 17$

4. $14x = 42$ $x = 3$ $75 = 15m$ $m = 5$ $3y = 100 - 28$ $y = 24$

Number Correct   LESSON FOLLOW-UP AND ERROR ANALYSIS

24–28: Have students check their answers as shown in the example.
18–23: Have students explain what a number and a letter together with no operation sign between them means. (to multiply)
Less than 18: Ask students to explain how to solve an equation. Check that they perform the same operation to both sides of the equal sign.

Lesson 3 Problem Solving   PRE-ALGEBRA

Study the first problem. Solve problems **2–5** in a similar way.

1. John bought several model kits for $9 each. He spent $36. How many kits did he buy?
   If x stands for the number of kits he bought, then ___9x___ stands for the cost of all the kits.

   Equation: ___9x = 36___ $x =$ ___4___

   John bought ___4___ model kits.

2. A train travels 70 km/h. How long does it take for this train to make a 630-km trip? If x stands for the number of hours for the trip, then ___70x___ stands for the total number of kilometres.

   Equation: ___70x = 630___ $x =$ ___9___

   It takes ___9___ h to make the trip.

3. 3 kg of apples cost $2.34 (234¢). How much does 1 kg of apples cost?
   If x stands for the cost of 1 kg, then ___3x___ stands for the cost of 3 kg.

   Equation: ___3x = 234___ $x =$ ___78___

   1 kg of apples costs ___78___ ¢.

4. Eight loaves of bread cost $7.84 (784¢). How much does one loaf of bread cost?
   If x stands for the cost of one loaf, then ___8x___ stands for the cost of eight loaves.

   Equation: ___8x = 784___ $x =$ ___98___

   One loaf of bread costs ___98___ ¢.

5. A board is 84 cm long. How many metres long is this board?
   If x stands for the number of metres, then ___12x___ stands for the number of centimetres.

   Equation: ___12x = 84___ $x =$ ___7___

   The board is ___7___ m long.

CHAPTER 3
Pre-Algebra Equations

Lesson 3
Solving Equations (division)

Prerequisite Skills: multiplication facts

Lesson Focus: solving equations using multiplication
Possible Score: 27
Time Frame: 10–15 minutes

Lesson 4 Solving Equations (multiplication) PRE-ALGEBRA

To solve an equation, you can multiply both sides of it by the same number.

$\dfrac{a}{5} = 35$

$5 \times \dfrac{a}{5} = 5 \times 35$

$\dfrac{\cancel{5} \times a}{\cancel{5}} = 175$

$a = 175$

Check
$\dfrac{a}{5} = 35$
$\dfrac{175}{5} = 35$
$35 = 35$

$r \div 3 = 11 + 34$
$(r \div 3) \times 3 = 45 \times \underline{3}$
$r = \underline{135}$

To change $\dfrac{a}{5}$ to a, both sides were multiplied by __5__.

To change $r \div 3$ to r, both sides were multiplied by __3__.

Solve each equation.

         a   b   c

1. $\dfrac{a}{8} = 7$ $a = 56$ $\dfrac{b}{13} = 9$ $b = 117$ $\dfrac{c}{4} = 6 + 12$ $c = 72$

2. $16 = \dfrac{r}{8}$ $r = 128$ $8 = s \div 7$ $s = 56$ $2 \times 9 = \dfrac{t}{5}$ $t = 90$

3. $g \div 17 = 9$ $g = 153$ $15 = \dfrac{h}{5}$ $h = 75$ $7 \times 6 = \dfrac{j}{3}$ $j = 126$

4. $\dfrac{m}{15} = 17$ $m = 255$ $23 = \dfrac{n}{28}$ $n = 644$ $p \div 19 = 3 \times 9$ $p = 513$

CHAPTER 3

Number Correct   **LESSON FOLLOW-UP AND ERROR ANALYSIS**

22–27: Have students check their answers as shown in the lesson.
17–21: Ask students to explain how $\dfrac{a}{8}$ and $a \div 8$ mean the same thing before they redo the pages.
Less than 17: Have students explain how to solve an equation. Check that they perform the same operation to both sides of the equal sign.

Lesson 4   Problem Solving   PRE-ALGEBRA

Study the first problem. Solve problems **2–5** in a similar way.

1. Joseph has $\frac{1}{4}$ the number of points he needs to win. He has 36 points. How many points does he need to win?
   If x stands for the number of points needed to win, then $\frac{1}{4}x$ or $\frac{x}{4}$ stands for the points he has now.
   Equation: $\frac{1}{4}x = 36$ or $\frac{x}{4} = 36$ $x =$ __144__
   Joseph needs __144__ points to win.

2. Mia has $\frac{1}{3}$ the number of points she needs to win. She has 48 points. How many points does she need to win?
   If x stands for the total number of points needed to win, then $\frac{x}{3}$ stands for the points she has now.
   Equation: $\frac{x}{3} = 48$ $x =$ __144__
   Mia needs __144__ points to win.

3. Three students are absent. This is $\frac{1}{6}$ of the entire class. How many students are in the class?
   If x stands for the total number of students, then $\frac{x}{6}$ stands for the number of students absent.
   Equation: $\frac{x}{6} = 3$ $x =$ __18__
   There are __18__ students in the class.

4. Alex drove 120 km and stopped for lunch. He had then travelled $\frac{1}{3}$ the total distance of his trip. What is the total distance of his trip?
   If x stands for the total distance, then $\frac{x}{3}$ stands for the distance he has already travelled.
   Equation: $\frac{x}{3} = 120$ $x =$ __360__
   The total distance of the trip is __360__ km.

5. Shea solved 12 problems. This was $\frac{1}{5}$ of all she has to solve. How many problems does she have to solve?
   Equation: $\frac{x}{5} = 12$ $x =$ __60__
   She has __60__ problems to solve in all.

CHAPTER 3
Pre-Algebra Equations

Lesson 4
Solving Equations (multiplication)

Prerequisite Skills: subtraction facts

Lesson Focus: solving equations using subtraction
Possible Score: 28
Time Frame: 15–20 minutes

Lesson 5 Solving Equations (subtraction)   PRE-ALGEBRA

To solve an equation, you can subtract the same number from both sides of it.

$v + 18 = 47$
$v + 18 - 18 = 47 - 18$
$v + 0 = 29$
$v = 29$

Check
$v + 18 = 47$
$29 + 18 = 47$
$47 = 47$

$c + 6 = 43 + 8$
$c + 6 - \underline{}6\underline{} = 51 - \underline{}6\underline{}$
$c + \underline{}0\underline{} = \underline{}45\underline{}$
$c = \underline{}45\underline{}$

To change $v + 18$ to v, __18__ was subtracted from both sides.

To change $c + 6$ to c, __6__ was subtracted from both sides.

CHAPTER 3

Solve each equation.

	a	b	c
1.	$d + 12 = 48$ $d = 36$	$36 + e = 84$ $e = 48$	$f + 14 = 18 + 18$ $f = 22$
2.	$38 = j + 13$ $j = 25$	$27 = 9 + h$ $h = 18$	$20 + 34 = 27 + l$ $l = 27$
3.	$12 + w = 76$ $w = 64$	$114 = x + 38$ $x = 76$	$300 - 30 = y + 50$ $y = 220$
4.	$200 + 50 = a + 212$ $a = 38$	$27 + b = 170 + 3$ $b = 146$	$100 - 2 = c + 43$ $c = 55$

Number Correct   LESSON FOLLOW-UP AND ERROR ANALYSIS
24–28: Have students circle the most difficult problems on the page and explain their choices.
18–23: Have students explain how to simplify an equation before they redo incorrect answers.
Less than 18: Have students explain how to solve an equation. Check that they perform the same operation to both sides of the equal sign.

55

Lesson 5 Problem Solving   PRE-ALGEBRA

Study the first problem. Solve problems **2–5** in a similar way.

1. A rectangle is 8 m longer than it is wide. If its length is 17 m, what is its width?
   If x stands for the number of metres wide, then ___$x + 8$___ stands for the number of metres long.

   Equation: ___$x + 8 = 17$___ $x =$ ___9___

   The width of the rectangle is ___9___ m.

2. A rectangle is 27 cm longer than it is wide. If its length is 45 cm, what is its width?
   If x stands for the number of centimetres wide, then ___$x + 27$___ stands for the number of centimetres long.

   Equation: ___$x + 27 = 45$___ $x =$ ___18___

   The width of the rectangle is ___18___ cm.

3. Maria's score of 94 is 8 points higher than Su-Lin's score. What is Su-Lin's score?
   If x stands for Su-Lin's score, then ___$x + 8$___ stands for Maria's score.

   Equation: ___$x + 8 = 94$___ $x =$ ___86___

   Su-Lin's score is ___86___.

4. The 17 men at work outnumber the women by 5. How many women are at work?
   If x stands for the number of women at work, then ___$x + 5$___ stands for the number of men at work.

   Equation: ___$x + 5 = 17$___ $x =$ ___12___

   There are ___12___ women at work.

5. The 48-min trip to work was 19 min longer than the trip home from work. How long did it take for the trip home?
   If x stands for the number of minutes for the trip home, then ___$x + 19$___ stands for the trip to work.

   Equation: ___$x + 19 = 48$___ $x =$ ___29___

   The trip home took ___29___ min.

CHAPTER 3
Pre-Algebra Equations

Lesson 5
Solving Equations (subtraction)

Prerequisite Skills: addition facts

Lesson Focus: solving equations using addition
Possible Score: 28
Time Frame: 15–20 minutes

Lesson 6 Solving Equations (addition)   PRE-ALGEBRA

To solve an equation, you can add the same number to both sides of it.

$t - 3 = 15$
$t - 3 + 3 = 15 + 3$
$t + 0 = 18$
$t = 18$

Check
$t - 3 = 15$
$18 - 3 = 15$
$15 = 15$

$b - 12 = 14 + 3$
$b - 12 + \underline{12} = 17 + \underline{12}$
$b + \underline{0} = \underline{29}$
$b = \underline{29}$

To change $t - 3$ to t, $\underline{3}$ was added to both sides.

To change $b - 12$ to b, $\underline{12}$ was added to both sides.

CHAPTER 3

Solve each equation.

             a   *b*   *c*

1. $b - 8 = 15$ $b = 23$ $x - 14 = 36$ $x = 50$ $c - 3 = 28 + 4$ $c = 35$

2. $42 = r - 12$ $r = 54$ $80 = e - 26$ $e = 106$ $20 + 11 = f - 14$ $f = 45$

3. $163 = a - 27$ $a = 190$ $9 \times 9 = m - 38$ $m = 119$ $t - 28 = 102$ $t = 130$

4. $117 = w - 83$ $w = 200$ $200 - 25 = g - 83$ $g = 258$ $h - 75 = 100 + 56$ $h = 231$

Number Correct

LESSON FOLLOW-UP AND ERROR ANALYSIS

24–28: Have students choose one problem on page 57 and write out the steps to solve it.
18–23: Have students explain how to simplify an equation before they redo incorrect answers.
Less than 18: Have students explain how to solve an equation. Check that they perform the same operation to both sides of the equal sign.

57

Lesson 6   Problem Solving   PRE-ALGEBRA

Study the first problem. Solve problems **2–5** in a similar way.

1. The temperature has fallen 12°C since noon. The present temperature is 17°C. What was the noon temperature?
   If x stands for the noon temperature, then ___$x - 12$___ stands for the present temperature.
   Equation: ___$x - 12 = 17$___ $x =$ ___29___
   The noon temperature was ___29___ °C.

2. The temperature has fallen 7°C since noon. The present temperature is 18°C. What was the noon temperature?
   If x stands for the noon temperature, then ___$x - 7$___ stands for the present temperature.
   Equation: ___$x - 7 = 18$___ $x =$ ___25___
   The noon temperature was ___25___ °C.

3. After selling 324 papers, Mr. Merk had 126 papers left. How many papers did he start with?
   If x stands for the number of papers he started with, then ___$x - 324$___ stands for the number left.
   Equation: ___$x - 324 = 126$___ $x =$ ___450___
   He had ___450___ papers to start with.

4. Andrew sold his football for $15.50 (1550¢). This was 95¢ less than the original cost. What was the original cost?
   If x stands for the original cost, then ___$x - 95$___ stands for the amount he sold the football for.
   Equation: ___$x - 95 = 1550$___ $x =$ ___1645___
   The original cost of the football was $___16.45___.

5. The width of a rectangle is 37 cm shorter than its length. The width is 75 cm. How long is the rectangle?
   If x stands for the measure of the length, then ___$x - 37$___ stands for the measure of the width.
   Equation: ___$x - 37 = 75$___ $x =$ ___112___
   The rectangle is ___112___ cm long.

CHAPTER 3
Pre-Algebra Equations

Lesson 6
Solving Equations (addition)

Prerequisite Skills: solving equations using all operations

Lesson Focus: solving equations
Possible Score: 33
Time Frame: 15–20 minutes

Lesson 7   Solving Equations Review   PRE-ALGEBRA

Solve each equation.

     a   b   c

1. $4b = 30 + 30$ $b = 15$ $13 + 26 = 3u$ $u = 13$ $7v = 42 + 42$ $v = 12$

2. $\dfrac{d}{5} = 100 - 40$ $d = 300$ $10 - 3 = \dfrac{y}{32}$ $y = 224$ $\dfrac{k}{37} = 10 - 8$ $k = 74$

3. $g + 27 = 49 - 4$ $g = 18$ $100 - 7 = 39 + x$ $x = 54$ $43 = n + 12$ $n = 31$

4. $p - 6 = 3 + 10$ $p = 19$ $56 - 3 = k - 42$ $k = 95$ $w - 39 = 90 + 3$ $w = 132$

5. $x + 16 = 33 + 12$ $x = 29$ $\dfrac{m}{14} = 14 - 4$ $m = 140$ $12 + n = 56 + 8$ $n = 52$

6. $7 \times 6 = 3t$ $t = 14$ $d + 291 = 400 + 26$ $d = 135$ $73 - 8 = n + 5$ $n = 60$

7. $4 + 8 = \dfrac{q}{12}$ $q = 144$ $g - 27 = 2 \times 45$ $g = 117$ $\dfrac{x}{15} = 20 - 5$ $x = 225$

CHAPTER 3

Number Correct   **LESSON FOLLOW-UP AND ERROR ANALYSIS**
28–33: Have students choose five equations and check their answers.
21–27: After simplifying, discuss that students perform the opposite of the indicated operation to solve each equation.
Less than 21: Before they redo page 60, have students underline the key words that indicate which operation they should use.

59

Lesson 7 Problem Solving   PRE-ALGEBRA

Safety Program	
Name	Points Earned
Tom	48
Susan	
Bob	
Mallory	
Al	

The bulletin-board chart was torn and some information is missing. Help complete the chart by using the information in the following problems. Write an equation for each problem. Solve the equation. Answer the problem.

1. Tom has earned three times as many points as Susan. How many points has Susan earned?

   Equation: __3x = 48__   x = __16__

   Susan has earned __16__ points.

2. Tom has one third the number of points that Bob has. How many points does Bob have?

   Equation: __$\frac{1}{3}x = 48$__   x = __144__

   Bob has __144__ points.

3. The number of points that Tom has is 27 less than the number of points that Mallory has. How many points does Mallory have?

   Equation: __x − 27 = 48__   x = __75__

   Mallory has __75__ points.

4. The number of points that Tom has earned is 27 more than the number of points that Al has earned. How many points has Al earned?

   Equation: __x + 27 = 48__   x = __21__

   Al has earned __21__ points.

CHAPTER 3
Pre-Algebra Equations

Lesson 7
Solving Equations Review

For further evaluation, copy the Chapter Test on page 267.
For maintaining skills, use the Cumulative Review on pages 217–218.

Possible Score: 20
Time Frame: 10–15 minutes

CHAPTER 3   PRACTICE TEST
Pre-Algebra Equations

Solve each equation.

	a	b	c
1.	$4m = 40$ $m = 10$	$90 = 6n$ $n = 15$	$42 - 20 = 2p$ $p = 11$
2.	$\frac{r}{6} = 7$ $r = 42$	$15 = \frac{s}{13}$ $s = 195$	$\frac{t}{2} = 40 + 3$ $t = 86$
3.	$a + 9 = 36$ $a = 27$	$27 = 6 + b$ $b = 21$	$14 + 20 = c + 4$ $c = 30$
4.	$x - 9 = 27$ $x = 36$	$36 = y - 14$ $y = 50$	$z - 6 = 30 + 12$ $z = 48$
5.	$42 + 18 = w + 20$ $w = 40$	$72 + 18 = \frac{x}{6}$ $x = 540$	$37 + 12 = y - 18$ $y = 67$
6.	$12d = 144$ $d = 12$	$17 = \frac{e}{3}$ $e = 51$	$30m = 3 \times 60$ $m = 6$

Write an equation for the problem. Solve.

7. Five workers are absent today. This is one fourth of all workers. How many workers are there?

   Equation: $\frac{1}{4}x = 5$ There are ___20___ workers.

Number Correct
LESSON FOLLOW-UP AND ERROR ANALYSIS
17–20: Have students choose five equations and check their answers.
13–16: Ask students to explain how after simplifying they perform the opposite of the indicated operation to solve each.
Less than 13: Have students choose an equation and explain the steps involved in solving it.

CHAPTER 4 PRETEST
Using Pre-Algebra

Complete the following.

	a	b	c
1.	$7x + 2x =$ __9x__	$9y + y =$ __10y__	$z + 2z =$ __3z__
2.	$6a + 2a =$ __8a__	$5b + b =$ __6b__	$c + 2c =$ __3c__

Solve each equation.

3. $3r + r = 36$ $5s + s = 42$ $t + 3t = 52$
     $r = 9$ $s = 7$ $t = 13$

4. $d + d + 8 = 48$ $e + e + 6 = 74$ $f + f - 5 = 95$
     $d = 20$ $e = 34$ $f = 50$

5. $u + 2u + 1 = 10$ $v + 3v + 4 = 24$ $w + 5w + 2 = 50$
     $u = 3$ $v = 5$ $w = 8$

Solve each problem.

6. Jenna made four times as many widgets as Carmen. They made a total of 60 widgets. How many widgets did Carmen make?

   Carmen made __12__ widgets.

7. A car averages 72 km per hour. At that rate, how far can the car travel in 3 hours?

   The car can travel __216__ km.

Prerequisite Skills: addition, subtraction, multiplication facts

Lesson 1 Combining Terms   PRE-ALGEBRA

Lesson Focus: combining like terms
Possible score: 28
Time Frame: 10–15 minutes

$3a + 2a = a + a + a + a + a$
$ = 5a$

$3a + 2a = (3 + 2)a$
$ = 5a$

$3b - 2b = b + b + b - b - b$
$ = \underline{\ 1b\ }$ or $\underline{\ b\ }$

$3b - 2b = (3 - 2)b$
$ = \underline{\ 1b\ }$ or $\underline{\ b\ }$

Complete the following:

	a	b	c
1.	$d + 3d = \underline{\ 4d\ }$	$5e + 2e = \underline{\ 7e\ }$	$7f + 2f = \underline{\ 9f\ }$
2.	$4g - 3g = \underline{\ g\ }$	$8h - 4h = \underline{\ 4h\ }$	$5j - j = \underline{\ 4j\ }$
3.	$2k + k = \underline{\ 3k\ }$	$5l - 3l = \underline{\ 2l\ }$	$3m + 2m = \underline{\ 5m\ }$
4.	$5n + 3n = \underline{\ 8n\ }$	$2p - p = \underline{\ p\ }$	$4q - q = \underline{\ 3q\ }$
5.	$8r - 2r = \underline{\ 6r\ }$	$5s + 4s = \underline{\ 9s\ }$	$5t + t = \underline{\ 6t\ }$
6.	$4u + 3u = \underline{\ 7u\ }$	$9v - v = \underline{\ 8v\ }$	$3w + w = \underline{\ 4w\ }$

Complete the following.

	a	b
7.	If $a = 5$, then $3a + 2a = \underline{\ 25\ }$.	If $b = 3$, then $5b - 2b = \underline{\ 9\ }$.
8.	If $c = 2$, then $3c + c = \underline{\ 8\ }$.	If $d = 1$, then $3d - d = \underline{\ 2\ }$.
9.	If $e = 5$, then $2e + 2e = \underline{\ 20\ }$.	If $f = 4$, then $5f - 4f = \underline{\ 4\ }$.
10.	If $g = 2$, then $g + 3g = \underline{\ 8\ }$.	If $h = 5$, then $2h - h = \underline{\ 5\ }$.
11.	If $j = 3$, then $2j + 4j = \underline{\ 18\ }$.	If $k = 3$, then $4k - 3k = \underline{\ 3\ }$.
12.	If $l = 5$, then $3l + 3l = \underline{\ 30\ }$.	If $m = 1$, then $6m - m = \underline{\ 5\ }$.

Number Correct   LESSON FOLLOW-UP AND ERROR ANALYSIS

24–28: Have students write a rule for adding and subtracting terms.
18–23: Ask students to explain what it means when no number precedes the letter. (The number 1 is understood and can be written as needed.)
Less than 18: Have students explain what a number and a letter together with no operation sign between them means. (to multiply)

Prerequisite Skills: combining like terms

Lesson Focus: solving equations involving two operations
Possible Score: 15
Time Frame: 10–15 minutes

Lesson 2 Solving Equations   PRE-ALGEBRA

Check

$x + 5x = 18$
$6x = 18$
$x = \dfrac{18}{6}$
$x = 3$

$x + 5x = 18$
$3 + (5 \times 3) = 18$
$3 + 15 = 18$
$18 = 18$

$y + y + 3 = 27$
$2y + 3 = 27$
$2y = 27 - 3$
$2y = 24$
$y = 24 \div 2$
$y = 12$

Check

$y + y + 3 = 27$
$12 + 12 + 3 = 27$
$27 = 27$

If $x + 5x = 18$, then $x = \underline{\;3\;}$ and $5x = \underline{\;15\;}$.

If $y + y + 3 = 27$, then $y = \underline{\;12\;}$

Solve each equation.

           a   b   c

1. $4a + a = 25$ $7b + b = 72$ $c + 6c = 49$
    $a = 5$ $b = 9$ $c = 7$

2. $d + d + 2 = 22$ $e + e + 8 = 28$ $f + f - 6 = 30$
    $d = 10$ $e = 10$ $f = 18$

3. $3g + g = 48$ $h + h - 5 = 25$ $5j + j = 54$
    $g = 12$ $h = 15$ $j = 9$

4. $k + k + 4 = 44$ $3l + l = 72$ $m + m - 7 = 19$
    $k = 20$ $l = 18$ $m = 13$

5. $n + 8n = 108$ $p + p + 12 = 60$ $2q + q = 72$
    $n = 12$ $p = 24$ $q = 24$

LESSON FOLLOW-UP AND ERROR ANALYSIS

Number Correct

64

- **13–15:** Have students check their answers as shown in the lesson example.
- **10–12:** When no number precedes the letter, discuss that the number 1 is understood and can be written as needed.
- **Less than 10:** Ask students to explain that a number and letter together means to multiply and to give an example.

Prerequisite Skills: solving equations

Lesson 3 Problem Solving PRE-ALGEBRA

Lesson Focus: solve problems using equations
Possible Score: 12
Time Frame: 10–15 minutes

Larry is twice as old as Marvin. Their combined age is 24 years. How old is each boy?

If x stands for Marvin's age, then __2x__ stands for Larry's age.

Equation: __$x + 2x = 24$__

Marvin is __8__ years old.

Larry is __16__ years old.

$x + 2x = 24$
$3x = 24$
$x = 24 \div 3$
$x = 8$
Since $x = 8$,
$2x = 2 \times 8$ or 16.

Check
$x + 2x = 24$
$8 + (2 \times 8) = 24$
$8 + 16 = 24$
$24 = 24$

Write an equation for each problem. Solve each problem.

1. An office has 28 workers. There are three times as many men as women. How many women are there? How many men are there?

   Equation: __$x + 3x = 28$__

   There are __7__ women and __21__ men.

2. During the summer Kim worked four times as many days as Lana. They worked a total of 75 days. How many days did each work?

   Equation: __$x + 4x = 75$__

   Lana worked __15__ days. Kim worked __60__ days.

3. A truck has a mass of 4200 kg. The mass of the truck body is six times that of the engine. What is the mass of the engine? What is the mass of the truck body?

   Equation: __$x + 6x = 4200$__

   The mass of the engine is __600__ kg and the mass of the truck body is __3600__ kg.

4. Jair is three times as old as Sue. Their combined age is 52. How old is each person?

   Equation: __$x + 3x = 52$__

   Sue is __13__ years old. Jair is __39__ years old.

LESSON FOLLOW-UP AND ERROR ANALYSIS

Number Correct
- **10–12:** Have students write a story problem that requires two answers, write an equation, and then solve the problem.
- **7–9:** Have students explain that once the value of the variable is found, it is used to find another answer. (If $x = 8$, then $2x = 16$.)
- **Less than 7:** Before students redo incorrect answers, remind them that they may choose any letter they wish to represent the unknown quantity.

Prerequisite Skills: solving equations

Lesson Focus: solve problems using equations
Possible Score: 12
Time Frame: 10–15 minutes

Lesson 4 Problem Solving   PRE-ALGEBRA

In an election between two girls, 75 votes were cast. Bianca received 5 more votes than Jaime. How many votes did each girl receive?

If x stands for the number of votes for Jaime, then __$x + 5$__ stands for the number of votes for Bianca.

Equation: __$x + (x + 5) = 75$__

Jaime received __35__ votes.

Bianca received __40__ votes.

$x + (x + 5) = 75$
$2x + 5 = 75$
$2x = 75 - 5$
$2x = 70$
$x = 35$
Since $x = 35$,
$x + 5 = 35 + 5$ or 40.

Check
$x + (x + 5) = 75$
$35 + 35 + 5 = 75$
$75 = 75$

Write an equation for each problem. Solve each problem.

1. Paul made 7 more gadgets than Jeremy. Together they made 55 gadgets. How many did each man make?

   Equation: __$x + (x + 7) = 55$__

   Paul made __31__ gadgets and Jeremy made __24__.

2. Two pairs of shoes cost $58. One pair costs $6 more than the other. How much did each pair cost?

   Equation: __$x + (x + 6) = 58$__

   One pair cost $__26__ and the other cost $__32__.

3. Yoko's mass is 8 kg more than Tara's mass. Their combined mass is 92 kg. What is each girl's mass?

   Equation: __$x + (x + 8) = 92$__

   Tara's mass is __42__ kg and Yoko's mass is __50__ kg.

4. Lisa has 12 more cases to unload than Mick does. They have a total of 150 cases to unload. How many cases does each have to unload?

   Equation: __$x + (x + 12) = 150$__

   Mick has __69__ cases and Lisa has __81__ cases.

LESSON FOLLOW-UP AND ERROR ANALYSIS

Number Correct

10–12: Have students write a story problem that requires two answers, write an equation, and then solve the problem.
7–9: Before students redo the page, discuss that when writing equations any letter can be used to represent the unknown quantity.
Less than 7: Ask students to explain how to find the value of the variable and how to use it to find another answer.

Prerequisite Skills: solving equations

Lesson 5 Problem Solving   PRE-ALGEBRA

Lesson Focus: solve problems using equations
Possible Score: 21
Time Frame: 10–15 minutes

Max has two boards that have a combined length of 16 m. One board is 1 m longer than twice the length of the other. What is the length of each board?

If x stands for the length of the shorter board, then __2x + 1__ stands for the length of the longer board.

Equation: __$x + (2x + 1) = 16$__

The shorter board is __5__ m long.

The longer board is __11__ m long.

$x + (2x + 1) = 16$
$3x + 1 = 16$
$3x = 16 - 1$
$3x = 15$
$x = 5$

Since $x = 5$,
$2x + 1 = (2 \times 5) + 1$ or 11.

Check

$x + (2x + 1) = 16$
$5 + (2 \times 5) + 1 = 16$
$5 + 10 + 1 = 16$
$16 = 16$

CHAPTER 4

Write an equation for each problem. Solve each problem.

1. Mark and Bill have a combined mass of 85 kg. Mark's mass is 20 kg less than twice Bill's mass. What is each boy's mass?

    Equation: __$x + (2x - 20) = 85$__

    Bill's mass is __35__ kg.

    Mark's mass is __50__ kg.

2. Cara and Amber have saved $43. Amber has saved $3 more than three times the amount Cara has saved. How much money has each girl saved?

    Equation: __$x + (3x + 3) = 43$__

    Cara has saved $__10__.

    Amber has saved $__33__.

3. A carpenter cut a board that was 5 m long into two pieces. The longer piece is 1 m longer than three times the length of the shorter piece. What is the length of each piece?

    Equation: __$x + (3x + 1) = 5$__

    The shorter piece is __1__ m long.

    The longer piece is __4__ m long.

Number Correct

LESSON FOLLOW-UP AND ERROR ANALYSIS

18–21: Have students write a story problem that requires two answers, write an equation, and then solve the problem.
14–17: Before students redo the pages, have them explain how to solve an equation and how one answer can help solve another.
Less than 14: Ask students to explain one problem—how they determined the equation, how they solved it, and so on.

67

Lesson 5 Problem Solving — PRE-ALGEBRA

Write an equation for each problem. Solve each problem.

1. Elise said that Box A is 2 kg heavier than Box D. She also said that together these boxes have a mass of 16 kg. What is the mass of each box?

   Equation: _____ $x + (x + 2) = 16$ _____

   Box A's mass is __9__ kg.
   Box D's mass is __7__ kg.

2. Box C is twice as heavy as Box A. Together their mass is 27 kg. What is the mass of each box?

   Equation: _____ $x + 2x = 27$ _____

   Box A's mass is __9__ kg.
   Box C's mass is __18__ kg.

3. Box B's mass is 1 kg more than twice the mass of Box D. They have a combined mass of 22 kg. What is the mass of each box?

   Equation: _____ $x + (2x + 1) = 22$ _____

   Box B's mass is __15__ kg.
   Box D's mass is __7__ kg.

4. Elise's mass is 1 kg more than Mark's. Their total mass is 97 kg. What is the mass of each person?

   Equation: _____ $x + (x + 1) = 97$ _____

   Mark's mass is __48__ kg.
   Elise's mass is __49__ kg.

CHAPTER 4
Using Pre-Algebra

Prerequisite Skills: solving equations

Lesson Focus: solving problems using formulas
Possible Score: 16
Time Frame: 20–25 minutes

Lesson 6 Problem Solving PRE-ALGEBRA

$$\text{distance} = \text{rate} \times \text{time}$$
$$d = r \times t$$

A robin flew 171 km in 3 hours. At what speed did the robin fly?

Equation: _____ $171 = r \times 3$ _____

The robin flew __57__ km per hour.

$d = r \times t$
$171 = r \times 3$
$\dfrac{171}{3} = r$
$57 = r$

Write an equation for each problem. Solve each problem.

1. At 450 km per hour, how far can a plane fly in 5 h?

   Equation: _____ $d = 450 \times 5$ _____

   The plane can fly __2250__ km in 5 h.

2. The Willards want to travel 744 km in 12 h. They plan to travel the same distance each hour. At what speed would they travel?

   Equation: _____ $744 = r \times 12$ _____

   They would travel __62__ km per hour.

3. A ship averages 25 knots per hour. How far can the ship travel in 2 days?

   Equation: _____ $d = 25 \times 48$ _____

   The ship can travel __1200__ knots in 2 days.

4. At what speed would a plane have to fly in order to travel 780 km in 2 h?

   Equation: _____ $780 = r \times 2$ _____

   It would fly at __390__ km per hour.

5. At 204.8 km per hour, how far can a race car travel in 4 h?

   Equation: _____ $d = 204.8 \times 4$ _____

   It can travel __819.2__ km in 4 h.

Number Correct   LESSON FOLLOW-UP AND ERROR ANALYSIS
- **13–16:** Have students write a story problem involving distance, rate, and time. Then have them write an equation and solve the problem.
- **10–12:** Ask students to explain which numbers represent distance, rate, and time before they redo incorrect answers.
- **Less than 10:** As they redo the pages, have students write formulas and write the appropriate numbers beneath each letter, then solve.

Lesson 6 Problem Solving   PRE-ALGEBRA

For all levers, $w \times d = W \times D$.

To balance the lever (or scale), how far from the fulcrum must the 12-gram mass be placed?

	Check
$w \times d = W \times D$	$w \times d = W \times D$
$10 \times 6 = 12 \times D$	$10 \times 6 = 12 \times 5$
$\frac{60}{12} = D$	$60 = 60$
$5 = D$	

The 12-g mass must be placed __5__ cm from the fulcrum.

Write an equation for each problem. Solve each problem.

1. A 60-kg boy sits 2 m from the fulcrum of a seesaw. How far from the fulcrum should a 40-kg girl sit so the seesaw is balanced?

   Equation: __$60 \times 2 = 40 \times D$__

   She should sit __3__ m from the fulcrum.

2. How much mass would have to be applied at point A so that the lever is balanced?

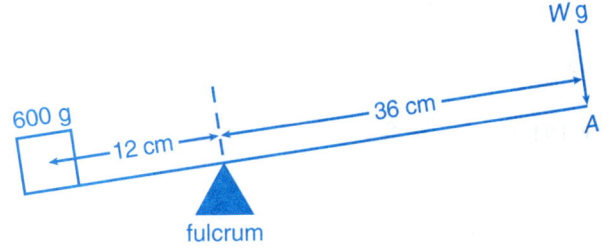

   Equation: __$600 \times 12 = W \times 36$__

   __200__ g would have to be applied at point A.

3. What mass is needed at point S on the scale so that the scale is balanced?

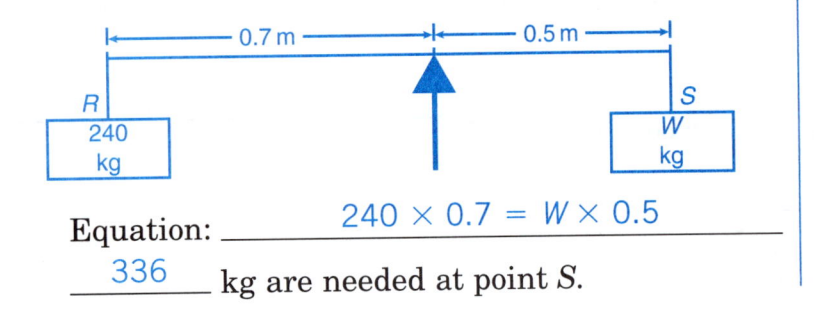

   Equation: __$240 \times 0.7 = W \times 0.5$__

   __336__ kg are needed at point S.

CHAPTER 4
Using Pre-Algebra

For further evaluation, copy the Chapter Test on page 268.
For maintaining skills, use the Cumulative Review on pages 219–220.

Possible Score: 10
Time Frame: 15–20 minutes

CHAPTER 4   PRACTICE TEST
Using Pre-Algebra

Solve each problem.

1. Maggie worked three times as many hours as Ann. They worked a total of 32 h. How many hours did Ann work?

   Equation: _____ $x + 3x = 32$ _____

   Ann worked __8__ h.

2. An eraser and a pencil cost 87¢. The pencil cost 9¢ more than the eraser. How much did the pencil cost?

   Equation: _____ $x + (x + 9) = 87$ _____

   The pencil cost __48__ ¢.

3. Dane and Jim earned 215 points in a contest. Dane earned 5 more than twice as many points as Jim. How many points did each boy earn?

   Equation: _____ $x + (2x + 5) = 215$ _____

   Jim earned __70__ points.

   Dane earned __145__ points.

4. Darla scored twice as many points as the combined scores of Gina and Hikaru. Darla scored 88 points. Gina scored 20 points. How many points did Hikaru score?

   Hikaru scored __24__ points.

5. At 51 km per hour, how far can a car travel in 3 h?

   It can travel __153__ km in 3 h.

6. To balance the lever, how far from the fulcrum must the 40-kg mass be placed?

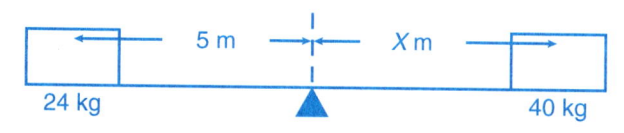

   5 m   X m
   24 kg   40 kg

   It must be __3__ m from the fulcrum.

LESSON FOLLOW-UP AND ERROR ANALYSIS

Number Correct
9–10: Have students indicate the most difficult problem to solve and explain their choice.
7–8: Have students look at each incorrect answer and explain the equation before they redo it.
Less than 7: As students redo the page, have them write formulas and write the appropriate numbers beneath each letter, then solve.

CHAPTER 5 PRETEST
Ratio, Rate, Proportion, and Percent

Circle each proportion below.

          a   b

1. $\frac{3}{16} = \frac{6}{24}$ ($\frac{7}{8} = \frac{28}{32}$)

2. $\frac{8}{20} = \frac{4}{5}$ ($\frac{2}{3} = \frac{10}{15}$)

3. ($\frac{7}{9} = \frac{21}{27}$)   ($\frac{24}{15} = \frac{8}{5}$)

Solve each of the following.

4. $\frac{n}{3} = \frac{9}{27}$ $\frac{3}{5} = \frac{15}{n}$

   $n = 1$ $n = 25$

5. $\frac{5}{6} = \frac{n}{36}$ $\frac{n}{8} = \frac{3}{6}$

   $n = 30$ $n = 4$

6. $\frac{8}{24} = \frac{n}{15}$ $\frac{n}{10} = \frac{9}{15}$

   $n = 5$ $n = 6$

7. $\frac{10}{25} = \frac{8}{n}$ $\frac{42}{n} = \frac{3}{4}$

   $n = 20$ $n = 56$

Complete the following.

               a   b

8. __4.32__ is 12% of 36.   7 is __$43\frac{3}{4}$__ % of 16.

9. $\frac{1}{2}$ is 50% of __1__.   45 is 75% of __60__.

10. $\frac{2}{5}$ is __80__ % of $\frac{1}{2}$.   __60__ is 30% of 200.

11. 3.6 is 80% of __4.5__.   1.8 is __75__ % of 2.4.

12. __5.561__ is 6.7% of 83.   135 is __150__ % of 90.

Prerequisite Skills: set notation

Lesson Focus: writing ratios
Possible Score: 20
Time Frame: 5–10 minutes

Lesson 1 Ratio

A **ratio** is a comparison of the numbers of two sets of like objects.
A **rate** is a comparison of the numbers of two sets of different objects.

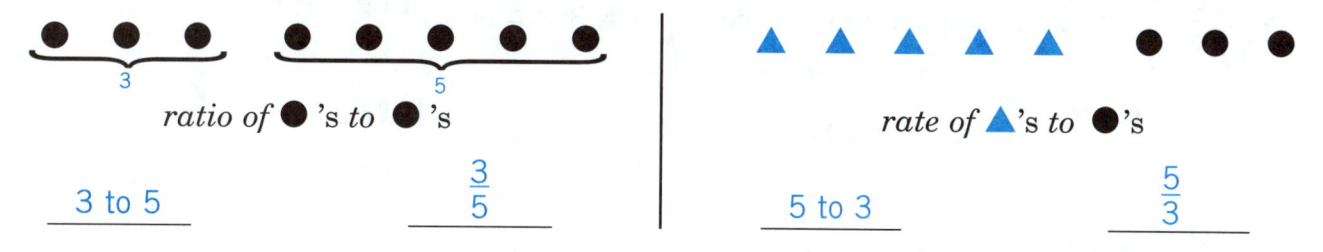

ratio of ●'s to ●'s

3 to 5 $\frac{3}{5}$

rate of ▲'s to ●'s

5 to 3 $\frac{5}{3}$

Express the ratio or the rate of the number of items in the first set to the number of items in the second set in two ways as shown.

			a	b
1.	{○, □}	{*, △, □}	2 to 3	$\frac{2}{3}$
2.	{Jim, John}	{Jo, Sue, Ann, Kay}	2 to 4	$\frac{2}{4}$
3.	{1, 2, 3, 4}	{a, b, c}	4 to 3	$\frac{4}{3}$
4.	{Bob, Dick, Al}	{1st, 2nd, 3rd}	3 to 3	$\frac{3}{3}$
5.	{m, n, o, p, q}	{w, x, y, z}	5 to 4	$\frac{5}{4}$

Express each of the following as a ratio or a rate in two ways as shown.

		a	b
6.	7 runs in 9 innings	7 to 9	$\frac{7}{9}$
7.	5 boys to 6 girls	5 to 6	$\frac{5}{6}$
8.	3 teachers for 72 students	3 to 72	$\frac{3}{72}$
9.	5 pages in 20 minutes	5 to 20	$\frac{5}{20}$
10.	5 touchdowns in 4 games	5 to 4	$\frac{5}{4}$
11.	11 chairs to 2 tables	11 to 2	$\frac{11}{2}$
12.	6 goals for 9 shots	6 to 9	$\frac{6}{9}$

Number Correct

LESSON FOLLOW-UP AND ERROR ANALYSIS

17–20: Have students create and explain five ratios they determine.
13–16: Have students explain the importance of writing the numbers in the ratio in the same order as they appear in the question.
Less than 13: Discuss that the first number in a ratio is the numerator and the second number is the denominator.

Prerequisite Skills: writing ratios

Lesson Focus: recognizing proportions
Possible Score: 20
Time Frame: 10–15 minutes

Lesson 2 Proportions (recognizing)

A **proportion** expresses the equality of two ratios.

$\frac{2}{3} = \frac{4}{6}$ ___is___ a proportion because $2 \times 6 = 3 \times 4$ is ___true___.

$\frac{5}{8} = \frac{3}{4}$ ___is not___ a proportion because $5 \times 4 = 8 \times 3$ is ___false___.

$\frac{4}{5} = \frac{7}{8}$ ___is not___ a proportion because $4 \times 8 = 5 \times 7$ is ___false___.

$\frac{3}{4} = \frac{9}{12}$ ___is___ a proportion because $3 \times 12 = 4 \times 9$ is ___true___.

Circle each proportion below.

	a	b
1.	(⎯) $\frac{2}{3} = \frac{8}{12}$ (⎯)	$\frac{1}{4} = \frac{2}{9}$
2.	$\frac{5}{8} = \frac{11}{16}$	(⎯) $\frac{5}{6} = \frac{20}{24}$ (⎯)
3.	(⎯) $\frac{1}{6} = \frac{2}{12}$ (⎯)	$\frac{14}{16} = \frac{7}{8}$
4.	(⎯) $\frac{5}{12} = \frac{15}{36}$ (⎯)	$\frac{8}{3} = \frac{15}{6}$
5.	$\frac{7}{20} = \frac{15}{40}$	$\frac{9}{24} = \frac{1}{3}$
6.	(⎯) $\frac{1}{3} = \frac{6}{18}$ (⎯)	(⎯) $\frac{15}{24} = \frac{5}{8}$ (⎯)
7.	$\frac{7}{12} = \frac{9}{16}$	$\frac{4}{5} = \frac{10}{12}$
8.	(⎯) $\frac{9}{10} = \frac{90}{100}$ (⎯)	(⎯) $\frac{8}{10} = \frac{4}{5}$ (⎯)
9.	$\frac{4}{12} = \frac{5}{16}$	$\frac{4}{3} = \frac{5}{4}$
10.	(⎯) $\frac{12}{25} = \frac{48}{100}$ (⎯)	(⎯) $\frac{125}{1000} = \frac{1}{8}$ (⎯)

Number Correct 74

LESSON FOLLOW-UP AND ERROR ANALYSIS

17–20: Have students express a rule for proving that two ratios are equal.
13–16: Check to see if students multiplied numerator by numerator and denominator by denominator instead of cross multiplying.
Less than 13: Have students explain how to cross multiply and compare products to determine proportions.

Prerequisite Skills: recognizing proportions

Lesson 3 Proportions PRE-ALGEBRA

Lesson Focus: solving proportions
Possible Score: 15
Time Frame: 10–15 minutes

Study how the proportions below are solved.

$\dfrac{5}{8} = \dfrac{15}{n}$
$5 \times n = 8 \times 15$
$5n = 120$
$n = \underline{\ 24\ }$

$\dfrac{2}{3} = \dfrac{n}{24}$
$2 \times 24 = 3 \times n$
$48 = 3n$
$\underline{\ 16\ } = n$

$\dfrac{5}{n} = \dfrac{6}{24}$
$5 \times 24 = n \times 6$
$120 = 6n$
$\underline{\ 20\ } = n$

$\dfrac{n}{6} = \dfrac{20}{24}$
$n \times 24 = 6 \times 20$
$24n = 120$
$n = \underline{\ 5\ }$

Solve each of the following.

	a	b	c
1.	$\dfrac{2}{3} = \dfrac{n}{18}$ $n = 12$	$\dfrac{3}{5} = \dfrac{n}{25}$ $n = 15$	$\dfrac{3}{4} = \dfrac{n}{100}$ $n = 75$
2.	$\dfrac{1}{4} = \dfrac{2}{n}$ $n = 8$	$\dfrac{5}{6} = \dfrac{10}{n}$ $n = 12$	$\dfrac{7}{8} = \dfrac{42}{n}$ $n = 48$
3.	$\dfrac{n}{6} = \dfrac{2}{3}$ $n = 4$	$\dfrac{n}{8} = \dfrac{21}{24}$ $n = 7$	$\dfrac{n}{3} = \dfrac{24}{36}$ $n = 2$
4.	$\dfrac{8}{n} = \dfrac{1}{2}$ $n = 16$	$\dfrac{5}{n} = \dfrac{20}{28}$ $n = 7$	$\dfrac{4}{n} = \dfrac{80}{100}$ $n = 5$
5.	$\dfrac{n}{2} = \dfrac{12}{8}$ $n = 3$	$\dfrac{5}{8} = \dfrac{n}{1000}$ $n = 625$	$\dfrac{3}{4} = \dfrac{36}{n}$ $n = 48$

CHAPTER 5

Number Correct **LESSON FOLLOW-UP AND ERROR ANALYSIS**
13–15: Have students tell why these proportions are equations.
10–12: Have students explain how to cross multiply and then how to solve each equation before they redo the page.
Less than 10: Check to see if students were confused by the fact that the variable is not always in the same place in these proportions.

Prerequisite Skills: solving proportions

Lesson Focus: solving story problems using proportions
Possible Score: 5
Time Frame: 10–15 minutes

Lesson 4 Proportions PRE-ALGEBRA

A train can travel 120 km in 2 h. At that rate, how far can the train travel in 3 hours?

Let n represent the number of kilometres travelled in 3 h. Then the following proportions can be obtained by thinking as follows.

Compare the number of hours to the number of kilometres travelled.	Compare the number of kilometres travelled to the number of hours.	Compare the first number of hours to the second and the first number of kilometres to the second.	Compare the second number of hours to the first and the second number of kilometres to the first.
$\frac{2}{120} = \frac{3}{n}$	$\frac{120}{2} = \frac{n}{3}$	$\frac{2}{3} = \frac{120}{n}$	$\frac{3}{2} = \frac{n}{120}$
$2n = 360$	$360 = 2n$	$2n = 360$	$360 = 2n$
$n = \underline{\ 180\ }$	$\underline{\ 180\ } = n$	$n = \underline{\ 180\ }$	$\underline{\ 180\ } = n$

Use a proportion to solve each problem.

1. If eight cases of merchandise cost $60, what would 12 cases cost?

   12 cases would cost $\underline{\ 90\ }$.

2. 2 kg of apples can be purchased for 98¢. At this rate, what would 1 kg of apples cost?

   1 kg of apples would cost $\underline{\ 49\ }$ ¢.

3. Caitlin delivered 450 flyers in 3 h. At this rate, how many flyers can she deliver in 4 h?

   She can deliver $\underline{\ 600\ }$ flyers in 4 h.

4. In his last game the Rams' quarterback threw 18 passes and completed 10. At this rate, how many passes will he complete if he throws 27 passes in a game?

   He will complete $\underline{\ 15\ }$ passes.

5. Mrs. Svage used 3 L of paint to cover 30 m². At this rate, how much paint will be needed to cover 40 m²?

   $\underline{\ 4\ }$ L will be needed.

Number Correct

76

4–5: Have students indicate the easiest problems to solve and explain their choices.
3: Before they redo incorrect answers, students may need help writing the proportions.
Less than 3: Before redoing the page, ask students to explain the examples at the top of the page for setting up the proportions.

LESSON FOLLOW-UP AND ERROR ANALYSIS

Prerequisite Skills: ratio and proportions

Lesson Focus: interpret scale drawings
Possible Score: 13
Time Frame: 15–20 minutes

Lesson 5   Scale Drawings

In a **scale drawing,** the dimensions of the object are in proportion to the actual object. The scale is the ratio of the drawing size to the actual size of the object.

Find the missing information.

scale: 3 cm: 1 km
drawing length: ?
actual length: 4.5 km

Set up a **proportion** to find the length.

scale → $\frac{3}{1} = \frac{n}{4.5}$ drawing length / actual length

$3 \times 4.5 = 1 \times n$
$13.5 = n$

The length of the drawing is 13.5 cm.

scale: ? cm: ? m
drawing length: 6 cm
actual length: 12 m

Set up a **ratio** to find the scale.

$\frac{6 \text{ cm}}{12 \text{ m}}$ drawing length / actual length

$\frac{1 \text{ cm}}{2 \text{ m}}$

The scale is 1 cm: 2 m.

Find the missing information.

a

1. scale: __1 cm: 4 m__
   drawing length: 5 cm
   actual length: 20 m

2. scale: 3 cm: 2 m
   drawing length: 13.5 cm
   actual length: __9 m__

3. scale: __1 cm: 16 m__
   drawing length: 2.5 cm
   actual length: 40 km

4. scale: 2 cm: 3 km
   drawing length: 3 cm
   actual length: __4.5 km__

b

scale: 2 cm: 5 m
drawing length: __14 cm__
actual length: 35 m

scale: 1.5 cm: 5 m
drawing length: __30 cm__
actual length: 100 m

scale: 1 m: 0.5 km
drawing length: 4 m
actual length: __2 km__

scale: __2 mm: 7 m__
drawing length: 1 mm
actual length: 3.5 m

Number Correct

LESSON FOLLOW-UP AND ERROR ANALYSIS

11–13: Have students write a story problem to go with one of the problems on page 77.
8–10: Have students name each numerator and denominator and tell what it represents.
Less than 8: Have students review how to solve a proportion.

Lesson 5 Problem Solving

Solve each problem.

1. Mr. Jonas made a scale drawing of an addition that he is making to his house. The scale he used is 1 cm: 3 m. The length of the addition is 12 m. What is the length of the addition on the drawing?

   The length of the addition on the drawing is ___4___ cm.

2. Elizabeth is designing a flower garden for her community. The garden will be 9 m wide and 13.5 m long. The drawing has a width of 5 cm. She includes in the drawing a key for the scale that is 1.25 cm long. What actual distance does the key for the scale represent?

   The drawing of Elizabeth's garden is scaled as ___1.25 cm: 2.25 m___.

3. Use the scale from problem 2 to find the length of Elizabeth's garden in the drawing.

   Elizabeth's drawing has a length of ___7.5___ cm.

4. Mrs. Finney is a sculptor of famous people. She uses a 1 cm: 2 cm scale. She is making a sculpture of the mayor of her city. The finished sculpture is 13.75 cm from the base of the neck to the tip of the head. What is the actual height of his head from the base of his neck to the tip of his head?

   The mayor's head from the base of his neck to the tip of his head is ___27.5___ cm.

5. The actual size of an artifact is 3 cm by 5.5 cm. In the archive files a drawing that measures 9 cm by 16.5 cm shows every detail of its design. Is the ratio that represents this scale drawing greater than 1, equal to 1, or less than 1?

   The scale for the archived drawing is ___greater than 1___.

Prerequisite Skills: solving proportions

Lesson 6 Problem Solving PRE-ALGEBRA

Lesson Focus: solving story problems using proportions
Possible Score: 14
Time Frame: 25–30 minutes

As **A** revolves twice, **B** revolves once.

As **C** revolves 4 times, **D** revolves 14 times.

As **E** revolves 4 times, **F** revolves 3 times.

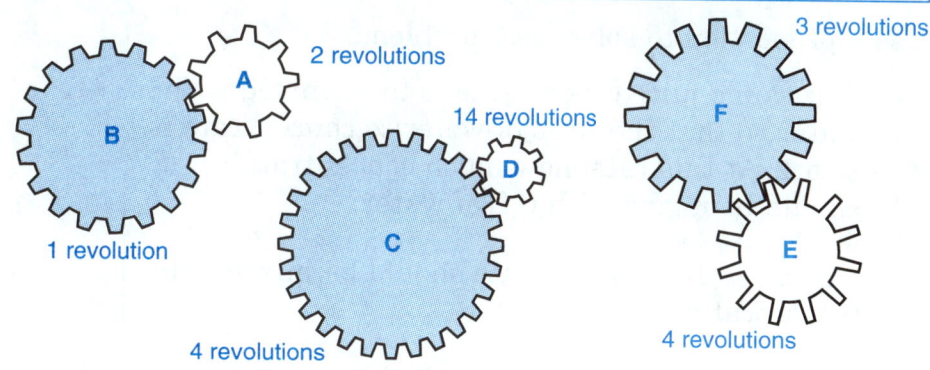

Gear A: 2 revolutions
Gear B: 1 revolution
Gear C: 4 revolutions
Gear D: 14 revolutions
Gear F: 3 revolutions
Gear E: 4 revolutions

Use a proportion to solve each problem.

1. When gear **A** has completed six revolutions, how many revolutions will gear **B** have made?

    Gear **B** will have made ___3___ revolutions.

2. While gear **B** is making 30 revolutions, how many revolutions will gear **A** make?

    Gear **A** will make ___60___ revolutions.

3. When gear **C** has completed 12 revolutions, how many revolutions will gear **D** have made?

    Gear **D** will have made ___42___ revolutions.

4. While gear **D** is making 84 revolutions, how many revolutions will gear **C** make?

    Gear **C** will make ___24___ revolutions.

5. When gear **E** has completed 56 revolutions, how many revolutions will gear **F** have made?

    Gear **F** will have made ___42___ revolutions.

6. While gear **F** is making 90 revolutions, how many revolutions will gear **E** make?

    Gear **E** will make ___120___ revolutions.

7. When gear **E** has completed 76 revolutions, how many revolutions will gear **F** have made?

    Gear **F** will have made ___57___ revolutions.

Number Correct

LESSON FOLLOW-UP AND ERROR ANALYSIS

12–14: Ask students to explain why it takes gear D 14 revolutions to equal gear C's 4 revolutions.
9–11: Before students redo incorrect answers, have them explain each proportion.
Less than 9: Have students refer to the example at the top of page 76 and use it to explain how to set up the proportions.

Lesson 6 Problem Solving   PRE-ALGEBRA

Use a proportion to solve each problem.

1. An orange-juice concentrate is to be mixed with water so that the ratio of water to concentrate is 3 to 1. At this rate, how much concentrate should be mixed with 6 L of water?

   ___2___ L of concentrate should be mixed with 6 L of water.

2. A shoe store sells 4 pairs of black shoes for every 7 pairs of brown shoes. There were 4900 pairs of brown shoes sold last year. How many pairs of black shoes were sold?

   ___2800___ pairs of black shoes were sold.

3. At the Kolbus Building, 3 out of every 7 employees use public transportation. There are 9800 employees at the building. How many use public transportation?

   ___4200___ use public transportation.

4. The ratio of box seats at the hockey arena to general-admission seats is 2 to 7. There are 2500 box seats. How many general-admission seats are there?

   There are ___8750___ general-admission seats.

5. At the snack bar, 7 hot dogs are sold for every 10 hamburgers sold. At this rate, how many hot dogs will be sold if 90 hamburgers are sold?

   ___63___ hot dogs will be sold.

6. Jacob delivered 90 flyers in 30 min. At this rate, how long will it take him to deliver 135 flyers?

   It will take him ___45___ min.

7. At the airport, four planes land every 8 min. At this rate, how many planes will land in 1 h?

   ___30___ planes will land in 1 h.

1.

2.

3.

4.

5.

6.

7.

CHAPTER 5
Ratio, Rate, Proportion, and Percent

Lesson 6
Problem Solving

Prerequisite Skills: writing fractions and decimals

Lesson Focus: changing percents to fractions and decimals
Possible Score: 28
Time Frame: 5–10 minutes

Lesson 7   Percent   PRE-ALGEBRA

If n stands for a number, then $n\%$ stands for the ratio of n to 100 or $\frac{n}{100}$.

$1\% = \frac{1}{100}$ or 0.01 | $37\% = \frac{37}{100}$ or 0.37 | $125\% = \frac{125}{100}$ or 1.25

$5\% = \frac{5}{100}$ or 0.05 | $53\% = \frac{53}{100}$ or 0.53 | $149\% = \frac{149}{100}$ or 1.49

Complete the following.

	percent	fraction	decimal
1.	3%	$\frac{3}{100}$	0.03
2.	27%	$\frac{27}{100}$	0.27
3.	121%	$\frac{121}{100}$	1.21
4.	7%	$\frac{7}{100}$	0.07
5.	39%	$\frac{39}{100}$	0.39
6.	141%	$\frac{141}{100}$	1.41
7.	9%	$\frac{9}{100}$	0.09
8.	11%	$\frac{11}{100}$	0.11
9.	167%	$\frac{167}{100}$	1.67
10.	57%	$\frac{57}{100}$	0.57
11.	251%	$\frac{251}{100}$	2.51
12.	69%	$\frac{69}{100}$	0.69
13.	391%	$\frac{391}{100}$	3.91
14.	87%	$\frac{87}{100}$	0.87

CHAPTER 5

Number Correct   **LESSON FOLLOW-UP AND ERROR ANALYSIS**

24–28: Have students give examples of percents greater than 100. (a store's markup on product; improvement in performance)

18–23: Ask students to explain why a zero must be inserted when changing a 1-digit percent to a decimal. (3% = 0.03)

Less than 18: Have students explain that a percent sign has the same value as a denominator of 100 in a fraction. (3% = $\frac{3}{100}$)

Prerequisite Skills: multiplication and division of whole numbers and fractions

Lesson Focus: changing fractions to percents and percents to fractions
Possible Score: 20
Time Frame: 20–25 minutes

Lesson 8 Fractions and Percent   PRE-ALGEBRA

Study how to change a fraction to a percent.

$\dfrac{4}{5} = \dfrac{n}{100}$

$400 = 5n$

$80 = n$

$\dfrac{4}{5} = \underline{80}\%$

$\dfrac{1}{8} = \dfrac{n}{100}$

$100 = 8n$

$12\dfrac{1}{2} = n$

$\dfrac{1}{8} = \underline{12\dfrac{1}{2}}\%$

Study how to change a percent to a fraction or mixed numeral.

$175\% = \dfrac{175}{100}$
$= \dfrac{7}{4}$ or $1\dfrac{3}{4}$

$3\dfrac{1}{2}\% = \dfrac{3\dfrac{1}{2}}{100}$
$= 3\dfrac{1}{2} \times \dfrac{1}{100}$
$= \dfrac{7}{2} \times \dfrac{1}{100}$
$= \dfrac{7}{200}$

Complete the following.

       a   b

1. $\dfrac{1}{4} = \underline{\ 25\ }\%$ $\dfrac{3}{8} = \underline{\ 37\dfrac{1}{2}\ }\%$

2. $\dfrac{1}{10} = \underline{\ 10\ }\%$ $\dfrac{3}{4} = \underline{\ 75\ }\%$

3. $\dfrac{1}{2} = \underline{\ 50\ }\%$ $\dfrac{5}{8} = \underline{\ 62\dfrac{1}{2}\ }\%$

4. $\dfrac{7}{10} = \underline{\ 70\ }\%$ $\dfrac{2}{5} = \underline{\ 40\ }\%$

5. $\dfrac{4}{5} = \underline{\ 80\ }\%$ $\dfrac{7}{8} = \underline{\ 87\dfrac{1}{2}\ }\%$

Change each of the following to a fraction or mixed numeral in simplest form.

6. $10\% = \underline{\ \dfrac{1}{10}\ }$ $80\% = \underline{\ \dfrac{4}{5}\ }$

7. $160\% = \underline{\ 1\dfrac{3}{5}\ }$ $12\dfrac{1}{2}\% = \underline{\ \dfrac{1}{8}\ }$

8. $250\% = \underline{\ 2\dfrac{1}{2}\ }$ $62\dfrac{1}{2}\% = \underline{\ \dfrac{5}{8}\ }$

9. $20\% = \underline{\ \dfrac{1}{5}\ }$ $16\% = \underline{\ \dfrac{4}{25}\ }$

10. $125\% = \underline{\ 1\dfrac{1}{4}\ }$ $37\dfrac{1}{2}\% = \underline{\ \dfrac{3}{8}\ }$

LESSON FOLLOW-UP AND ERROR ANALYSIS

Number Correct

82

17–20: Have students write three mixed numerals and change each to a percent.
13–16: Have students explain how to change a fraction to a percent. (first change fractions to equivalent fractions with a denominator of 100)
Less than 13: As students redo problems 6–10, ask them to explain what a percent sign means. (hundredths)

Prerequisite Skills: decimal place values

Lesson Focus: changing decimals to percents and percents to decimals
Possible Score: 16
Time Frame: 5–10 minutes

Lesson 9   Decimals and Percent

Study how to change a decimal to a percent.

$0.3 = 0.30 = \frac{30}{100} = \underline{30}\%$

$1.24 = \frac{124}{100} = \underline{124}\%$

$0.375 = \frac{37.5}{100} = \underline{37.5}\%$

$1.6 = 1.60 = \frac{160}{100} = \underline{160}\%$

$0.59 = \frac{59}{100} = \underline{59}\%$

$2.125 = \frac{212.5}{100} = \underline{212.5}\%$

Study how to change a percent to a decimal.

$17.6\% = \frac{17.6}{100} = \underline{0.176}$

$7.25\% = \frac{7.25}{100} = \underline{0.0725}$

$16\frac{3}{4}\% = 16.75\% = \frac{16.75}{100} = \underline{0.1675}$

$8.4\% = \frac{8.4}{100} = \underline{0.084}$

$9.69\% = \frac{9.69}{100} = \underline{0.0969}$

$37\frac{1}{2}\% = 37.5\% = \frac{37.5}{100} = \underline{0.375}$

CHAPTER 5

Complete the following.

a

	decimal	percent
1.	0.2	20%
2.	1.9	190%
3.	0.02	2%
4.	0.36	36%
5.	1.47	147%
6.	0.067	6.7%
7.	0.123	12.3%
8.	1.625	162.5%

b

	percent	decimal
1.	52%	0.52
2.	148%	1.48
3.	5.4%	0.054
4.	8.75%	0.0875
5.	183.75%	1.8374
6.	$9\frac{1}{2}\%$	0.095
7.	$7\frac{1}{4}\%$	0.0725
8.	$8\frac{3}{4}\%$	0.0875

Number Correct

LESSON FOLLOW-UP AND ERROR ANALYSIS

13–16: Have students choose a column and list the percents from least to greatest.
10–12: Have students explain how to change fractions to decimals ($\frac{1}{2}$ = 0.5) before changing to percents.
Less than 10: Ask students to explain how and why the decimal point is moved two places when converting decimals and percents.

Prerequisite Skills: changing fractions, decimals, and percents

Lesson Focus: changing fractions, decimals, and percents
Possible Score: 32
Time Frame: 25–30 minutes

Lesson 10 Fractions, Decimals, and Percent

Change each fraction to a percent. Change each percent to a fraction or mixed numeral in simplest form.

a | b

1. $\frac{1}{8} = \underline{12\frac{1}{2}}\%$ | $30\% = \underline{\frac{3}{10}}$

2. $80\% = \underline{\frac{4}{5}}$ | $\frac{1}{5} = \underline{20}\%$

3. $\frac{3}{5} = \underline{60}\%$ | $120\% = \underline{1\frac{1}{5}}$

4. $87\frac{1}{2}\% = \underline{\frac{7}{8}}$ | $\frac{3}{4} = \underline{75}\%$

5. $\frac{1}{10} = \underline{10}\%$ | $150\% = \underline{1\frac{1}{2}}$

6. $31\frac{1}{4}\% = \underline{\frac{5}{16}}$ | $\frac{9}{10} = \underline{90}\%$

7. $\frac{4}{25} = \underline{16}\%$ | $64\% = \underline{\frac{16}{25}}$

8. $110\% = \underline{1\frac{1}{10}}$ | $\frac{9}{20} = \underline{45}\%$

Change each decimal to a percent. Change each percent to a decimal.

9. $0.5 = \underline{50}\%$ | $4\% = \underline{0.04}$

10. $17.7\% = \underline{0.177}$ | $1.1 = \underline{110}\%$

11. $0.67 = \underline{67}\%$ | $6.625\% = \underline{0.06625}$

12. $8.46\% = \underline{0.0846}$ | $1.58 = \underline{158}\%$

13. $0.125 = \underline{12.5}\%$ | $4.075\% = \underline{0.04075}$

14. $6.007\% = \underline{0.06007}$ | $0.312 = \underline{31.2}\%$

15. $6\frac{1}{4}\% = \underline{0.0625}$ | $9\frac{3}{4}\% = \underline{0.0975}$

16. $7\frac{3}{4}\% = \underline{0.0775}$ | $5\frac{1}{2}\% = \underline{0.055}$

LESSON FOLLOW-UP AND ERROR ANALYSIS

Number Correct: 84

27–32: Have students write decimals for problems 1–8 and fractions for problems 9–16.
20–26: Before students redo the page, ask them to explain how to change a fraction or decimal to a percent or vice versa.
Less than 20: Have students explain the value of the percent sign when changing percents to fractions or decimals. (value of hundredths)

Prerequisite Skills: percents, fractions, and decimals; inequalities, ordering

Lesson Focus: use inequality symbols to compare and order percents, fractions, and decimals
Possible Score: 32
Time Frame: 15–20 minutes

Lesson 11 Comparing and Ordering

When comparing percents, fractions, and decimals, use inequality symbols or the equal symbol to show the relationship.

> **The inequality symbols that show order are**
> \> greater than   ≥ greater than or equal to
> < less than   ≤ less than or equal to

Replace the _____ with <, >, or = in the following number sentences.

0.60 _____ $\frac{6}{100}$ $\frac{1}{3}$ _____ 30%   15% _____ 0.165

To make comparing easier, rewrite the numbers so that each is in the same format.

0.60 __>__ 0.06 $33\frac{1}{3}$% __>__ 30%   0.15 __<__ 0.165

Replace the _____ with <, >, or = in the following number sentences.

	a	b	c
1.	0.07 __<__ $\frac{2}{3}$	$\frac{1}{3}$ __>__ 0.33	85% __<__ 8.5
2.	$\frac{4}{5}$ __>__ 20%	12% __>__ 0.012	15% __>__ $\frac{1}{15}$
3.	80% __>__ $\frac{3}{5}$	$\frac{6}{7}$ __>__ 85%	2.5 __=__ $\frac{5}{2}$
4.	0.14% __<__ $\frac{14}{10}$	0.082 __<__ 80%	$\frac{7}{11}$ __<__ 0.64

Write the numbers from least to greatest.

5. $\frac{1}{2}$, 0.55, 45%; __45%__ < __$\frac{1}{2}$__ < __0.55__   90%, 0.99, 1; __90%__ < __0.99__ < __1__

Write the numbers from greatest to least.

6. $\frac{3}{4}$, 7.5, 68%; __7.5__ > __$\frac{3}{4}$__ > __68%__   0.115, 11%, $\frac{1}{11}$; __0.115__ > __11%__ > __$\frac{1}{11}$__

Number Correct
LESSON FOLLOW-UP AND ERROR ANALYSIS
26–32: Have students make a list from least to greatest of the commonly used fractions. Then write them in decimal and percent form.
19–25: Remind students that the point of an inequality symbol points to the smaller of two numbers.
Less than 19: Show students how they can use a number line to help them compare numbers.

85

Lesson 11 Problem Solving

Solve each problem.

1. Jim ate $\frac{3}{5}$ of a large pizza from a pizzeria. Sonny ate 75% of a large pizza from the same pizzeria. Who ate the largest portion of pizza?

   __Sonny__ ate the most pizza.

2. Sarah, Johannah, and Trista collect postcards. They each have the same number of postcards in their collections. They decided to compare how many each had from New Brunswick. Sarah said, "half of my collection is from New Brunswick." Trista said, "$\frac{5}{8}$ of my collection is from New Brunswick." Then Johannah proclaimed, "I have you both beat, 62% of my collection is from New Brunswick." Is Johannah's statement true or false? List the girls in order from the one who has the greatest number of New Brunswick postcards to the one who has the fewest New Brunswick postcards.

   Johannah's statement is __false__.

   From greatest to fewest, the number of New Brunswick postcards is __Trista, Johannah, Sarah__.

3. Mr. Morrison owned 5 hectares (ha) of land. He decided to distribute the property among his three children. The oldest child received the deed to $\frac{3}{8}$ of his land. The middle child was given $\frac{1}{3}$. The youngest child received about 30% of the land. Write the size of each child's land as a decimal. Decide if the size of land received matched each child's order in the family.

   The oldest child received __0.375__ of the land.

   The middle child received about __0.33__ of the land.

   The youngest child received __0.3__ of the land.

   The largest parcel of land was given to the __oldest__ child and the smallest parcel of land was given to the __youngest__ child.

1.

2.

3.

CHAPTER 5
Ratio, Rate, Proportion, and Percent

Lesson 11
Comparing and Ordering

Prerequisite Skills: changing percents to decimals

Lesson Focus: finding a percent of a number
Possible Score: 27
Time Frame: 25–30 minutes

Lesson 12   Percent of a Number   PRE-ALGEBRA

What number is $16\frac{1}{2}\%$ of 90?

$n = 16\frac{1}{2}\% \times 90$
$= 0.165 \times 90$
$= \underline{14.85}$

$\underline{14.85}$ is $16\frac{1}{2}\%$ of 90.

What number is 135% of 83?

$n = 135\% \times 83$
$= 1.35 \times 83$
$= \underline{112.05}$

$\underline{112.05}$ is 135% of 83.

Complete the following.

	a		b	
1.	__8__ is 40% of 20.		__4.8__ is 32% of 15.
2.	__96__ is 120% of 80.		__29.76__ is 62% of 48.
3.	__22.77__ is 33% of 69.		__57__ is 150% of 38.
4.	__525__ is $62\frac{1}{2}\%$ of 840.		__5.561__ is 6.7% of 83.
5.	__$\frac{3}{16}$__ is 50% of $\frac{3}{8}$.		__5.07__ is 7.8% of 65.
6.	__408__ is 85% of 480.		__5.9__ is 25% of 23.6.
7.	__24__ is $37\frac{1}{2}\%$ of 64.		__70__ is 175% of 40.
8.	__6.72__ is 6% of 112.		__46.08__ is 9.6% of 480.
9.	__432__ is 80% of 540.		__6.225__ is 12.5% of 49.8.
10.	__14.4__ is 8% of 180.		__124.8__ is 130% of 96.

CHAPTER 5

Number Correct   LESSON FOLLOW-UP AND ERROR ANALYSIS
23–27:   Have students give examples where problems like these are actually used. (taxes, sale prices, increased performance)
18–22:   Have students explain how to change each percent to a decimal and then multiply.
Less than 18:   Before students redo the pages, change a percent to a decimal and ask them to explain each step.

Lesson 12   Problem Solving   PRE-ALGEBRA

Solve each problem.

1. Of the building permits issued, 85% were for single-family dwellings. There were 760 permits issued. How many were for single-family dwellings?

   __646__ were for single-family dwellings.

2. Leona answered all the questions on a test. She had 90% of them correct. There were 40 questions in all. How many did she have correct?

   She had __36__ correct.

3. Of the 45 seats on the bus, 60% are filled. How many seats are filled?

   __27__ seats are filled.

4. An oil tank will hold 250 L. The tank is 80% full. How many litres of oil are in the tank?

   __200__ L of oil are in the tank.

5. A contractor is to remove 600 m³ of earth. So far, 70% of the work has been done. How many cubic metres of earth have been removed?

   __420__ m³ of earth have been removed.

6. Mrs. Hughes bought a mixture of grass seed that contained 75% bluegrass seed. She purchased 2.5 kg of grass seed in all. How many kilograms of bluegrass seed did she get?

   She got __1.875__ kg of bluegrass seed.

7. Mr. Jones' car gets 6.6 km per litre of fuel efficiency. He can improve his fuel efficiency by 15% by getting a tune-up. By how much will his fuel efficiency improve with a tune-up?

   His fuel efficiency will improve by __0.99__ km per litre.

CHAPTER 5
Ratio, Rate, Proportion, and Percent

Lesson 12
Percent of a Number

Prerequisite Skills: solving proportions

Lesson Focus: finding what percent one number is of another
Possible Score: 27
Time Frame: 25–30 minutes

Lesson 13   Percent of a Number   PRE-ALGEBRA

25 is what percent of 40?	$\frac{3}{8}$ is what percent of $\frac{1}{2}$?
$25 = n\% \times 40$	$\frac{3}{8} = n\% \times \frac{1}{2}$
$25 = \frac{n}{100} \times 40$	$\frac{3}{8} = \frac{n}{100} \times \frac{1}{2}$
$25 = \frac{40n}{100}$	$\frac{3}{8} = \frac{n}{200}$
$2500 = 40n$	$600 = 8n$
$\underline{\ 62.5\ } = n$	$\underline{\ 75\ } = n$
25 is __62.5__% of 40.	$\frac{3}{8}$ is __75__% of $\frac{1}{2}$.

Complete the following.

    a   *b*

1. 32 is __50__% of 64.   40 is __80__% of 50.

2. 88 is __110__% of 80.   67 is __100__% of 67.

3. $\frac{3}{8}$ is __50__% of $\frac{3}{4}$.   0.8 is __25__% of 3.2.

4. $18\frac{3}{4}$ is __25__% of 75.   96 is __80__% of 120.

5. 50 is __$62\frac{1}{2}$__% of 80.   1.6 is __25__% of 6.4.

6. $\frac{2}{3}$ is __80__% of $\frac{5}{6}$.   19 is __25__% of 76.

7. 78 is __75__% of 104.   19 is __20__% of 95.

8. 0.72 is __150__% of 0.48.   24 is __60__% of 40.

9. $8\frac{1}{3}$ is __25__% of $33\frac{1}{3}$.   64 is __80__% of 80.

10. $6\frac{1}{4}$ is __$12\frac{1}{2}$__% of 50.   0.69 is __25__% of 2.76.

Number Correct   **LESSON FOLLOW-UP AND ERROR ANALYSIS**
23–27: Have students choose a problem and write a story problem it would solve.
18–22: Before students redo the pages, review that they write a 1 beneath each whole number, then solve the resulting proportion. $(\frac{25}{1} = \frac{40n}{100})$
Less than 18: Have students explain what *is* and *of* mean in an equation. (*Is* means "equals" and *of* means "times.")

Lesson 13 Problem Solving   PRE-ALGEBRA

Solve each problem.

1. Last season a baseball player hit 48 home runs. So far this season he has hit 30 home runs. The number of home runs he has hit so far this season is what percent of the number of home runs he hit last season?

   The number he has hit this season is ___$62\frac{1}{2}$___ % of the number he hit last season.

2. In April, 175 cases of toy cars were sold. In May, 125 cases were sold. April's sales were what percent of May's sales?

   April's sales were ___140___ % of May's sales.

3. The down payment on a bike is $15. The bike costs $75. The down payment is what percent of the cost?

   The down payment is ___20___ % of the cost.

4. On a spelling test, Janice spelled 17 words correctly. There were 20 words on the test. What percent of the words did she spell correctly?

   She spelled ___85___ % correctly.

5. The Andersons are planning to take a 960-km trip. They will travel 840 km by car. What percent of the distance will they travel by car?

   They will travel ___$87\frac{1}{2}$___ % of the distance by car.

6. During hockey practice, Lea attempted 30 penalty shots and made 21. What percent of these penalty shots did she make?

   She made ___70___ % of the shots.

7. Emily's mass is 54 kg, and Marta's mass is 36 kg. Emily's mass is what percent of Marta's mass?

   Emily's mass is ___150___ % of Marta's mass.

CHAPTER 5
Ratio, Rate, Proportion, and Percent

Lesson 13
Percent of a Number

Prerequisite Skills: changing percents to fractions

Lesson Focus: finding a number when a percent of it is known
Possible Score: 27
Time Frame: 25–30 minutes

Lesson 14 Percent of a Number   PRE-ALGEBRA

32 is 16% of what number?

$32 = 16\% \times n$
$32 = \frac{16}{100} \times n$
$32 = \frac{16n}{100}$
$3200 = 16n$
$\underline{200} = n$

32 is 16% of __200__.

1.4 is 5.6% of what number?

$1.4 = 5.6\% \times n$
$1.4 = \frac{5.6}{100} \times n$
$1.4 = \frac{5.6n}{100}$
$140 = 5.6n$
$\underline{25} = n$

1.4 is 5.6% of __25__.

Complete the following.

a   b

1. 37 is 20% of __185__.   92 is 50% of __184__.

2. 3.4 is 25% of __13.6__.   60 is 150% of __40__.

3. 60 is 60% of __100__.   9 is 30% of __30__.

4. 50 is 40% of __125__.   78 is 60% of __130__.

5. 264 is 6% of __4400__.   84 is 12% of __700__.

6. 18 is 75% of __24__.   2.6 is 50% of __5.2__.

7. 8.7 is 30% of __29__.   72 is 80% of __90__.

8. 9 is 100% of __9__.   1.3 is 65% of __2__.

9. 144 is 24% of __600__.   2.16 is 3.6% of __60__.

10. 192 is 75% of __256__.   12.8 is 6.4% of __200__.

Number Correct   LESSON FOLLOW-UP AND ERROR ANALYSIS
23–27: Have students choose a problem and write a story problem using the numbers in that problem.
18–22: Have students show that *what number* becomes *n* in the equation before they redo incorrect answers.
Less than 18: Have students explain what *is* and *of* mean in an equation. (*Is* means "equals" and *of* means "times.")

CHAPTER 5

Lesson 14   Problem Solving   PRE-ALGEBRA

Solve each problem.

1. Mr. Buccola has a tree that is 15 m tall. He estimates that the tree is 75% as tall now as it will be when fully grown. How tall will the tree be when fully grown?

   The tree will be ___20___ m tall.

2. There are 35 boys on the school football team. This number represents 5% of the school's total enrollment. What is the school's total enrollment?

   The school's total enrollment is ___700___.

3. Jessica has read 120 pages of a library book. This is 40% of the book. How many pages are there in the book?

   There are ___300___ pages in the book.

4. When operating at 75% capacity, a factory can produce 360 cases of nails each day. How many cases of nails can be produced each day when the factory is operating at full capacity?

   ___480___ cases can be produced each day.

5. Brianna received 212 votes for class secretary. This was 53% of the total number of votes cast. How many votes were cast?

   ___400___ votes were cast.

6. Kristen has earned 75% of the points she needs for a prize. She has earned 660 points. How many points are needed to win a prize?

   ___880___ points are needed.

7. Emma can throw a baseball 8 m. This is 80% as far as her older brother can throw the ball. How far can her older brother throw the ball?

   Her older brother can throw the ball ___100___ m.

CHAPTER 5
Ratio, Rate, Proportion, and Percent

Lesson 14
Percent of a Number

Prerequisite Skills: changing fractions, decimals, and percents

Lesson 15 Percent   PRE-ALGEBRA

Lesson Focus: finding percents, parts, and wholes
Possible Score: 39
Time Frame: 25–30 minutes

Complete the following.

a   *b*

1. __12__ is 40% of 30.   73 is __20__% of 365.

2. 26 is __52__% of 50.   24 is 60% of __40__.

3. 39 is 52% of __75__.   __19__ is 25% of 76.

4. 37 is __74__% of 50.   __216__ is 60% of 360.

5. 18 is 25% of __72__.   69 is __25__% of 276.

6. __23.04__ is 24% of 96.   8 is 16% of __50__.

7. 0.7 is __50__% of 1.4.   __49.2__ is 50% of 98.4.

8. 3.9 is 75% of __5.2__.   0.09 is __36__% of 0.25.

9. __48.96__ is 6.8% of 720.   0.95 is 1.9% of __50__.

10. 64 is 125% of __51.2__.   __986__ is 100% of 986.

11. 175 is __140__% of 125.   98 is 150% of $65\frac{1}{3}$.

12. __864__ is 120% of 720.   275 is __220__% of 125.

13. $\frac{1}{3}$ is __40__% of $\frac{5}{6}$.   30 is 75% of __40__.

14. __600__ is 60% of 1000.   1 is __100__% of 1.

15. 15 is 50% of __30__. $1\frac{1}{2}$ is 75% of 2.

16. __1__ is 25% of 4.   73 is 25% of __292__.

Number Correct
LESSON FOLLOW-UP AND ERROR ANALYSIS
33–39: Have students choose a problem and illustrate the three types of percent problems.
25–32: Ask students to choose a percent problem on page 93 and explain how to solve it.
Less than 25: As they redo incorrect answers, help students write each equation as a proportion and then solve.

Lesson 15 Problem Solving   PRE-ALGEBRA

Solve each problem.

1. There are 850 students at a school. Of these, 36% are in grade 8. How many students are in the grade 8?

   __306__ students are in grade 8.

2. Mail was delivered to 171 out of the 180 houses on Saylor Street. To what percent of the houses on Saylor Street was mail delivered?

   Mail was delivered to __95__% of the houses.

3. A savings bond costs $18.75 and can be redeemed at maturity for $25. The cost of the bond is what percent of its value at maturity?

   Its cost is __75__% of its value at maturity.

4. Mrs. Wilson sold merchandise to 25% of the clients she contacted. She sold to six clients. How many clients did she contact?

   She contacted __24__ clients.

5. A store sold 185 bicycles last month. Of those, 60% were girls' bicycles. How many girls' bicycles were sold?

   __111__ girls' bicycles were sold.

6. Of the library books turned in today, 95% were turned in on time. There were 285 books turned in on time. What was the total number of books turned in?

   There were __300__ books turned in.

7. The enrollment at Pleasant Street School is 110% of last year's enrollment. The enrollment last year was 390. What is the enrollment this year?

   The enrollment this year is __429__.

For further evaluation, copy the Chapter Test on page 269.
For maintaining skills, use the Cumulative Review on pages 221–222.

Possible Score: 20
Time Frame: 20–25 minutes

CHAPTER 5   PRACTICE TEST
Ratio, Rate, Proportion, and Percent

Express each of the following as a ratio or a rate in two ways as shown.

	a	b
1. 8 goals in 3 games	8 to 3	$\frac{8}{3}$
2. 5 policemen to 4 firemen	5 to 4	$\frac{5}{4}$
3. 4 planes in 30 minutes	4 to 30	$\frac{4}{30}$
4. 5 quarts for 9 boys	5 to 9	$\frac{5}{9}$

Solve the following.

a

b

5. $\frac{n}{3} = \frac{12}{36}$ $n = 1$ $\frac{4}{n} = \frac{16}{20}$ $n = 5$

6. $\frac{8}{9} = \frac{n}{45}$ $n = 40$ $\frac{7}{8} = \frac{49}{n}$ $n = 56$

7. $\frac{18}{24} = \frac{n}{16}$ $n = 12$ $\frac{n}{12} = \frac{4}{16}$ $n = 3$

Complete the following.

a

b

8. __22.4__ is 35% of 64.   11 is __20__ % of 55.

9. 18 is __72__ % of 25.   30 is 7.5% of __400__.

10. $1\frac{3}{4}$ is __70__ % of $2\frac{1}{2}$.   __$3\frac{3}{8}$__ is $12\frac{1}{2}$% of 27.

11. __270__ is 150% of 180.   2.4 is __25__ % of 9.6.

Number
Correct   LESSON FOLLOW-UP AND ERROR ANALYSIS
17–20: Have students choose a problem from problems 8–11 and write a story problem it would solve.
13–16: Check to see if students forgot to cross multiply when solving proportions. Review solving proportions.
Less than 13: Before students redo incorrect answers, ask them to explain how to solve proportions.

95

CHAPTER 6 PRETEST
Simple/Compound Interest

Complete the following for simple interest.

	principal	rate	time	interest
1.	$320	7%	1 year	$22.40
2.	$300	$5\frac{1}{2}$%	$\frac{1}{2}$ year	$8.25
3.	$800	10%	1 year	$80
4.	$500	8%	$\frac{1}{4}$ year	$10
5.	$600	16%	2 years	$192
6.	$26 000	9%	4 years	$9360

Interest is to be compounded in each account below. Find the total amount that will be in each account after the period of time indicated.

	principal	rate	time	compounded	total amount
7.	$200	6%	2 years	annually	$224.72
8.	$100	5%	3 years	annually	$115.76
9.	$300	8%	$1\frac{1}{2}$ years	semiannually	$337.46
10.	$400	5%	1 year	quarterly	$420.38

Prerequisite Skills: finding percent of a number

Lesson Focus: finding simple interest
Possible Score: 18
Time Frame: 20–25 minutes

Lesson 1 Simple Interest PRE-ALGEBRA

If the rate of interest is 12% a year, what will the interest be on a $300 loan for $1\frac{1}{2}$ years?

$interest = principal \times rate \times time$ (in years)

$$i = p \times r \times t$$
$$= 300 \times 0.12 \times \frac{3}{2}$$
$$= 36 \times \frac{3}{2}$$
$$= 54$$

The interest will be $ __54.00__ .

If the rate of interest is $9\frac{1}{2}$% a year, what will the interest be on a $100 loan for 2 years?

$$i = p \times r \times t$$
$$= \underline{100} \times \underline{0.095} \times \underline{2}$$
$$= \underline{9.50} \times \underline{2}$$
$$= \underline{19}$$

The interest will be $ __19.00__ .

Find the interest for each loan described below.

	principal	rate	time	interest
1.	$250	10%	2 years	$50
2.	$400	12%	2 years	$96
3.	$550	8%	$1\frac{1}{4}$ years	$55
4.	$650	$11\frac{1}{2}$%	3 years	$224.25
5.	$600	16%	3 years	$288
6.	$500	$11\frac{1}{4}$%	1 year	$56.25
7.	$1500	15%	$1\frac{1}{3}$ years	$300
8.	$1000	$12\frac{1}{2}$%	3 years	$375
9.	$2890	14%	$2\frac{1}{2}$ years	$1011.50
10.	$2600	9%	$2\frac{1}{2}$ years	$585

Number Correct — LESSON FOLLOW-UP AND ERROR ANALYSIS

15–18: Ask students how much it would take to totally repay each loan in problems 1–10. (principal + interest = total repayment)
12–14: Check to see if students changed each percent to a decimal before they multiplied.
Less than 12: Have students explain how to change mixed numerals to fractions or decimals before they multiply. ($1\frac{1}{2} = \frac{3}{2}$ or 1.5)

CHAPTER 6

97

Lesson 1 Problem Solving PRE-ALGEBRA

Solve each problem.

1. Mr. Wilkinson borrowed $600 for $1\frac{1}{2}$ years. He is to pay 9% annual interest. How much interest is he to pay?

   He will pay $ __81.00__ interest.

2. Hillary had $350 in a savings account for $\frac{1}{2}$ year. Interest was paid at an annual rate of 5%. How much interest did she receive?

   She received $ __8.75__ interest.

3. Suppose you deposit $700 in a savings account at $5\frac{1}{2}$% interest. How much interest will you receive in one year?

   You will receive $ __38.50__ .

4. The Tremco Company borrowed $10 000 at 12% annual interest for a 1-year period. How much interest did the company have to pay? What was the total amount (principal + interest) the company needed to repay the loan?

   The company had to pay $ __1200.00__ interest.

   The total amount needed was $ __11 200.00__ .

5. Ian borrowed $700 for 1 year. Interest on the first $300 of the loan was 18%, and interest on the remainder of the loan was 12%. How much interest did he pay?

   He paid $ __102.00__ interest.

6. Molly's mother borrowed $460 at 10% annual interest. What would be the interest if the loan were repaid after $\frac{1}{2}$ year? What would the interest be if the loan were repaid after $\frac{3}{4}$ year?

   The interest would be $ __23.00__ for $\frac{1}{2}$ year.

   The interest would be $ __34.50__ for $\frac{3}{4}$ year.

CHAPTER 6
Simple/Compound Interest

Lesson 1
Simple Interest

Prerequisite Skills: finding percent of a number

Lesson Focus: finding principal, rate, time, and interest
Possible Score: 18
Time Frame: 25–30 minutes

Lesson 2 Simple Interest   PRE-ALGEBRA

$36 interest is paid in 2 years at a flat rate of 9%. Find the principal.

$$i = p \times r \times t$$
$$36 = p \times 0.09 \times 2$$
$$36 = 0.18p$$
$$\frac{36}{0.18} = p$$
$$\underline{200} = p$$

The principal is $ __200.00__ .

$16 interest is paid in 2 years on $100 principal. Find the rate.

$$i = p \times r \times t$$
$$16 = 100 \times r \times 2$$
$$16 = 200r$$
$$\frac{16}{200} = r$$
$$\underline{0.08} = r$$

The rate is __8__ %.

$50 interest is paid on $200 principal at a rate of 10%. Find the time.

$$i = p \times r \times t$$
$$50 = 200 \times 0.10 \times t$$
$$50 = 20t$$
$$\frac{50}{20} = t$$
$$\underline{2.5} = t$$

The time is __2.5__ years.

Complete the following.

	principal	rate	time	interest
1.	$100	7%	3 years	$21
2.	$325	8%	$1\frac{1}{2}$ years	$39
3.	$375	10%	$\frac{1}{2}$ year	$18.75
4.	$780	15%	2 years	$234
5.	$1200	9%	2 years	$216
6.	$1400	8%	$1\frac{1}{2}$ years	$168
7.	$3500	$8\frac{1}{2}$%	$1\frac{1}{2}$ years	$446.25
8.	$8000	$7\frac{1}{2}$%	$2\frac{1}{2}$ years	$1500
9.	$18 050	12%	3 years	$6498
10.	$25 000	15%	3 years	$11 250

CHAPTER 6

Number Correct

LESSON FOLLOW-UP AND ERROR ANALYSIS

15–18: Ask students how much it would take to totally repay each loan in problems 1–10. (principal + interest = total repayment)
12–14: Have students explain how to use the formula ($i = p \times r \times t$) at the top of the page to solve each problem.
Less than 12: Before students redo page 100, ask them to explain what is given and what is asked for in each problem.

Lesson 2 Problem Solving   PRE-ALGEBRA

Solve each problem.

1. Mrs. Vernon paid $72 interest for a 2-year loan at 9% annual interest. How much money did she borrow?

   She borrowed $ __400.00__ .

2. Matthew paid $63 interest for a $350 loan for $1\frac{1}{2}$ years. What was the rate of interest?

   The rate of interest was __12__ %.

3. Suppose you borrow $600 at 10% interest. What period of time would you have the money if the interest is $30?

   The period of time would be __$\frac{1}{2}$__ year.

4. Albertito had $740 in a savings account at 5% interest. The money was in the account for $\frac{1}{4}$ year. How much interest did he receive? Suppose he withdrew the principal and interest after $\frac{1}{4}$ year. How much money would he withdraw from the account?

   He received $ __9.25__ interest.

   He would withdraw $ __749.25__ from the account.

5. How much must you deposit at $5\frac{1}{2}$% annual interest in order to earn $33 in 1 year?

   You would need $ __600.00__ in the account.

6. The interest on a $300 loan for 2 years is $90. What rate of interest is charged?

   The rate of interest is __15__ %.

7. How much must you have on deposit at $6\frac{1}{2}$% of annual interest in order to earn $221 in a year?

   You would have to deposit $ __3400.00__ .

CHAPTER 6
Simple/Compound Interest

Prerequisite Skills: finding percent of a number

Lesson Focus: finding interest compounded annually
Possible Score: 14
Time Frame: 20–25 minutes

Lesson 3 Compound Interest

Interest paid on the original principal and the interest already earned is called **compound interest**.

Bev had $400 in a savings account for 3 years that paid 6% interest compounded annually. What was the total amount in her account at the end of the third year?

At the end of 1 year:

interest = 400 × 0.06 × 1 = 24.00 or $24

new principal = 400 + 24 = 424 or $424

At the end of 2 years:

interest = 424 × 0.06 × 1 = 25.44 or $25.44

new principal = 424 + 25.44 = 449.44 or $449.44

At the end of 3 years:

interest = 449.44 × 0.06 × 1 = 26.9664 or $26.97

total amount = __449.44__ + __26.97__ = __476.41__ or $__476.41__

Assume interest is compounded annually. Find the total amount for each of the following.

	principal	rate	time	total amount
1.	$500	6%	2 years	$561.80
2.	$700	$5\frac{1}{2}$%	2 years	$779.12
3.	$800	5%	3 years	$926.10
4.	$800	$6\frac{1}{2}$%	3 years	$966.36
5.	$200	9%	3 years	$259.01
6.	$1000	8%	2 years	$1166.40

Number Correct

LESSON FOLLOW-UP AND ERROR ANALYSIS

12–14: Have students compare each total amount to its principal and circle the total amount that shows the greatest increase and explain why.

9–11: Check to see if students forgot to add the interest when determining the new principal.

Less than 9: Have students explain how to insert a zero when changing 1-digit percents to decimals. (6% = 0.06)

101

Lesson 3 Problem Solving

Solve each problem.

1. Heidi had $600 in a savings account for 2 years. Interest was paid at the rate of 6% compounded annually. What was the total amount in her account at the end of 2 years?

   The total amount was $___674.16___.

2. Travis deposited $400 in an account that pays 5% interest compounded annually. What will be the total amount in his account after 2 years? After 3 years?

   It will be $___441.00___ after 2 years.

   It will be $___463.05___ after 3 years.

3. Aubrey deposited $300 in an account that pays 7% interest compounded annually. Tori deposited $300 in an account at an annual rate of 7% (simple interest). After 3 years what will be the total amount in Aubrey's account? In Tori's account?

   Aubrey's account will have $___367.51___.

   Tori's account will have $___363.00___.

4. Ms. Sanchez has $500 in her savings account, which pays 5% interest compounded annually. What will be the value of the account after 3 years?

   The value will be $___578.81___.

5. Landon deposited $300 at 6% interest compounded annually. Elisa deposited $200 at $6\frac{1}{2}$% interest compounded annually. Who will have the greater account after 3 years? How much greater will it be?

   ___Landon___ will have the greater account.

   It will be $___115.70___ greater.

CHAPTER 6
Simple/Compound Interest

Lesson 3
Compound Interest

Prerequisite Skills: finding percent of a number

Lesson Focus: finding compound interest
Possible Score: 14
Time Frame: 25–30 minutes

Lesson 4 Compound Interest

Interest may be paid annually (each year), semiannually (twice a year), quarterly (four times a year), monthly (every month), or daily (every day).

Ed had $100 in an account for $1\frac{1}{2}$ years that paid 6% interest compounded semiannually. What was the total amount in his account at the end of $1\frac{1}{2}$ years?

At the end of $\frac{1}{2}$ year:

interest = $100 \times 0.06 \times \frac{1}{2}$ = 3.00 or $3

new principal = 100 + 3 = 103 or $103

At the end of 1 year:

interest = $103 \times 0.06 \times \frac{1}{2}$ = 3.09 or $3.09

new principal = 103 + 3.09 = 106.09 or $106.09

At the end of $1\frac{1}{2}$ years:

interest = $106.09 \times 0.06 \times \frac{1}{2}$ = 3.1827 or $3.18

total amount = __106.09__ + __3.18__ = __109.27__ or $__109.27__

Find the total amount for each of the following.

	principal	rate	time	compounded	total amount
1.	$200	6%	$1\frac{1}{2}$ years	semiannually	$218.55
2.	$300	5%	2 years	semiannually	$331.15
3.	$100	5%	1 year	quarterly	$105.10
4.	$400	7%	$\frac{3}{4}$ year	quarterly	$421.37
5.	$500	8%	4 months	monthly	$513.47
6.	$600	9%	$\frac{1}{4}$ year	monthly	$613.60

Number Correct LESSON FOLLOW-UP AND ERROR ANALYSIS
- **12–14:** Have students choose a problem and find the total interest for that problem.
- **9–11:** Discuss that the *time* and *compounded* columns together indicate how many times to multiply and add.
- **Less than 9:** Before students redo the pages, have them explain how to convert *semiannually, quarterly,* and *monthly* to fractions.

Lesson 4 Problem Solving

Solve each problem.

1. Mrs. Fauler had $600 in a savings account that paid 5% interest compounded semiannually. What was the value of her account after $1\frac{1}{2}$ years?

   The value was $ __646.14__ .

2. How much interest would $3000 earn in two years at 7% interest compounded semiannually?

   It would earn $ __442.58__ interest.

3. Suppose $100 were deposited in each savings account with rates of interest as follows:
   Account **A**—6% compounded annually
   Account **B**—6% compounded semiannually
   Account **C**—6% compounded quarterly
   What would be the value of each account after 1 year?

   $ __106.00__ will be in account **A**.

   $ __106.09__ will be in account **B**.

   $ __106.14__ will be in account **C**.

4. Assume $200 was deposited in a 2-year account at 9%. How much more interest would be in the account if the interest were compounded annually rather than computed as simple interest?

   There would be $ __1.62__ more in the account.

5. Account **A** has $500 at 8% interest compounded quarterly. Account **B** has $500 at 8% interest compounded semiannually. Which account will have a greater amount of money after 1 year? How much more?

   Account __A__ will have more money.

   It will have $ __0.41__ more.

For further evaluation, copy the Chapter Test on page 270.
For maintaining skills, use the Cumulative Review on pages 223–224.
For assessment of Chapters 1–6, use the Mid-Test on pages 207–208.

Possible Score: 10
Time Frame: 15–20 minutes

CHAPTER 6   PRACTICE TEST
Simple/Compound Interest

Complete the following for simple interest.

	principal	rate	time	interest
1.	$150	15%	3 years	$67.50
2.	$700	$8\frac{1}{2}$%	2 years	$119.00
3.	$645	12%	$\frac{1}{4}$ year	$19.35
4.	$540	10%	$2\frac{1}{2}$ years	$135.00
5.	$3840	$9\frac{1}{2}$%	2 years	$729.60
6.	$1800	15%	2 years	$540.00

Interest is to be compounded in each account below. Find the total amount that will be in each account after the period of time indicated.

	principal	rate	time	compounded	total amount
7.	$300	7%	2 years	annually	$343.47
8.	$600	5%	3 years	annually	$694.58
9.	$500	6%	2 years	semiannually	$562.75
10.	$400	9%	$\frac{1}{4}$ year	monthly	$409.07

Number Correct   LESSON FOLLOW-UP AND ERROR ANALYSIS
  9–10: Have students choose a problem from problems 7–10 and find the total interest for that problem.
   7–8: Check to see if students had difficulty with problems 7–10. If so, have them explain how to find compound interest.
         (Review Lessons 3 and 4.)
Less than 7: Check to see if students had difficulty with problems 1–6. If so, have them explain how to find simple interest.
         (Review Lesson 2)

CHAPTER 7 PRETEST
Metric Measurement

Circle the unit you would use to measure each of the following:

1. capacity of a tank   metre   (litre)   gram

2. length of a string   (centimetre)   centilitre   centigram

3. weight of an ant   millimetre   millilitre   (milligram)

Write *1000; 0.01;* or *0.001* to make each sentence true.

4. The prefix *milli* means __0.001__.

5. The prefix *kilo* means __1000__.

6. The prefix *centi* means __0.01__.

Measure each line segment to the nearest unit indicated.

7. __5__ cm

8. __65__ mm

Complete the following:

a	b
9. 1 m = __0.001__ km	100 mm = __10__ cm
10. 2 L = __2000__ mL	2 kg = __2000__ g
11. 0.5 g = __500__ mg	300 cm = __3__ m
12. 1.4 kL = __1400__ L	0.05 km = __50__ m

CHAPTER 7
Metric Measurement
106

CHAPTER 7 PRETEST

Prerequisite Skills: whole number and decimal place values

Lesson Focus: recognizing metric units of length, capacity, and mass
Possible Score: 18
Time Frame: 5–10 minutes

Lesson 1   Metric Prefixes

A **metre** is a unit of *length*.

A **litre** is a unit of *capacity*.

A **gram** is a unit of *mass*.

Kilo means 1000.   *Kilometre* means __1000__ m.

Hecto means 100.   *Hectolitre* means __100__ L.

Deca means 10.   *Decagram* means __10__ g.

Deci means 0.1.   *Decimetre* means __0.1__ m.

Centi means 0.01.   *Centilitre* means __0.01__ L.

Milli means 0.001.   *Milligram* means __0.001__ g.

The most commonly used prefixes are *kilo, centi,* and *milli*.

Tell whether the following would be measured in *metres, litres,* or *grams*.

           a   *b*

1. amount of juice in a glass __litres__   mass of a pencil __grams__

2. distance a baseball is thrown __metres__   length of a bus __metres__

3. amount of fuel in a truck __litres__   mass of a bird __grams__

Complete the following as shown.

4. **Kilo**litre means __1000 litres__.   **Kilo**gram means __1000 grams__.

5. **Centi**gram means __0.01 gram__.   **Centi**metre means __0.01 metre__.

6. **Milli**litre means __0.001 litre__.   **Milli**metre means __0.001 metre__.

Name two things that could be measured with each of the following.

7. metres   __Answers will vary.__

8. litres

9. grams

Number Correct

LESSON FOLLOW-UP AND ERROR ANALYSIS

- **14–18:** Have students circle the longest unit of length and ☐ the smallest unit of capacity.
- **11–13:** Have students explain the relationship between units and that all units of measure in the metric system use the same prefixes.
- **Less than 11:** Ask students to explain the meaning of the prefixes and give examples for length, capacity, and mass.

Prerequisite Skills: recognizing metric prefixes

Lesson Focus: converting metric units of length
Possible Score: 15
Time Frame: 5–10 minutes

Lesson 2   Length

To name a unit of length other than the metre, a *prefix* is attached to the word *metre*. This prefix denotes the relationship of that particular unit to the metre.

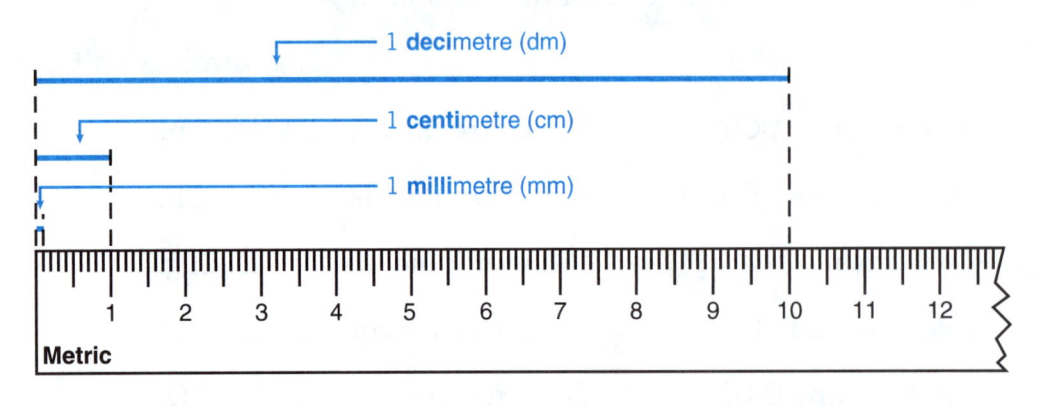

1 mm = __0.001__ m   1 dm = __0.1__ m

1 cm = __0.01__ m   1 **kilo**metre (km) = 1000 m

In each pair of measurements below, circle the measurement for the greater length.

	a	b	c
1.	(1 km); 1 dm	(1 dm); 1 cm	(1 km); 1 mm
2.	1 mm ;(1 dm)	1 cm ;(1 km)	(1 cm); 1 mm

Complete the following.

	a	b	c
3.	1 m = __100__ cm	1 m = __1000__ mm	1 m = __10__ dm
4.	0.01 m = __1__ cm	0.001 m = __1__ mm	0.1 m = __1__ dm
5.	1000 m = __1__ km	1 m = __0.001__ km	0.001 km = __1__ m

Number Correct
108

LESSON FOLLOW-UP AND ERROR ANALYSIS

13–15: Have students measure their heights in centimetres and millimetres.
10–12: Before students redo the page, ask them to choose two units of measure and explain the relationship between them.
Less than 10: Discuss that when changing from a larger unit of measure to a smaller unit, the answer is larger. (1 m = 100 cm)

Prerequisite Skills: recognizing metric prefixes

Lesson Focus: converting metric units of length
Possible score: 19
Time Frame: 5–10 minutes

Lesson 3 Units of Length

To change from	to millimetres, multiply by	to centimetres, multiply by	to metres, multiply by	to kilometres, multiply by
millimetres		0.1	0.001	0.000 001
centimetres	10		0.01	0.00 001
metres	1000	100		0.001
kilometres	1 000 000	100 000	1000	

Using the table makes it easy to complete exercises like the following.

  8.43 km = _____?_____ m   75 mm = _____?_____ cm
  1 km = 1000 m   1 mm = 0.1 cm
  8.43 km = (8.43 × 1000) m   75 mm = (75 × 0.1) cm
  8.43 km = __8430__ m   75 mm = __7.5__ cm

Complete.

          a   b

1. 5 km = __5000__ m   0.452 km = __452__ m

2. 38 m = __0.038__ km   948 m = __0.948__ km

3. 7.5 m = __750__ cm   80 m = __8000__ cm

4. 4 cm = __0.04__ m   75 cm = __0.75__ m

5. 92 cm = __920__ mm   4.86 cm = __48.6__ mm

6. 92 mm = __9.2__ cm   7 mm = __0.7__ cm

7. 0.5 m = __500__ mm   0.003 m = __3__ mm

8. 92 mm = __0.092__ m   3600 mm = __3.6__ m

9. A city block is about 200 m long.
    How long is a city block in kilometres? __0.2 km__

10. How long would five city blocks be in metres? __1000 m__   In kilometres? __1 km__

Number Correct   LESSON FOLLOW-UP AND ERROR ANALYSIS
  16–19: Have students select several objects and give their approximate measurements.
  12–15: Have students explain how to use the chart at the top of the page to find the number by which to multiply.
Less than 12: Ask students to explain how an answer changes as you convert from a larger unit to a smaller unit or vice versa.

Prerequsite Skills: recognizing metric prefixes

Lesson Focus: converting metric units of capacity
Possible Score: 16
Time Frame: 5–10 minutes

Lesson 4 Capacity

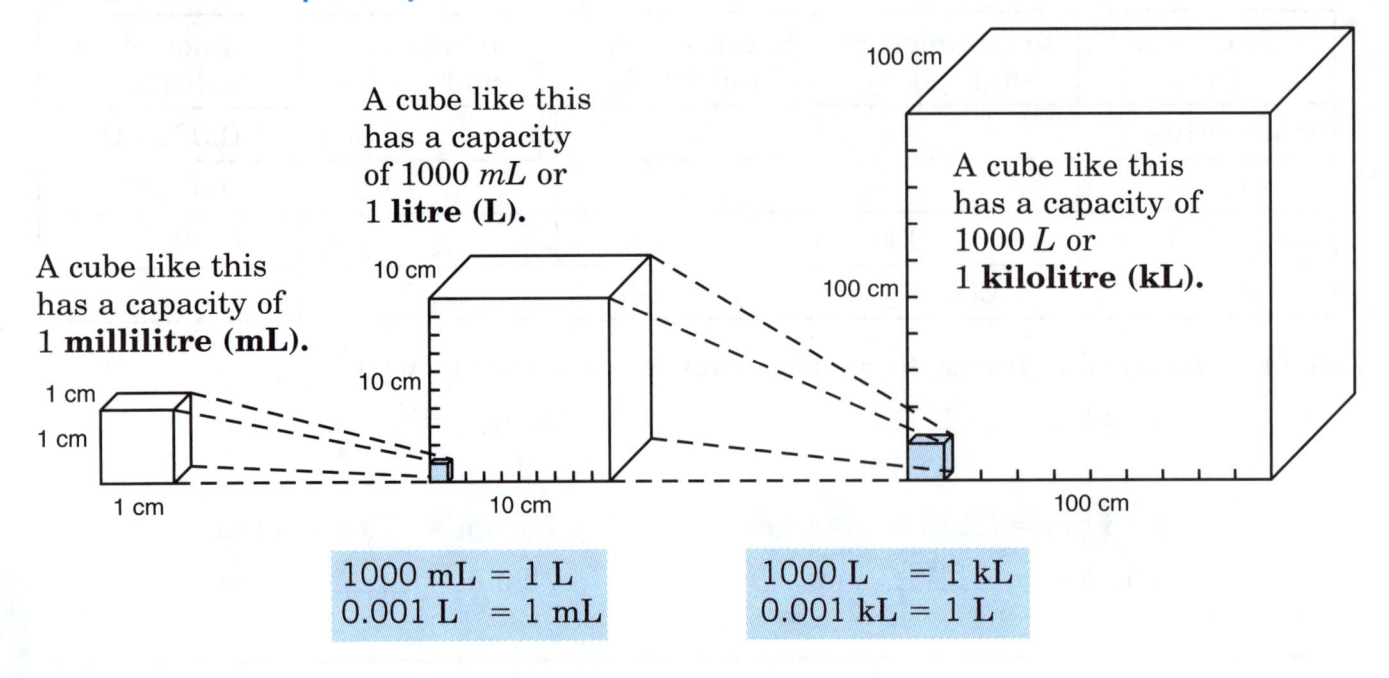

A cube like this has a capacity of 1 **millilitre (mL)**.

A cube like this has a capacity of 1000 mL or 1 **litre (L)**.

A cube like this has a capacity of 1000 L or 1 **kilolitre (kL)**.

1000 mL = 1 L
0.001 L = 1 mL

1000 L = 1 kL
0.001 kL = 1 L

Underline the measurement for the greater amount.

　　　　　　a　　　　　　　　　　　　　　　　b

1. 10 L, <u>10 kL</u>　　　　　　　　　　　　100 mL, <u>1 kL</u>

2. 0.1 kL, <u>1000 L</u>　　　　　　　　　　　10 mL, <u>1 L</u>

3. <u>1000 L</u>, 10 000 mL　　　　　　　　　<u>0.001 kL</u>, 1 mL

4. 500 L, <u>1 kL</u>　　　　　　　　　　　　700 mL, <u>1 L</u>

Complete the following.

5. 1 L = __1000__ mL　　　　　　　　　　0.1 L = __100__ mL

6. 1 kL = __1000__ L　　　　　　　　　　0.01 kL = __10__ L

7. 0.001 kL = __1__ L　　　　　　　　　1000 mL = __1__ L

8. 100 L = __0.1__ kL　　　　　　　　　10 kL = __10 000__ L

Number Correct

LESSON FOLLOW-UP AND ERROR ANALYSIS

13–16: Have students name an object that might come in (or could be placed in) each of the boxes pictured at the top of the page.
10–12: Before students redo incorrect answers, have them explain the metric prefixes at the top of Lesson 1 and the relationship between each.
Less than 10: Help students use a unit of length (cm) to explain capacity. You may want to make a model of each box.

Prerequisite Skills: recognizing metric prefixes

Lesson Focus: converting metric units of capacity
Possible Score: 16
Time Frame: 10–15 minutes

Lesson 5 Units of Capacity

1.2 kL = __?__ L
1 kL = 1000 L
1.2 kL = (1.2 × 1000) L
1.2 kL = __1200__ L

54 L = __?__ kL
1 L = 0.001 kL
54 L = (54 × 0.001) kL
54 L = __0.054__ kL

Complete the following.

　　　　　　　a　　　　　　　　　　　　　　　　　b

1. 6.4 L = __6400__ mL　　　　　　6000 mL = __6__ L
2. 25 kL = __25 000__ L　　　　　　752 L = __0.752__ kL
3. 78 L = __78 000__ mL　　　　　　529 mL = __0.529__ L
4. 0.986 kL = __986__ L　　　　　　42 L = __0.042__ kL
5. 7.5 L = __7500__ mL　　　　　　7.5 mL = __0.0075__ L
6. 7.5 kL = __7500__ L　　　　　　7.5 L = __0.0075__ kL

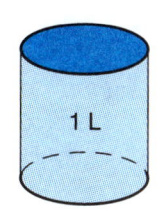

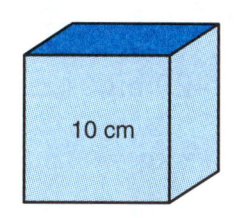

1 L of water will fill a cube with side length 10 cm.

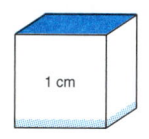

1 mL of water will fill a cube with side length 1 cm.

Would you use millilitres or litres to measure each of the following?

7. a dose of cough medicine　　　　L　　　(mL)

8. water in an aquarium　　　　　　(L)　　　mL

9. perfume in a bottle　　　　　　　L　　　(mL)

Number Correct

LESSON FOLLOW-UP AND ERROR ANALYSIS

13–16: Have students choose a measure in problems 1–6 and write a story problem using that measure.
10–12: Have students explain how they determine whether to multiply by 1000 or by 0.001 when converting between units.
Less than 10: Check to see if students understand that moving a decimal point to the right increases the answer and to the left decreases it.

111

Prerequisite Skills: recognizing metric prefixes

Lesson Focus: converting metric units of mass
Possible Score: 14
Time Frame: 5–10 minutes

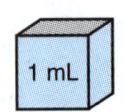

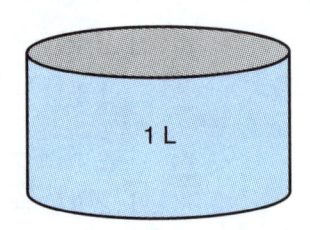

Lesson 6   Units of Mass

An aspirin tablet has a mass of about 350 **milligrams (mg)**.

1 mL of water has a mass of 1 **gram (g)**.

1 L of water has a mass of 1 **kilogram (kg)**.

1000 mg = 1 g
0.001 g = 1 mg

1000 g = 1 kg
0.001 kg = 1 g

65 g = __?__ mg
1 g = 1000 mg
65 g = (65 × 1000) mg
65 g = __65 000__ mg

250 g = __?__ kg
1 g = 0.001 kg
250 g = (250 × 0.001) kg
250 g = __0.250__ kg

Complete the following.

                a   b

1. 26 g = __26 000__ mg   6.2 g = __6200__ mg
2. 75.2 mg = __0.0752__ g   2420 mg = __2.42__ g
3. 89 kg = __89 000__ g   7.5 kg = __7500__ g
4. 835 g = __0.835__ kg   5.6 g = __0.0056__ kg
5. 60.5 g = __60 500__ mg   60.5 g = __0.0605__ kg

6. A teaspoon holds about 5 mL of water. What is the mass of 5 mL in grams? In milligrams?

   It's mass is __5__ g.
   It's mass is __5000__ mg.

6.

7. A nickel has a mass of about 5 g. What is the mass of 200 nickels in grams? In kilograms?

   200 nickels have a mass of about __1000__ g.
   200 nickels have a mass of about __1__ kg.

7.

LESSON FOLLOW-UP AND ERROR ANALYSIS

Number Correct
12–14: Have students choose a measure in problems 1–5 and write a story problem using that measure.
9–11: Have students explain how they determine whether to multiply by 1000 or by 0.001 when converting between units of mass.
Less than 9: Check to see if students understand that moving a decimal point to the right increases the answer and to the left decreases it.

Prerequisite Skills: metric units of length, capacity, and mass

Lesson Focus: problem solving with metric units
Possible Score: 9
Time Frame: 10–15 minutes

Lesson 7 Problem Solving

Solve each problem.

1. A pitcher contained 1.2 L of milk. You used 250 mL of milk from the pitcher. How many millilitres of milk are left in the pitcher?

    __950__ mL are left.

2. Megan says she is 1.6 m tall. Nicole says she is 162 cm tall. Who is taller? How many centimetres taller?

    __Nicole__ is __2__ cm taller.

3. A jet flew 1 km on 15 L of fuel. How many kilolitres of fuel are needed for the jet to fly 3000 km?

    The jet would need __45__ kL of fuel.

4. 2 L of grape juice will fill eight glasses of the same size. What is the capacity of each glass in millilitres?

    Each glass has a capacity of __250__ mL.

5. Tim bought 6 kg of meat for $25.20. What was the cost per kilogram?

    The cost was $__4.20__ per kilogram.

6. During a contest, Frog A jumped 59.3 cm. Frog B jumped 590 mm. Which frog jumped farther? How many centimetres farther?

    Frog __A__ jumped __0.3__ cm farther.

7. Ben drove 158 km. Ali drove 230 km. How much farther than Ben did Ali drive?

    She drove __72__ km farther.

CHAPTER 7

Number Correct

LESSON FOLLOW-UP AND ERROR ANALYSIS

8–9: Have students choose the most difficult problem on the page and explain their choices.
6–7: Ask students to explain that metric prefixes have the same meaning whether converting units of length, capacity, or mass.
Less than 6: Before students redo incorrect answers, have them explain how they determine which operation to use.

Prerequisite Skills: reading a thermometer

Lesson Focus: understanding Celsius
Possible Score: 14
Time Frame: 5–10 minutes

Lesson 8 Temperature

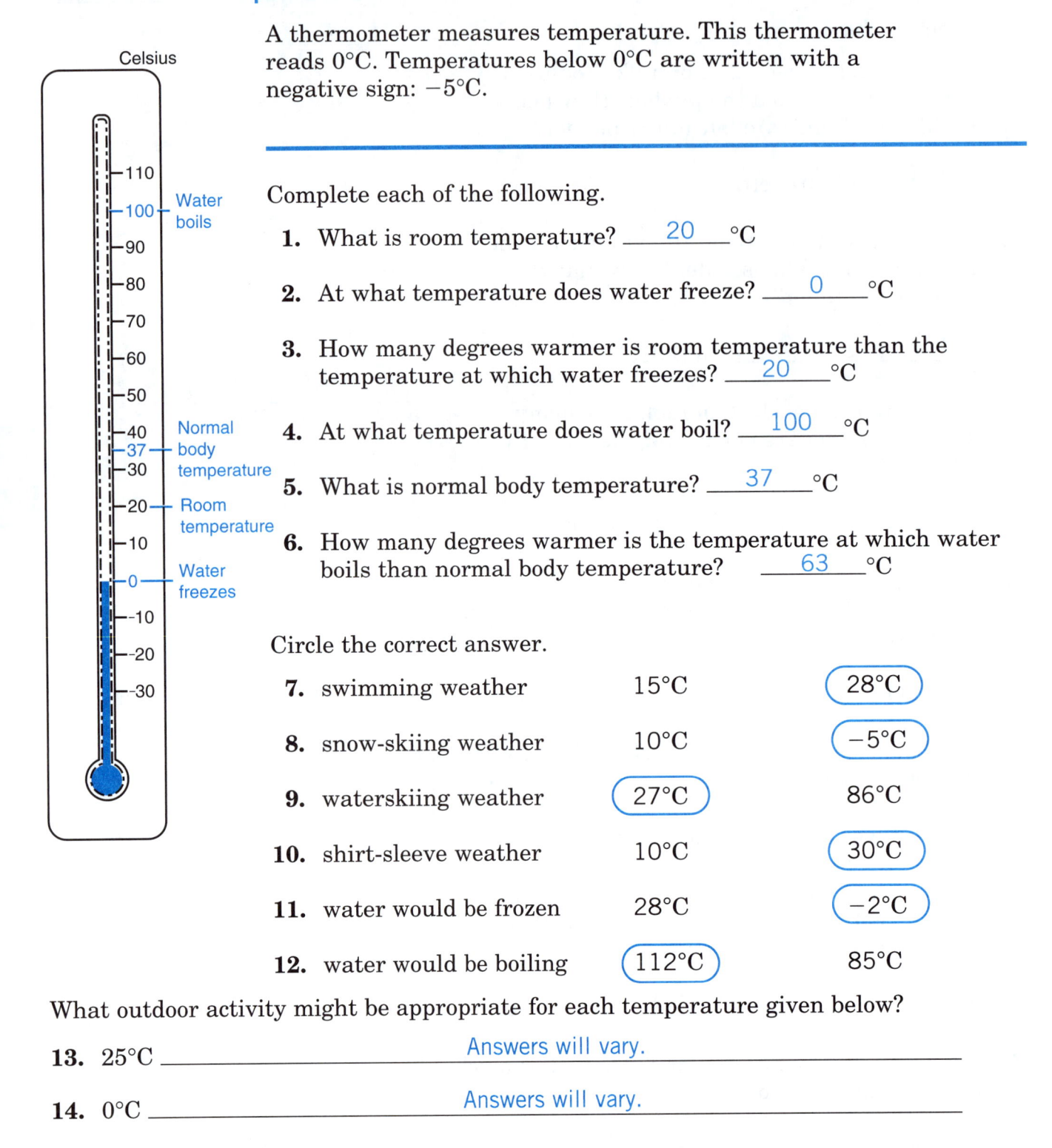

A thermometer measures temperature. This thermometer reads 0°C. Temperatures below 0°C are written with a negative sign: −5°C.

Celsius thermometer showing:
- 110
- 100 — Water boils
- 90
- 80
- 70
- 60
- 50
- 40
- 37 — Normal body temperature
- 30
- 20 — Room temperature
- 10
- 0 — Water freezes
- −10
- −20
- −30

Complete each of the following.

1. What is room temperature? __20__ °C

2. At what temperature does water freeze? __0__ °C

3. How many degrees warmer is room temperature than the temperature at which water freezes? __20__ °C

4. At what temperature does water boil? __100__ °C

5. What is normal body temperature? __37__ °C

6. How many degrees warmer is the temperature at which water boils than normal body temperature? __63__ °C

Circle the correct answer.

7. swimming weather   15°C   (28°C)

8. snow-skiing weather   10°C   (−5°C)

9. waterskiing weather   (27°C)   86°C

10. shirt-sleeve weather   10°C   (30°C)

11. water would be frozen   28°C   (−2°C)

12. water would be boiling   (112°C)   85°C

What outdoor activity might be appropriate for each temperature given below?

13. 25°C _____Answers will vary._____

14. 0°C _____Answers will vary._____

Number Correct
9–14: Have students give the high (or low) temperature of the day in Celsius.
Less than 9: Ask students to explain how to use the thermometer on the page.

For further evaluation, copy the Chapter Test on page 271.
For maintaining skills, use the Cumulative Review on pages 225–226.

Possible Score: 30
Time Frame 10–15 minutes

CHAPTER 7   PRACTICE TEST
Metric Measurement

Measure each line segment to the nearest unit as indicated.

1. __5__ cm
2. __35__ mm

Complete the following.

	a		*b*
3. 25 cm = __250__ mm		35 m = __3500__ cm	
4. 6 km = __6000__ m		7 m = __0.007__ km	
5. 260 cm = __2.6__ m		600 mm = __0.6__ m	
6. 7.5 m = __7500__ mm		2.5 mm = __0.25__ cm	
7. 12 L = __12 000__ mL		13.5 mL = __0.0135__ L	
8. 5.4 kL = __5400__ L		1200 L = __1.2__ kL	
9. 0.045 L = __45__ mL		260 L = __0.26__ kL	
10. 58 kg = __58 000__ g		400 g = __0.4__ kg	
11. 3000 mg = __3__ g		3.8 kg = __3800__ g	
12. 0.6 g = __600__ mg		50 mg = __0.05__ g	

13. Water freezes at __0__ °C.
14. Water boils at __100__ °C.

Solve each problem.

15. There are 5.5 kL of water in a tank. If 3200 L of water are used, how many litres will be in the tank? How many kilolitres is that?

    There will be __2300__ L in the tank.
    That is __2.3__ kL.

16. Chloe jumped 1.45 m. Evan jumped 138 cm. Who jumped farther? How much farther?

    __Chloe__ jumped __7__ cm farther.

Number Correct

LESSON FOLLOW-UP AND ERROR ANALYSIS

26–30: Have students indicate the longest, the largest, and the heaviest measures in problems 3–14.
20–25: Discuss that metric prefixes have the same meaning whether converting units of length, capacity, or mass.
Less than 20: Have students explain how they determine whether to multiply by 1000 or by 0.001 when converting units of measure.

CHAPTER 8 PRETEST
More Metric Measurement and Estimation

Complete the following.

	a	b
1.	40 mm = __4__ cm	4 cm = __40__ mm
2.	48 h = __2__ days	5 m = __500__ cm
3.	3 h = __180__ min	5 kg = __5000__ g
4.	2000 L = __2__ kL	5 min 4 s = __304__ s
5.	4000 g = __4__ kg	3 kL = __3000__ L
6.	8000 mL = __8__ L	7 g = __7000__ mg

Add, subtract, or multiply.

	a	b	c
7.	8 h 3 min +2 h 4 min ――――― 10 h 7 min	3 min 49 s −1 min 27 s ――――― 2 min 22 s	2 kg ×4 ――― 8 kg
8.	5 min 2 s +2 min 2 s ――――― 7 min 4 s	5 h 2 min −2 h 3 min ――――― 2 h 59 min	1 kg ×7 ――― 7 kg

Round as indicated.

	a nearest ten	b nearest hundred	c nearest thousand
9. 8324	8320	8300	8000
10. 74 485	74 490	74 500	74 000

Prerequisite Skills: operations with whole numbers and fractions

Lesson Focus: metric units of length
Possible Score: 31
Time Frame: 15–20 minutes

Lesson 1 Units of Length

1 m = 100 cm
1 m = 1000 mm
1 cm = 10 mm

6 cm = __?__ mm
1 cm = 10 mm
6 cm = (6 × 10) mm
6 m = __60__ mm

40 mm = __?__ cm
10 mm = 1 cm
40 mm = (40 ÷ 10) cm
40 mm = _____ cm

4 m = __?__ cm
1 m = 100 cm
4 m = (4 × 100) cm
4 m = __400__ cm

Complete the following.

                a   b

1. 50 mm = __5__ cm   5 cm = __50__ mm

2. 7 cm = __70__ mm   6 m = __600__ cm

3. 400 cm = __4000__ mm   2 m = __2000__ mm

4. 5 m = __500__ cm   12 cm = __120__ mm

5. 7000 mm = __7__ m   5 m = __500__ cm

6. 7 m = __7000__ mm   4 cm = __40__ mm

7. 100 mm = __10__ cm   6 m = __6000__ mm

8. 3000 mm = __3__ m   3 m = __300__ cm

9. 3 cm = __30__ mm   8 cm = __80__ mm

10. 20 mm = __2__ cm   2 cm = __20__ mm

11. 5 m = __5000__ mm   1 m = __1000__ mm

Number Correct   LESSON FOLLOW-UP AND ERROR ANALYSIS

26–31: Have students measure their height in metres and centimetres.
20–25: Before students redo incorrect answers, ask them to explain the relationship between millimetres, centimetres and metres.
Less than 20: Have students explain how to convert from a larger unit to a smaller unit of measure (7 cm = 70 mm) or vice versa.

Lesson 1 Problem Solving

Solve each problem.

1. The Higgins' fence is 200 cm high. What is the height of the fence in metres?

   The height of the fence is ___2___ m.

2. Mr. Baxter is 2 m tall. How many centimetres tall is he?

   He is ___200___ cm tall.

3. Kasey purchased 2 m of ribbon. How many millimetres of ribbon did she purchase?

   She purchased ___2000___ mm of ribbon.

4. A football field is 59 m wide. What is the width of the field in centimetres?

   The width of the field is ___5900___ cm.

5. Zach threw a baseball 5000 cm. How many metres did he throw the baseball?

   He threw the baseball ___50___ m.

6. Mr. Kelly's lot is 20 m wide. What is the width of the lot in centimetres?

   The width of the lot is ___2000___ cm.

7. Logan has a piece of wire 3000 mm long. What is the length of the wire in metres?

   The length of the wire is ___3___ m.

8. When the high-jump bar is set at 200 cm, what is the height of the bar in millimetres?

   The height is ___2000___ mm.

9. Mrs. Avilla's garage door is 3 m wide. What is the width of the door in millimetres?

   The width of the door is ___3000___ mm.

CHAPTER 8
More Metric Measurement and Estimation

Lesson 1
Units of Length

Prerequisite Skills: operations with whole numbers and fractions

Lesson 2 Units of Capacity, Time, Mass

Lesson Focus: metric units of capacity, time, and mass
Possible Score: 30
Time Frame: 15–20 minutes

1000 mL = 1 L	60 s = 1 min	1000 mg = 1 g
1000 L = 1 kL	60 min = 1 h	1000 g = 1 kg
	24 h = 1 day	1000 kg = 1 t

300 s = __?__ min

1 min = 60 s
300 s = (300 ÷ 60) min
300 s = __5__ min

5 kg = __?__ g

1 kg = 1000 g
5 kg = (5 × 1000) g
5 kg = __5000__ g

Complete the following.

            a   *b*

1. $5\frac{1}{2}$ min = __330__ s   2 min 14 s = __134__ s

2. 3000 g = __3__ kg   3 h 40 min = __220__ min

3. 5000 L = __5__ kL   5 kL = __5000__ L

4. $3\frac{1}{2}$ h = __210__ min   3 L = __3000__ mL

5. 3 t = __3000__ kg   2 L = __2000__ mL

6. 3000 mL = __3__ L   6 kg = __6000__ g

7. 72 min = __$1\frac{1}{5}$__ h   3 min 51 s = __231__ s

8. 2000 mL = __2__ L   10 g = __10 000__ mg

9. 96 h = __4__ days   5 h 16 min = __316__ min

10. 4 g = __4000__ mg   2 h 15 min = __135__ min

11. 3 days = __72__ h   3 kL = __3000__ L

Number Correct   LESSON FOLLOW-UP AND ERROR ANALYSIS

26–30: Next to each problem, have students label *l* for liquid measure, *t* for time, and *m* for mass, as appropriate.
20–25: Before students redo incorrect answers, have them explain the examples at the top of the page.
Less than 20: Ask students to choose two units of measure and explain how to convert between them.

CHAPTER 8

Lesson 2 Problem Solving

Solve each problem.

1. It took Jeremy 3 min 25 s to run around the block. How many seconds did it take him to run around the block?

    It took him ____205____ s.

2. The capacity of a container is 2 kL. What is the capacity of the container in litres?

    The capacity is ____2000____ L.

3. Last month 64 000 mL of milk were delivered to the Collins' house. How many litres of milk was that?

    That was ____64____ L of milk.

4. Mrs. Johnson purchased a 2-kg can of shortening. How many grams of shortening did she purchase?

    She purchased ____2000____ g of shortening.

5. Mr. Singer showed his class a film that lasted 120 min. How many hours did the film last?

    The film lasted ____2____ h.

6. The cooling system of Mr. Bigg's car has a capacity of 8 L. What is the capacity of the cooling system in millilitres?

    The capacity is ____8000____ mL.

7. The Smith's baby had a mass of 4 kg at birth. What was the mass of the baby in grams?

    The baby's mass was ____4000____ g.

8. An elephant at the zoo has a mass of 2 t. What is the mass of the elephant in kilograms?

    The elephant has a mass of ____2000____ kg.

CHAPTER 8
More Metric Measurement and Estimation

Lesson 2
Units of Capacity, Time, Mass

Prerequisite Skills: measurement equivalents; addition of whole numbers

Lesson Focus: adding units of time
Possible Score: 35
Time Frame: 15–20 minutes

Lesson 3 Adding Time

$$\begin{array}{r}3\text{ h }45\text{ min}\\+5\text{ h }35\text{ min}\\\hline 80\text{ min}\end{array}$$

$(45 + 35)$ min = __80__ min

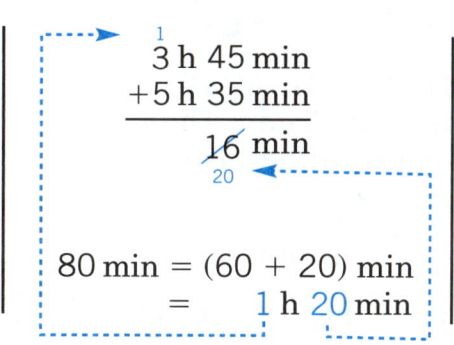

$$\begin{array}{r}\overset{1}{3}\text{ h }45\text{ min}\\+5\text{ h }35\text{ min}\\\hline \cancel{16}\text{ min}\\{}_{20}\end{array}$$

80 min = (60 + 20) min
     = 1 h 20 min

$$\begin{array}{r}\overset{1}{3}\text{ h }45\text{ min}\\+5\text{ h }35\text{ min}\\\hline 9\text{ h }\cancel{16}\text{ min}\\{}_{20}\end{array}$$

$(1 + 3 + 5)$ h = __9__ h

Complete the following.

	a		b	
1.	75 min = 1 h __15__ min		73 s = 1 min __13__ s
2.	90 s = 1 min __30__ s		71 s = 1 min __11__ s
3.	79 min = 1 h __19__ min		100 min = 1 h __40__ min
4.	75 min = 1 h __15__ min		95 s = 1 min __35__ s

Find each sum.

	a	b	c	d
5.	7 min 1 s +2 min 1 s ――――― 9 min 2 s	3 min 14 s + 2 min 29 s ――――― 5 min 43 s	3 h 1 min +2 h 2 min ――――― 5 h 3 min	7 h 20 min +2 h 15 min ――――― 9 h 35 min
6.	3 h 4 min +2 h 7 min ――――― 5 h 11 min	8 min 7 s +2 min 3 s ――――― 10 min 10 s	3 h 21 min +2 h 16 min ――――― 5 h 37 min	5 h 30 min +3 h 30 min ――――― 9 h
7.	3 h 40 min +2 h 40 min ――――― 6 h 20 min	7 h 34 min +5 h 49 min ――――― 13 h 23 min	9 min 7 s +3 min 6 s ――――― 12 min 13 s	5 h 32 min +2 h 45 min ――――― 8 h 17 min
8.	3 h 35 min +6 h 30 min ――――― 10 h 5 min	5 h 40 min +1 h 25 min ――――― 7 h 5 min	1 min 17 s +4 min 53 s ――――― 6 min 10 s	3 h 30 min +2 h 45 min ――――― 6 h 15 min
9.	9 min 35 s +2 min 25 s ――――― 12 min	7 h 11 min +1 h 49 min ――――― 9 h	3 h 30 min +2 h 30 min ――――― 6 h	9 min 20 s +3 min 40 s ――――― 13 min

Number Correct
LESSON FOLLOW-UP AND ERROR ANALYSIS

30–35: Have students choose one problem and write a story problem it would solve.
23–29: Check to see if students added correctly but forgot to convert the smaller unit when possible.
Less than 23: Have students explain how to solve problems 1–4 where two different names are given for the same time. Relate to problems 5–9.

121

Lesson 3 Problem Solving

Solve each problem.

1. The first feature at the Rex Theatre lasts 1 h 45 min. The second lasts 1 h 36 min. The features are shown one after the other. How long will the double feature last?

   It will last ___3___ h ___21___ min.

2. Aaron took 3 min 45 s to solve the first part of a puzzle and 6 min 25 min to solve the second part. How long did it take him to solve the whole puzzle?

   It took Aaron ___10___ min ___10___ s to solve the whole puzzle.

3. Lee ran the first leg of a two-person relay in 6 min 20 s, and Elmer ran the second leg in 5 min 55 s. How long did it take the boys to run the relay?

   It took them ___12___ min ___15___ s.

4. It took Del 1 h 45 min to clean the garage and 4 h 45 min to clean the house. How much time did Del spend cleaning in all?

   Del spent ___6___ h ___30___ min cleaning in all.

5. Margaret did the crossword puzzle in 6 min 35 s and the word-search puzzle in 4 min 50 s. How much time did Margaret spend on the two puzzles?

   Margaret spent ___11___ min ___25___ s on the puzzles.

6. Mrs. Little took 3 min 30 s to fill her car with gasoline and 2 min 55 s to pay for the gasoline. How long did she spend at the gas station?

   Mrs. Little spent ___6___ min ___25___ s at the gas station.

7. Tyler's sister ran the first kilometre of a 2-km race in 4 min 35 s. She ran the second kilometre in 4 min 53 s. What was her time for the race?

   Her time was ___9___ min ___28___ s.

Prerequisite Skills: measurement equivalents; subtraction of whole numbers

Lesson Focus: subtracting units of time
Possible Score: 35
Time Frame: 15–20 minutes

Lesson 4 Subtracting Time

$\begin{array}{r} 664 \\ 7\!\!\!/\text{min } 4\!\!\!/\text{ s} \\ -2\text{ min } 7\text{ s} \\ \hline \end{array}$

7 min 4 s = (6 + 1) min + 4 s
= 6 min (60 + 4) s
= 6 min 64 s

$\begin{array}{r} 664 \\ 7\!\!\!/\text{min } 4\!\!\!/\text{ s} \\ -2\text{ min } 7\text{ s} \\ \hline 57 \text{ s} \end{array}$

(64 − 7) s = __57__ s

$\begin{array}{r} 664 \\ 7\!\!\!/\text{min } 4\!\!\!/\text{ s} \\ -2\text{ min } 7\text{ s} \\ \hline 4\text{ min } 57\text{ s} \end{array}$

(6 − 2) min = __4__ min

Complete the following.

	a	b
1.	15 min 4 s = 14 min __64__ s	4 h 3 min = 3 h __63__ min
2.	7 h 2 min = 6 h __62__ min	6 min 2 s = 5 min __62__ s
3.	5 h 1 min = 4 h __61__ min	2 min 45 s = 1 min __105__ s
4.	4 h 3 min = 3 h __63__ min	3 h 1 min = 2 h __61__ min

Find each difference.

	a	b	c	d
5.	9 min 7 s − 3 min 6 s = 6 min 1 s	7 h 14 min − 3 h 9 min = 4 h 5 min	3 h 45 min − 2 h 19 min = 1 h 26 min	5 min 3 s − 1 min 2 s = 4 min 1 s
6.	16 h 9 min − 9 h 9 min = 7 h	8 min 2 s − 3 min 1 s = 5 min 1 s	6 h 1 min − 3 h 1 min = 3 h	8 min 27 s − 5 min 16 s = 3 min 11 s
7.	15 min 7 s − 12 min 9 s = 2 min 58 s	2 h 1 min − 1 h 1 min = 1 h	5 h 19 min − 3 h 45 min = 1 h 34 min	8 min 4 s − 2 min 9 s = 5 min 5 s
8.	3 h 15 min − 1 h 30 min = 1 h 45 min	8 min 20 s − 5 min 40 s = 2 min 40 s	16 h 25 min − 8 h 55 min = 7 h 30 min	9 min 6 s − 2 min 8 s = 6 min 58 s
9.	6 min 0 s − 2 min 51 s = 3 min 9 s	8 h − 4 h 15 min = 3 h 45 min	8 min − 4 min 11 s = 3 min 49 s	7 h − 2 h 30 min = 4 h 30 min

Number Correct

LESSON FOLLOW-UP AND ERROR ANALYSIS

30–35: Have students choose one problem and write a story problem it would solve.
23–29: Have students explain how to rename when the top measure is smaller than the bottom measure.
(See 7a: 7 s−9 s; 23 s−9 s)
Less than 23: Ask students to explain what is being done in rows 1–4 and how that relates to the remainder of page 123.

123

Lesson 4 Problem Solving

Solve each problem.

1. A frozen ham took 7 h 30 min to thaw in the fridge, and a small frozen turkey took 16 h 15 min. How much longer did the turkey take to thaw than the ham?

   The turkey took __8__ h __45__ min longer to thaw.

2. The first game of a doubleheader lasted 2 h 48 min. The second game lasted 3 h 19 min. How much longer did the second game last than the first?

   The second game lasted __31__ min longer.

3. Abby ran four laps of the track in 4 min 45 s. Renée ran four laps in 5 min 10 s. How much longer did it take Renée to run the four laps?

   It took Renée __0__ min __25__ s longer.

4. Of Heather's 9-h workday, 5 h 45 min were spent packing crates. How much time did she have left for other tasks?

   She had __3__ h __15__ min left.

5. It took Sonya 4 h 30 min to paint the living room and 1 h 45 min to paint the bathroom. How long did she paint in all?

   She painted for __6__ h __15__ min.

6. The Ace Factory operates 3 h 30 min in the morning and 4 h 15 min in the afternoon. How much longer does the factory operate in the afternoon than in the morning?

   It operates __45__ min longer in the afternoon.

7. Katie took 4 min 35 s to run 1 km. Her boyfriend took 5 min 45 s to run 1 km. How much longer did Katie's boyfriend take to run 1 km?

   It took Katie's boyfriend __1__ h __10__ min longer.

CHAPTER 8
More Metric Measurement and Estimation

Lesson 4
Subtracting Time

Prerequisite Skills: measurement equivalents, multiplication of whole numbers

Lesson Focus: multiplying units of measure
Possible Score: 28
Time Frame: 15–20 minutes

Lesson 5   Multiplying Measures

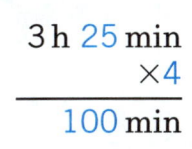

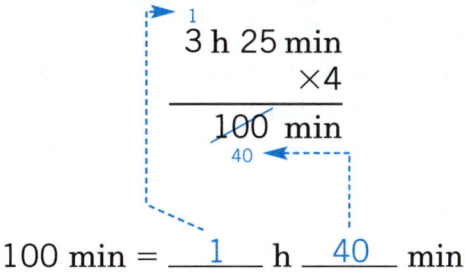

$$\begin{array}{r} 3\text{ h }25\text{ min} \\ \times 4 \\ \hline 100\text{ min} \end{array}$$

$$\begin{array}{r} \overset{1}{3}\text{ h }25\text{ min} \\ \times 4 \\ \hline 100\text{ min} \\ 40 \end{array}$$

$$\begin{array}{r} \overset{1}{3}\text{ h }25\text{ min} \\ \times 4 \\ \hline 13\text{ h }100\text{ min} \\ 40 \end{array}$$

(4 × 25) min = __100__ min | 100 min = __1__ h __40__ min | [(4 × 3) + 1] h = __13__ h

---

Find each product.

	a	b	c
1.	4 cm ×4 = **16 cm**	3 min 12 s ×4 = **12 min 48 s**	3 kL ×3 = **9 kL**
2.	3 h 20 min ×3 = **10 h**	5 kg ×5 = **25 kg**	7 m ×2 = **14 m**
3.	55 cm ×9 = **495 cm**	65 L ×7 = **455 L**	2.5 kg ×6 = **15 kg**
4.	37 g ×9 = **333 g**	106 m ×6 = **636 m**	9 min 36 s ×5 = **48 min**
5.	3 m ×8 = **24 m**	5 min 20 s ×6 = **32 min**	300 mg ×3 = **900 mg**
6.	2 kg ×5 = **10 kg**	1 h 10 min ×6 = **7 h**	60 L ×3 = **180 L**
7.	2.5 km ×8 = **420 km**	4.5 m ×6 = **27 m**	7.1 kg ×9 = **63.9 kg**

Number Correct
LESSON FOLLOW-UP AND ERROR ANALYSIS
24–28: Have students choose one problem and write a story problem it would solve.
18–23: Check to see if students multiplied correctly but forgot to convert the smaller unit as needed.
Less than 18: Have students explain how to multiply the whole numbers and then convert the smaller units if needed.

Lesson 5   Problem Solving

Solve each problem.

1. Six tables are to be placed end to end. Each table is 128 cm. What will be the total length of the tables?

   The total length will be ___768___ cm.

2. It takes 2 min 15 s to assemble a doodad. How long will it take to assemble three doodads?

   It will take ___6___ min ___45___ s.

3. Each doodad has a mass of 1.6 kg. There are six doodads in each case. How much will the mass of the doodads in a case be?

   They will have a mass of ___9.6___ kg.

4. A laundry purchased five large bottles of detergent. Each bottle contained 2 L. How much detergent was purchased?

   ___10___ L of detergent were purchased.

5. Mr. Mitchell purchased eight boards. Each board was 2.5 m long. What was the total length of the boards he purchased?

   The total length was ___20___ m.

6. Nine cases of art supplies are to be shipped. Each case has a mass of 4 kg. How much will the mass of the shipment be?

   It will have a mass of ___36___ kg.

7. Each shift at the Kempf Factory lasts 7 h 30 min. There are nine shifts each week. How long does the factory operate each week?

   The factory operates ___67___ h ___30___ min each week.

CHAPTER 8
More Metric Measurement and Estimation

126

Lesson 5
Multiplying Measures

Prerequisite Skills: measurement equivalents, whole number operations

Lesson Focus: converting, adding, subtracting, and multiplying units of measure
Possible Score: 18
Time Frame: 10–15 minutes

Lesson 6 Measurement

Complete the following.

	a		b
1.	2 h = __120__ min	3 cm = __30__ mm
2.	5 kg = __5000__ g	2 kg = __2000__ g
3.	700 L = __0.7__ kL	3 min 38 s = __218__ s
4.	70 mm = __7__ cm	4 L = __4000__ mL
5.	240 s = __4__ min	5 m = __500__ cm

Add, subtract, or multiply.

	a	b	c
6.	3 h 7 min +5 h 4 min ———— 8 h 11 min	4 min 14 s −2 min 7 s ———— 2 min 7 s	7.2 L ×3 ——— 21.6 L
7.	16 h 25 min +4 h 35 min ———— 21 h	6.2 cm ×5 ——— 31 cm	6 h −2 h 3 min ———— 3 h 57 min

Solve each of the following.

8. Freda slept for 8 h 30 min last night. Francis slept for 11 h 15 min last night. How much longer did Francis sleep than Freda?

   Francis slept for __2__ h __45__ min longer than Freda.

8.

9. It is 112 m from home plate to the right-field fence at the foul pole. What is this distance in centimetres?

   It is __11 200__ cm.

9.

CHAPTER 8

Number Correct

LESSON FOLLOW-UP AND ERROR ANALYSIS

15–18: Have students give the answers to problem 8 in minutes and 9 in metres.
12–14: Have students write the measurement equivalents they need to redo in incorrect answers.
Less than 12: In problems 6 and 7, ask students to explain how to convert the smaller units in the answers.

Prerequisite Skills: whole number place values

Lesson Focus: rounding whole numbers
Possible Score: 27
Time Frame: 5–10 minutes

Lesson 7   Rounding Numbers

Round 32 to the nearest ten.

32 is nearer 30 than 40.
32 rounded to the nearest ten is __30__.

Round 75 to the nearest ten.

75 is as near 70 as 80. In such cases, use the *greater multiple* of ten.
75 rounded to the nearest ten is __80__.

Round 4769 to the nearest hundred.

4769 is nearer 4800 than 4700.
4769 rounded to the nearest hundred is __4800__.

Round 4500 to the nearest thousand.

4500 is as near 4000 as 5000. In such cases, use the *greater multiple* of one thousand.
4500 rounded to the nearest thousand is __5000__.

Round to the nearest ten.

	a	b	c
1.	28 __30__	73 __70__	85 __90__
2.	244 __240__	477 __480__	655 __660__
3.	1696 __1700__	2792 __2790__	8245 __8250__

Round to the nearest hundred.

	a	b	c
4.	321 __300__	479 __500__	550 __600__
5.	1459 __1500__	2628 __2600__	1650 __1700__
6.	24 136 __24 100__	35 282 __35 300__	47 350 __47 400__

Round to the nearest thousand.

	a	b	c
7.	4325 __4000__	6782 __7000__	7500 __8000__
8.	5943 __6000__	8399 __8000__	8500 __9000__
9.	16 482 __16 000__	27 501 __28 000__	43 500 __44 000__

Number Correct

LESSON FOLLOW-UP AND ERROR ANALYSIS

23–27: Have students give an example of when a rounded number might be used instead of the exact number.
18–22: Before students redo the page, have them draw a caret (^) beneath the place value to which they are to round.
Less than 18: In 3-, 4-, and 5-digit numbers check to see if students always rounded to the nearest hundred or thousand.

Prerequisite Skills: rounding; addition and subtraction of whole numbers

Lesson Focus: estimating; adding and subtracting whole numbers
Possible Score: 24
Time Frame: 10–15 minutes

Lesson 8   Estimating Sums and Differences

Estimate the sum of 744 and 378.

		estimated sum	actual sum
744	to the nearest hundred →	700	744
+378	to the nearest hundred →	+400	+378
		1100	1122

To estimate the sum of 6375 and 8678, round 6375 to __6000__ and 8678 to __9000__. The estimated sum would be 6000 + 9000 or __15 000__.

Estimate the difference between 6232 and 2948.

		estimated difference	actual difference
6232	to the nearest thousand →	6000	6232
−2948	to the nearest thousand →	−3000	−2948
		3000	3284

To estimate the difference between 38 735 and 12 675, round 38 735 to __40 000__ and 12 675 to __10 000__. The estimated difference would be 40 000 − 10 000 or __30 000__.

Estimate each sum or difference. Then find each sum or difference.

	a	estimate	b	estimate	c	estimate
1.	739 +435 1174	700 +400 1100	678 −245 433	700 −200 500	743 +825 1568	700 +800 1500
2.	7254 −1326 5928	7000 −1000 6000	1375 +6427 7802	1000 +6000 7000	2795 −1246 1549	3000 −1000 2000
3.	7524 +3542 11 066	8000 +4000 12 000	6852 −4526 2326	7000 −5000 2000	7689 +3824 11 513	8000 +4000 12 000
4.	25 243 −12 675 12 568	25 000 −13 000 12 000	76 425 +23 142 99 567	80 000 +20 000 100 000	95 245 −58 624 36 621	100 000 −60 000 40 000

Number Correct   **LESSON FOLLOW-UP AND ERROR ANALYSIS**
20–24: Have students circle each estimated answer (sum) that is greater than the actual sum.
16–19: Check to see if students performed the wrong operation after rounding the numbers.
Less than 16: Have students explain how to circle numbers to estimate a sum or difference.

Prerequisite Skills: rounding; multiplication of whole numbers

Lesson Focus: estimating, multiplying whole numbers
Possible Score: 30
Time Frame: 10–15 minutes

Lesson 9 Estimating Products

Study how to estimate the product of 187 and 63.

estimated product

187 — to the nearest hundred → 200
×63 — to the nearest ten → ×60
———
12 000

actual product

187
×63
———
561
11 220
———
11 781

To estimate 86 × 224, round 86 to __90__ and 224 to __200__.

The estimated product would be 90 × 200 or __18 000__.

Write the estimated product on each _____. Then find each product.

	a		b		c	
1.	72 ×38 = 2736	2800	91 ×57 = 5187	5400	55 ×65 = 3575	4200
2.	69 ×48 = 3312	3500	56 ×78 = 4368	4800	75 ×66 = 4950	5600
3.	84 ×63 = 5292	4800	93 ×43 = 3999	3600	74 ×45 = 3330	3500
4.	125 ×78 = 9750	8000	469 ×36 = 16 884	20 000	724 ×63 = 45 612	42 000
5.	427 ×43 = 18 361	16 000	825 ×73 = 60 225	56 000	974 ×47 = 45 778	50 000

Number Correct

130

LESSON FOLLOW-UP AND ERROR ANALYSIS

26–30: Have students circle each estimated answer (product) that is less than the actual product.
20–25: Have students explain how to show renamed digits when multiplying.
Less than 20: Ask students to explain how to circle numbers to the greatest place value and estimate answers (products).

For further evaluation, copy the Chapter Test on page 272.
For maintaining skills, use the Cumulative Review on pages 227–228.

Possible Score: 28
Time Frame: 10–15 minutes

CHAPTER 8 PRACTICE TEST
More Metric Measurement and Estimation

Complete the following.

	a	b
1.	3 cm = __30__ mm	3 m = __300__ cm
2.	90 min = __$1\frac{1}{2}$__ h	4 kg = __4000__ g
3.	5 kg = __5000__ g	2 h 27 min = __147__ min
4.	96 h = __4__ days	3 cm = __30__ mm
5.	3 kL = __3000__ L	3 L = __3000__ mL

Add, subtract, or multiply.

6.
```
  3 min 6 s        8 h 6 min        2 min 14 s
 +2 min 8 s       -2 h 4 min              ×3
 ──────────       ─────────        ──────────
  5 min 14 s       6 h 2 min        6 min 42 s
```

7.
```
  6 h 30 min         3.3 m          6 h 50 min
 -4 h 45 min           ×4          +2 h 48 min
 ───────────        ──────         ───────────
  1 h 45 min        13.2 m          9 h 38 min
```

Round as indicated.

		a nearest ten	b nearest hundred	c nearest thousand
8.	4773	4770	4800	5000
9.	63 575	63 580	63 600	64 000

Write an estimate for each exercise. Then find the answer.

10.
```
   7129   7000          9046   9000          296    300
  +4516  +5000         -3978  -4000          ×78    ×80
  ─────  ─────         ─────  ─────        ─────  ─────
  11 645 12 000         5068   5000        23 088 24 000
```

Number Correct   LESSON FOLLOW-UP AND ERROR ANALYSIS

24–28: Have students circle the numbers in row 10 to the nearest tens and find the answers.
18–23: In rows 6 and 7, check to see if students forgot to convert the smaller unit when possible.
Less than 18: Have students explain how to circle a number to a given place value before they redo incorrect answers.

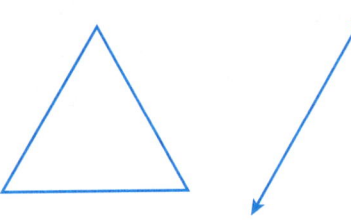

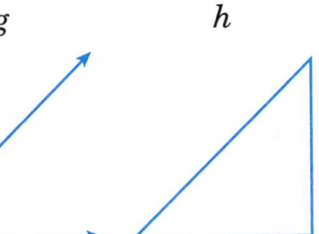

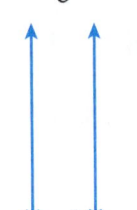

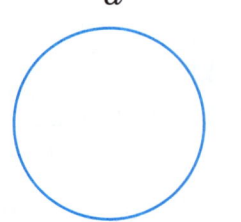

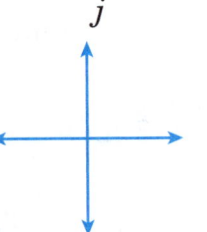

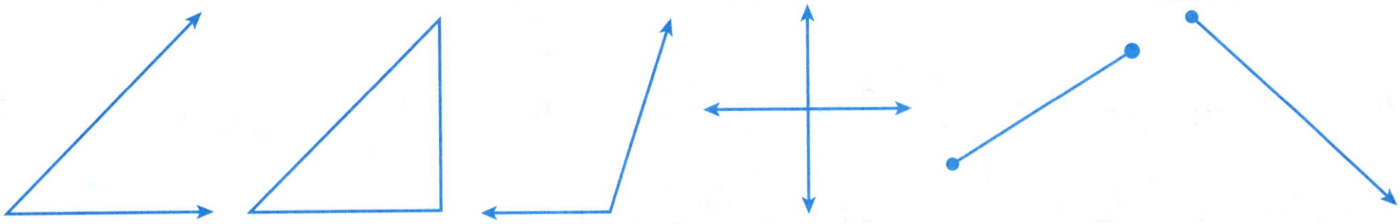

CHAPTER 9   PRETEST
Geometry

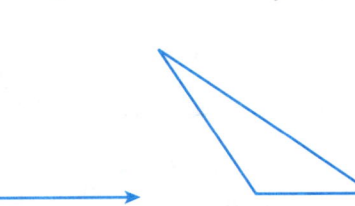

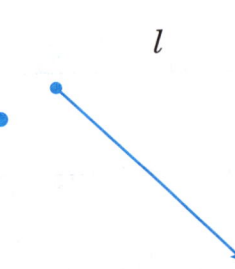

On the _____ before each name below, write the letter(s) of the figures(s) it describes above.

	a		b		c
1.	__l__ ray	__k__ line segment	__a,h__ isosceles triangle
2.	__b__ line	__i__ obtuse angle	__f__ obtuse triangle
3.	__d__ circle	__h__ right triangle	__j__ perpendicular lines
4.	__g__ acute angle	__c__ parallel lines	__a__ equilateral triangle
5.	__e__ right angle	__a__ acute triangle	__f__ scalene triangle

Use a protractor to find the measure of each angle below.

6. a. __60__ °   b. __90__ °

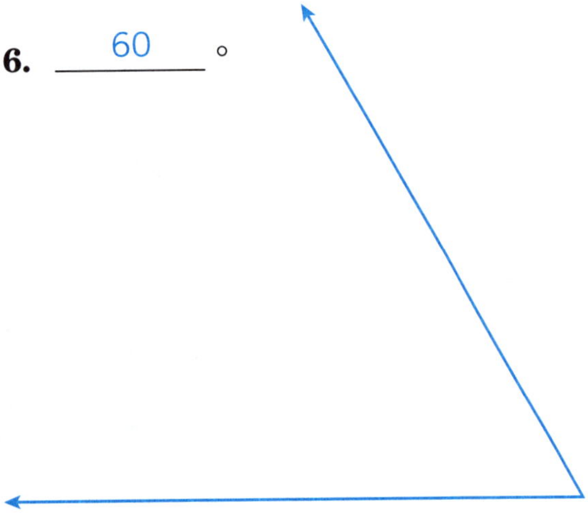

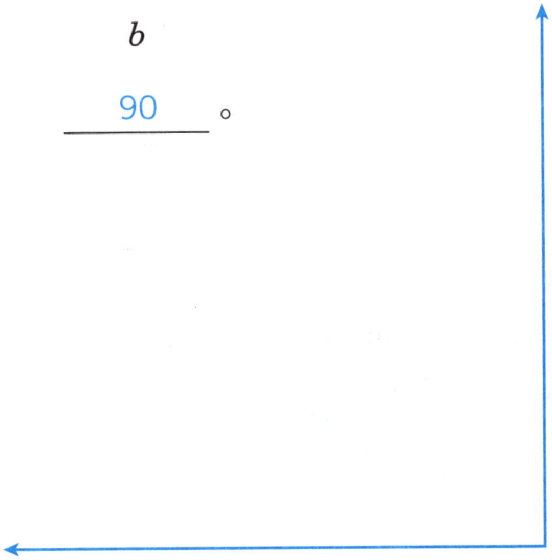

Prerequisite Skills: identifying points

Lesson Focus: recognizing and identifying lines, line segments, and rays
Possible Score: 18
Time Frame: 5–10 minutes

Lesson 1 Lines, Line Segments, and Rays

Line BC or $\overleftrightarrow{BC}$

Any two points on a line can be used to name that line. Do $\overleftrightarrow{BC}$ and $\overleftrightarrow{CB}$ name the same line? __yes__

Line segment JK or $\overline{JK}$

$\overline{JK}$ consists of all points on the line between and including *endpoints* J and K. Do $\overline{JK}$ and $\overline{JK}$ name the same line segment? __yes__

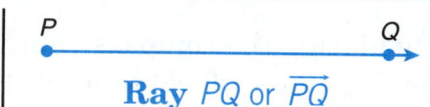

Ray PQ or $\overrightarrow{PQ}$

$\overrightarrow{PQ}$ consists of endpoint P and all points on $\overleftrightarrow{PQ}$ that are on the same side of P as Q. Do $\overrightarrow{PQ}$ and $\overrightarrow{QP}$ name the same ray? __no__

Complete the following as shown.

1. E———D   line ED or DE $\overleftrightarrow{ED}$ or $\overleftrightarrow{DE}$
   Endpoints: __None__

2. G———F   ray FG $\overrightarrow{FG}$
   Endpoint: __F__

3. L———M   line segment LM or ML $\overline{LM}$ or $\overline{ML}$
   Endpoints: __L and M__

4. R———S   line RS or SR $\overleftrightarrow{RS}$ or $\overleftrightarrow{SR}$
   Endpoint(s): __None__

5. X———Y   line segment XY or YX $\overline{XY}$ or $\overline{YX}$
   Endpoint(s): __X and Y__

6. T———V   ray TV $\overrightarrow{TV}$
   Endpoint(s): __T__

7. A———C   ray CA $\overrightarrow{CA}$
   Endpoint(s): __C__

8. W———Z   line WZ or ZW $\overleftrightarrow{WZ}$ or $\overleftrightarrow{ZW}$
   Endpoint(s): __None__

9. H———N   line segment HN or NH $\overline{HN}$ or $\overline{NH}$
   Endpoint(s): __H and N__

Number Correct

LESSON FOLLOW-UP AND ERROR ANALYSIS

15–18: Have students give examples of objects that are suggested by lines, line segments, and rays.
12–14: Have students explain the difference between a *line* (no endpoints) and a *ray* (one endpoint).
Less than 12: Ask students to explain how to name lines and line segments. (the order of the letters makes no difference with lines but it does with rays)

Prerequisite Skills: identifying points

Lesson Focus: naming parts of a circle
Possible Score: 9
Time Frame: less than 5 minutes

Lesson 2 Circles

By placing the compass point at point *P*, you can locate all the points in a plane (never-ending flat surface) that are the same distance from point *P*.

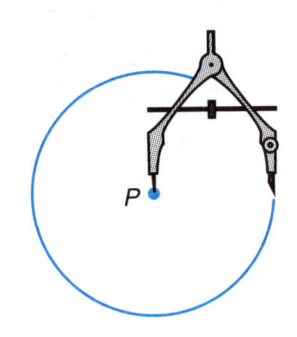

A **circle** is a set of points in a plane such that each point is the same distance from some given point called the *centre*.

You can name a circle by naming its centre. Circle *P* is shown at the left.

A **radius** of a circle is a line segment from the centre of the circle to a point on the circle.

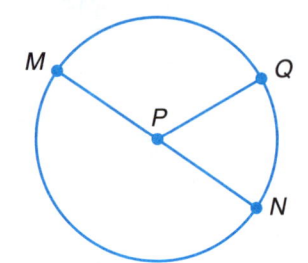

$\overline{PM}$ is a radius of circle *P*. Name two more radii of circle *P*. __$\overline{PQ}, \overline{PN}$__

A **diameter** of a circle is a line segment that has its endpoints on the circle and passes through the centre of the circle.

Name a diameter of circle *P*. __$\overline{MN}, \overline{NM}$__

Name the centre, a radius, and a diameter of each circle.

	centre	radius	diameter
1.	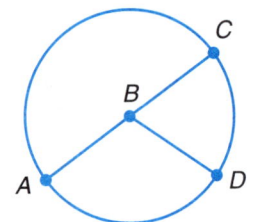 point B	$\overline{BD}, \overline{BA}$ or $\overline{BC}$	$\overline{AC}$ or $\overline{CA}$
2.	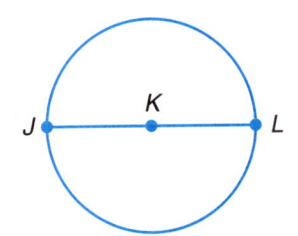 point K	$\overline{KJ}$ or $\overline{KL}$	$\overline{JL}$ or $\overline{LJ}$

Write *True* or *False* after each statement.

3. All radii of the same circle have the same length. __True__

4. All diameters of the same circle have the same length. __True__

5. The length of a diameter of a circle is twice the length of a radius. __True__

Number Correct
- 8–9: Have students construct a circle, draw a diameter and a radius, and label all points.
- 6–7: Before students redo the page, ask them to define *radius* and *diameter* in their own words. (radius is one-half the length of a diameter)
- Less than 6: Have students explain how a radius and a diameter are named. (both by their endpoints)

LESSON FOLLOW-UP AND ERROR ANALYSIS

134

Prerequisite Skills: naming angles

Lesson Focus: identifying types of angles
Possible Score: 6
Time Frame: 5–10 minutes

Lesson 3   Angles

An **angle** is formed by two rays that have a common endpoint.

Study how angle ACB (denoted ∠ACB) is constructed below.

Step 1	*Step 2*	*Step 3*	
Draw circle Q and diameter AB.	Select point C anywhere on circle Q.	Draw $\overrightarrow{CA}$ and $\overrightarrow{CB}$.	Compare ∠ACB with a corner of a page of this book.

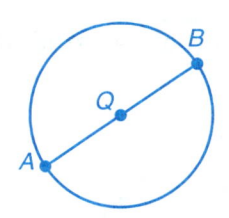

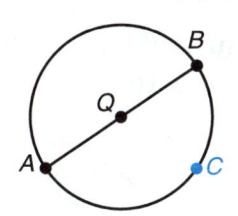

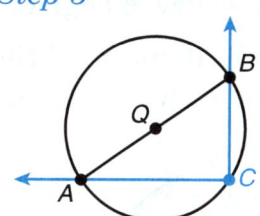

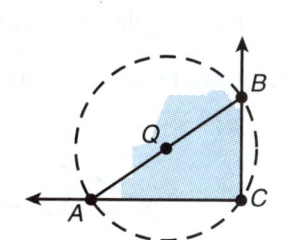

Angles such as ∠ACB are called **right angles**.

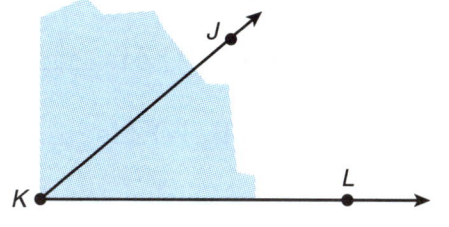

Does ∠JKL appear to be larger or smaller than a right angle? __smaller__

Angles like ∠JKL are called **acute angles**.

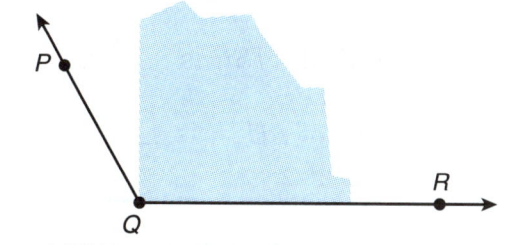

Does ∠PQR appear to be larger or smaller than a right angle? __larger__

Angles like ∠PQR are called **obtuse angles**.

Compare each angle with a model of a right angle. Then describe each angle by writing either *acute*, *obtuse*, or *right* on each _____.

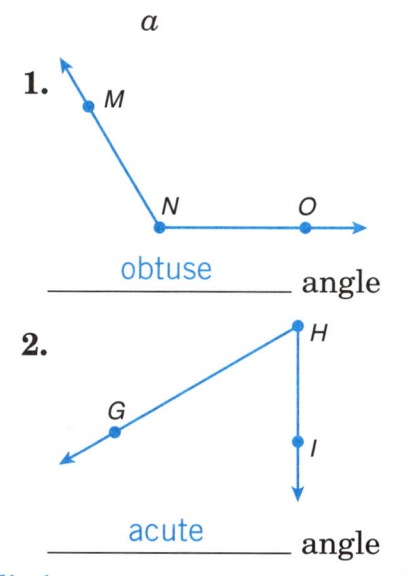

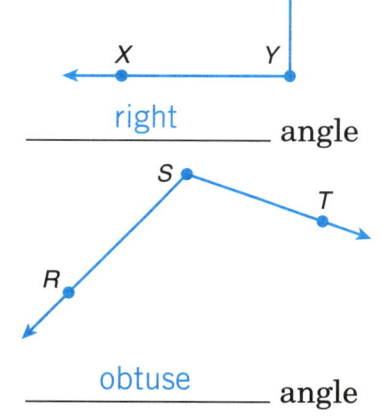

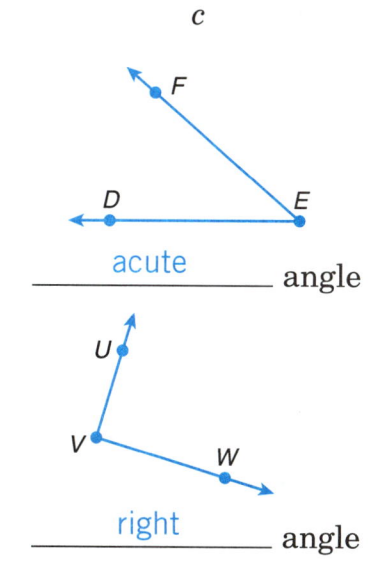

	a	b	c
1.	__obtuse__ angle	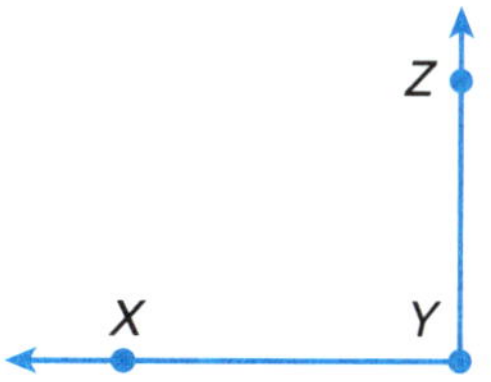 __right__ angle	__acute__ angle
2.	__acute__ angle	__obtuse__ angle	__right__ angle

Number Correct   LESSON FOLLOW-UP AND ERROR ANALYSIS
 5–6: Have students name each angle in rows 1 and 2.
 3–4: Check to see if the positions of the angles in row 2 confused students. If so, remind them to compare them to right angles.
 Less than 3: Have students explain how angles are named. (by three points with the common endpoint, the *vertex*, as the middle letter)

Prerequisite Skills: naming angles

Lesson Focus: measuring angles with a protractor
Possible Score: 6
Time Frame: 5–10 minutes

Lesson 4 Angle Measurement

To use a protractor to measure an angle:

a. Place the centre of the protractor at the vertex of the angle.

b. Align one side of the angle with the base of the protractor so that the other side of the angle intersects the curved edge of the protractor.

c. Use the scale starting at 0 and read the measure of the angle where the other side of the angle intersects the curved edge of the protractor.

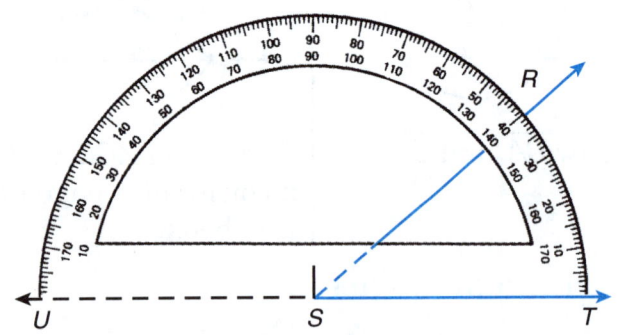

The measurement of ∠TSR is __40°__.
40° is read 40 *degrees*.
The measurement of ∠USR is __140°__.

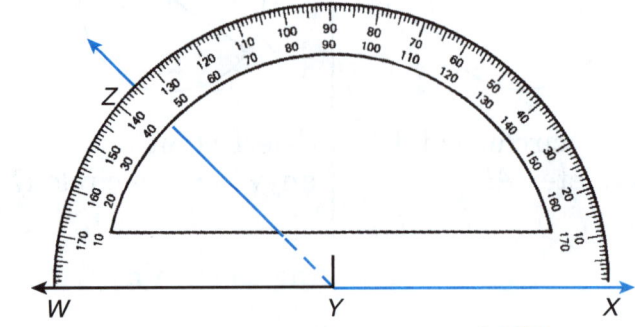

The measurement of ∠XYZ is __135°__.
The measurement of ∠WYZ is __45°__.

Use a protractor to measure each angle below.

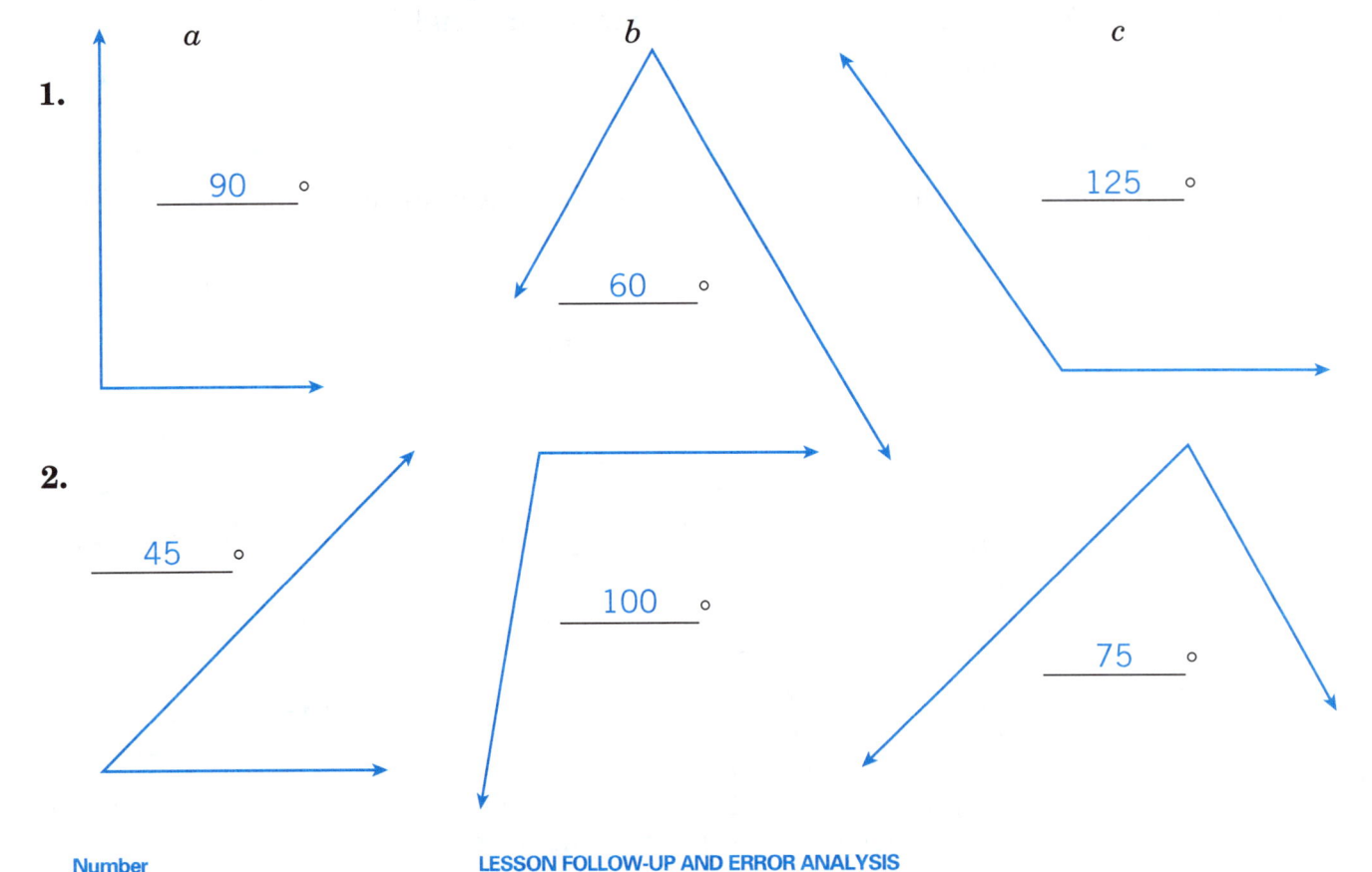

   a   b   c

1.   __90__°   __60__°   __125__°

2.   __45__°   __100__°   __75__°

Number Correct

136

LESSON FOLLOW-UP AND ERROR ANALYSIS

5–6: Have students identify each angle as acute, right, or obtuse.
3–4: Discuss how to use the larger numbers on the protractor when the angle is greater than a right angle and vice versa.
Less than 3: Have students demonstrate how to use a protractor before they redo the page. (Align the bottom of the protractor with the angle.)

Prerequisite Skills: measuring angles with a protractor

Lesson Focus: measuring and drawing angles
Possible Score: 16
Time Frame: 10–15 minutes

Lesson 5 Angle Measurement

Use a protractor to measure each angle in the figure below.

	a angle	measurement		b angle	measurement
1.	∠AFB	30°		∠BFD	90°
2.	∠BFC	50°		∠CFE	70°
3.	∠CFD	40°		∠AFD	120°
4.	∠DFE	30°		∠BFE	120°
5.	∠AFC	80°		∠AFE	150°

Use a protractor to draw angles having the measurements given below.

	a	b	c
6.	75°	90°	30°

See students' work

7.	120°	60°	135°

Number Correct

LESSON FOLLOW-UP AND ERROR ANALYSIS

13–16: Have students label and name every angle in rows 6 and 7.
10–12: In rows 1–5, check to see if students were confused as to which angle to measure because all the angles have the same vertex (common endpoint).
Less than 10: Ask students to demonstrate how to draw an angle. (First, draw the bottom ray and then mark where the other ray should go.)

CHAPTER 9

137

Prerequisite Skills: measuring angles with a protractor

Lesson Focus: measuring, identifying, and drawing congruent angles
Possible Score: 6
Time Frame: 10–15 minutes

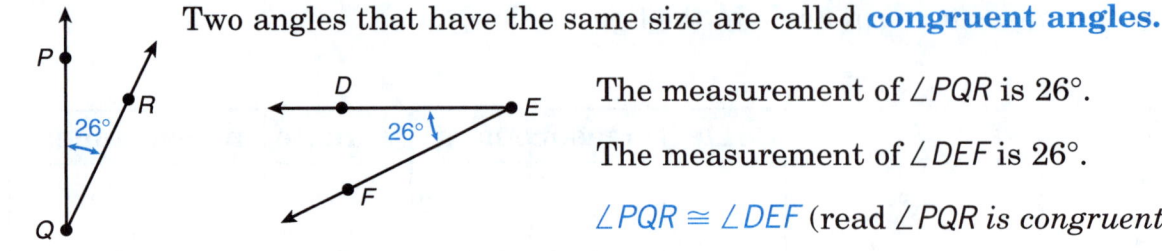

Lesson 6 Congruent Angles

Two angles that have the same size are called **congruent angles.**

The measurement of ∠PQR is 26°.

The measurement of ∠DEF is 26°.

∠PQR ≅ ∠DEF (read ∠PQR is congruent to ∠DEF)

For each exercise, measure both angles. Write *congruent* if the angles are congruent. Write *not congruent* if the angles are not congruent.

                                      a   b

1. _____congruent_____   _____congruent_____

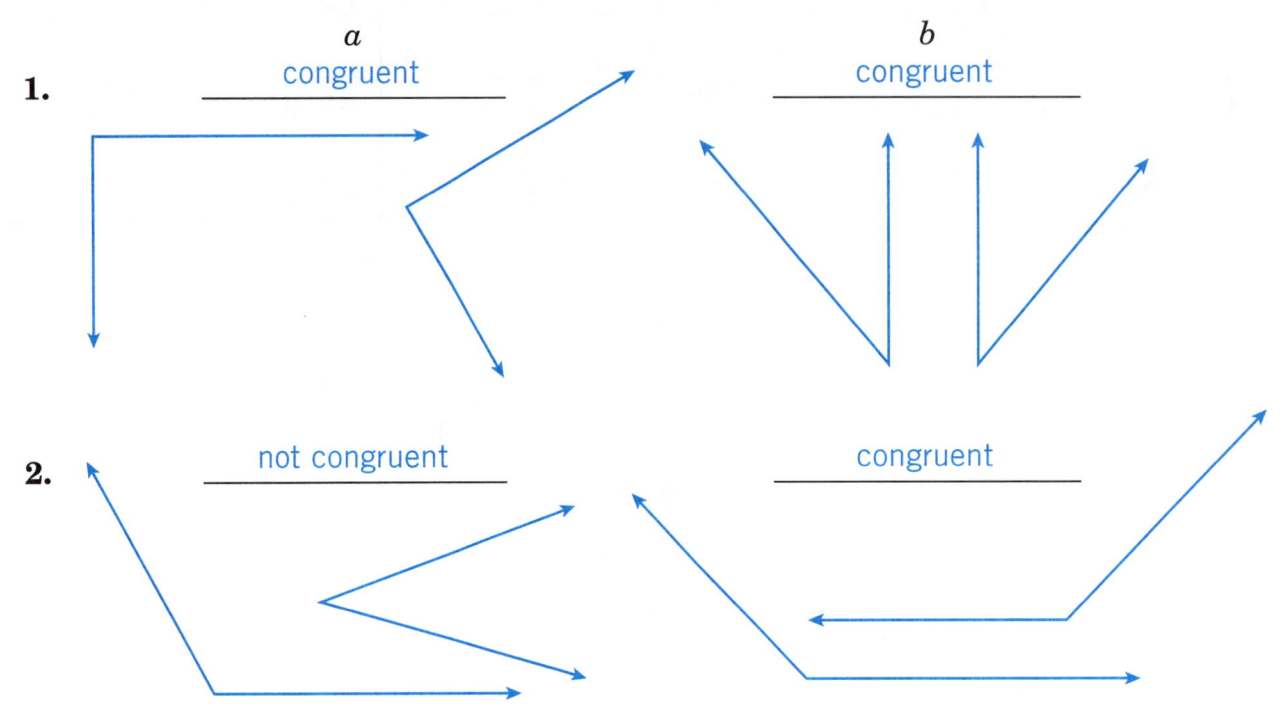

2. ___not congruent___   _____congruent_____

Find the measurement for each angle below. Then draw an angle congruent to each angle.

3. ___25___°   ___150___°

See students' work.

Number Correct

LESSON FOLLOW-UP AND ERROR ANALYSIS

5–6: Have students circle the larger angle in any pair that is not congruent.

3–4: Check to see if students can properly use a protractor. Discuss that congruent angles cannot be identified if they are not measured accurately.

Less than 3: Have students explain how to construct an angle congruent to another angle. (Draw a ray first, then measure for the second ray.)

138

Prerequisite Skills: identifying lines

Lesson Focus: parallel and perpendicular lines
Possible Score: 9
Time Frame: less than 5 minutes

Lesson 7 Parallel and Perpendicular Lines

parallel lines

perpendicular lines

Parallel lines are always the same distance apart. They will never intersect, even if extended.

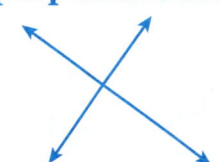

Perpendicular lines form right angles.

Write *parallel* if the lines are parallel. Write *perpendicular* if the lines are perpendicular. Write *neither* if the lines are neither parallel nor perpendicular.

　　　　　　　　　　a　　　　　　　　　　　　b　　　　　　　　　　　　c

1.　　　perpendicular　　　　　　　parallel　　　　　　　　neither

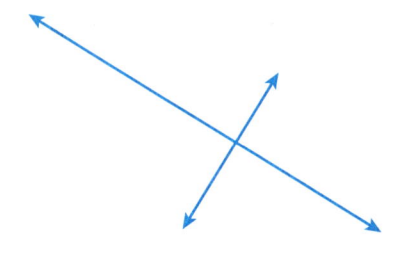

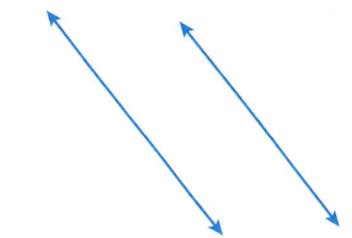

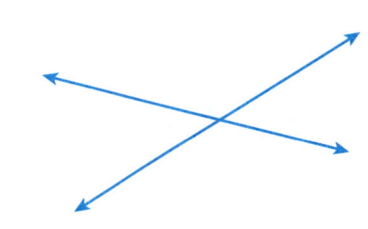

2.　　　　parallel　　　　　　　　　parallel　　　　　　　perpendicular

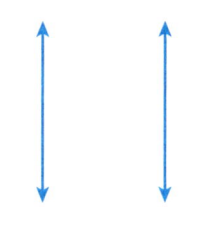

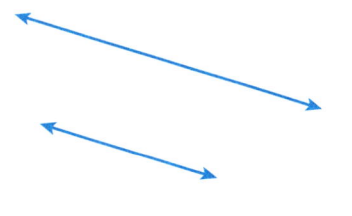

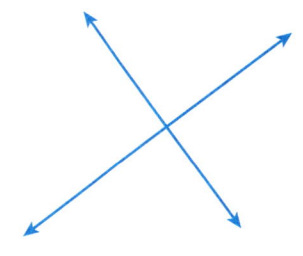

3.　　　　parallel　　　　　　　　　neither　　　　　　　perpendicular

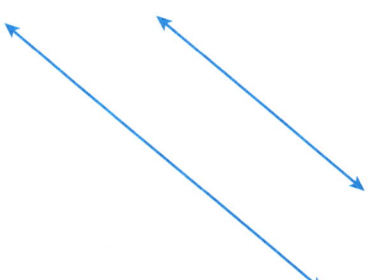

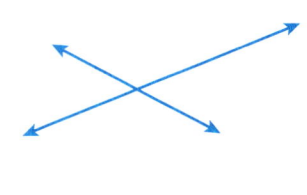

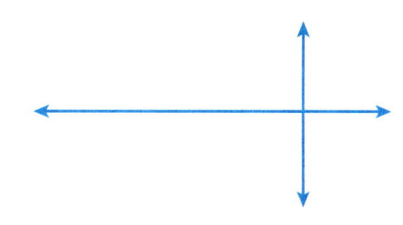

Number Correct

LESSON FOLLOW-UP AND ERROR ANALYSIS

8–9: Have students give a real-life example that suggests parallel and perpendicular lines.
6–7: Have students define the terms *parallel* and *perpendicular* in their own words.
Less than 6: Ask students to explain how a corner of a piece of paper (right angle) can be used to check that lines are perpendicular.

Prerequisite Skills: recognizing types of angles; measuring length

Lesson Focus: identifying triangles
Possible Score: 6
Time Frame: 5–10 minutes

Lesson 8 Triangles

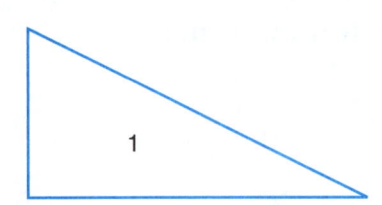

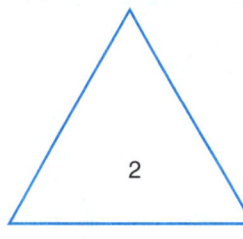

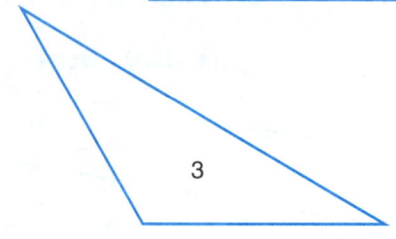

1.   2.   3.

Compare the angles of each triangle with a model of a right angle.

| An **acute triangle** contains all acute angles. |

Which triangle above is an acute triangle? __2__

| A **right triangle** contains one right angle. |

Which triangle above is a right triangle? __1__

| An **obtuse triangle** contains one obtuse angle. |

Which triangle above is an obtuse triangle? __3__

Use a ruler to compare the lengths of the sides of each triangle.

| A **scalene triangle** has no sides the same length. |

Which triangle above is a scalene triangle? __3__

| An **isosceles triangle** has two or more sides the same length. |

Which triangles above are isosceles triangles? __1, 2__

| An **equilateral triangle** has all three sides the same length. |

Which triangle above is an equilateral triangle? __2__

Compare the angles of each triangle below with a model of a right angle. Then describe each triangle as being either *acute, obtuse,* or *right.*

1.   a   b   c

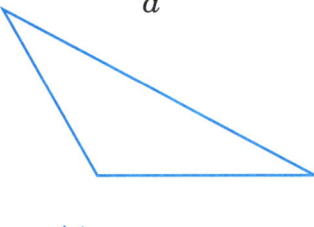

   __obtuse__ triangle   __right__ triangle   __acute__ triangle

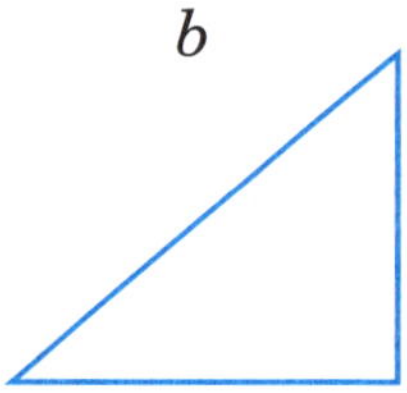

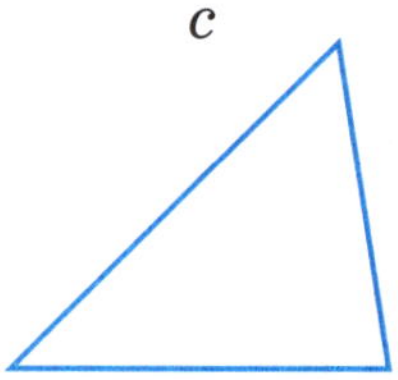

2. Compare the lengths of the sides of each triangle. Then describe each triangle as being either *scalene, isosceles,* or *equilateral.*

   __scalene__ triangle 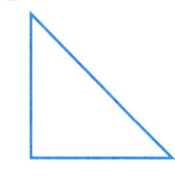   __isosceles__ triangle 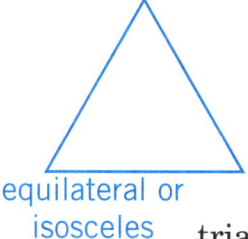   __equilateral or isosceles__ triangle

Number Correct

140

LESSON FOLLOW-UP AND ERROR ANALYSIS

5–6: Have students interchange the directions and then complete.
3–4: Ask students to explain how triangles are identified by their angles in row 1 and by their sides in row 2.
Less than 3: Have students explain how the same triangle can be identified by its angles and by its sides.

For further evaluation, copy the Chapter Test on page 273.
For maintaining skills, use the Cumulative Review on pages 229–230.

Possible Score: 15
Time Frame: 5–10 minutes

CHAPTER 9   PRACTICE TEST
Geometry

Use the figures below to answer the questions that follow.

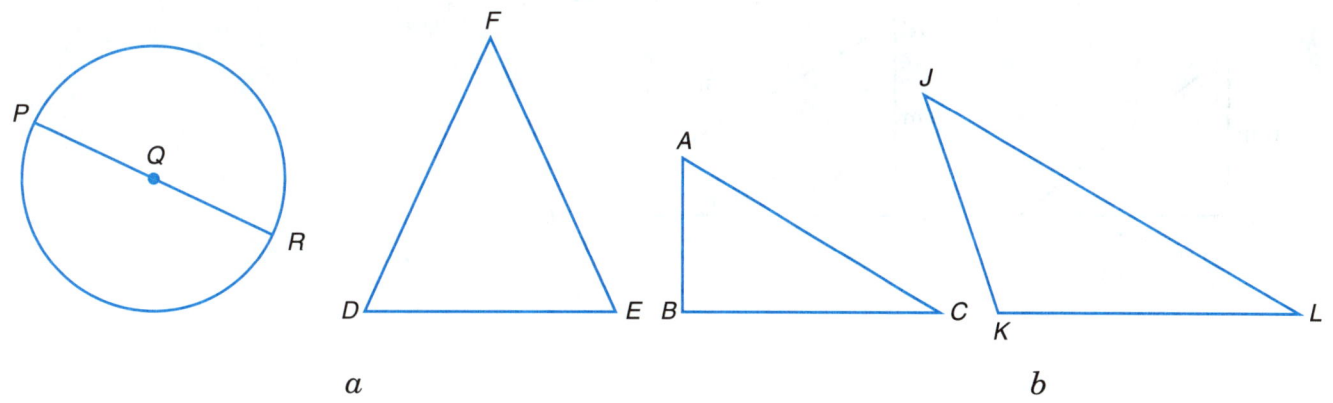

a *b*

1. Name a radius of circle Q. __QP or QR__   Name a diameter of circle Q. __PR or RP__
2. Which figure is a right triangle?   Which figure is an isosceles triangle?
   __△ABC__   __△DEF__

3. Which figures are scalene triangles?   Which figure is an obtuse triangle?
   __△ABC or △JKL__   __△JKL__

Use a protractor to measure each angle. Then describe each angle by writing *acute*, *obtuse*, or *right*.

   a   *b*   *c*
4. __90__°, __right__   __40__°, __acute__   __105__°, __obtuse__

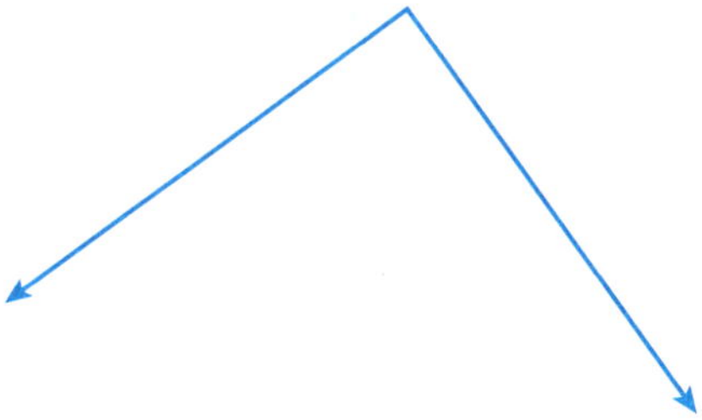

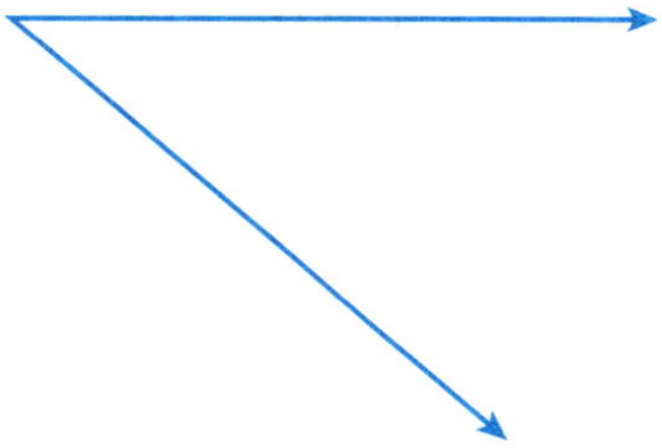

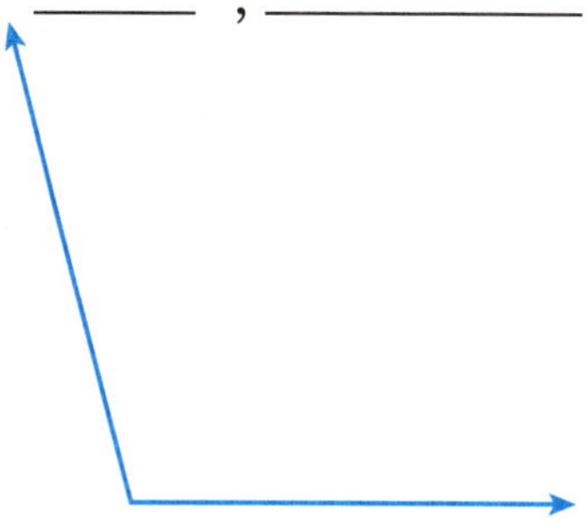

Tell whether each pair of lines is *parallel* or *perpendicular*.

5. __parallel__   __perpendicular__   __perpendicular__

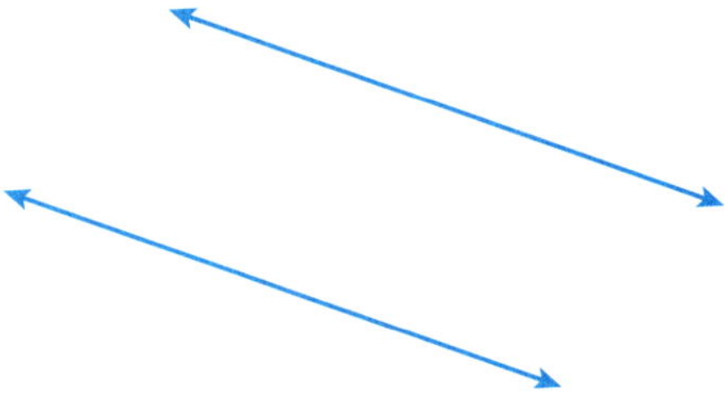

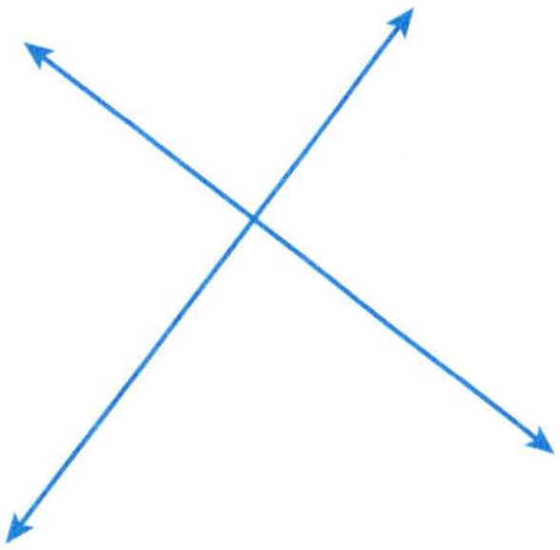

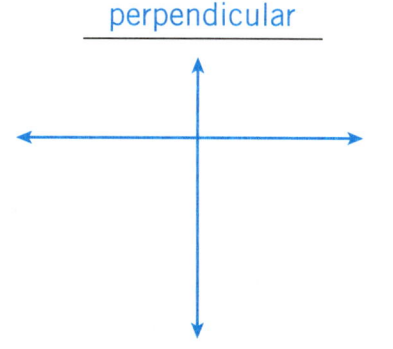

Number Correct   LESSON FOLLOW-UP AND ERROR ANALYSIS
13–15:   Ask students if it is possible to construct a right equilateral triangle and have them show why or why not.
10–12:   Check to see if students are confused by the different kinds of triangles. If so, review Lesson 8.
Less than 10:   In row 4, check to see if students described each angle correctly but did not get an accurate measurement with a protractor.

CHAPTER 10 PRETEST
Similar Triangles

Find the length of the side shown in colour in each pair of similar triangles below.

1. 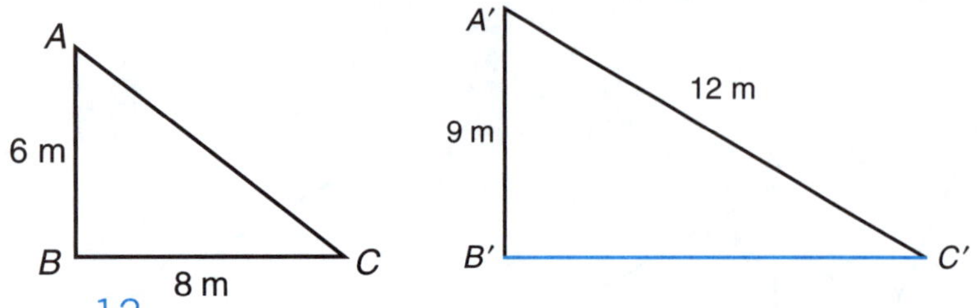
   - AB = 6 m, BC = 8 m
   - A'B' = 9 m, A'C' = 12 m
   - __12__ m

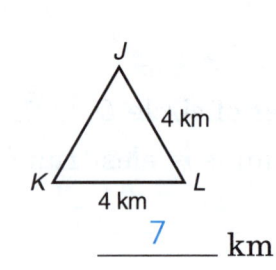

2. 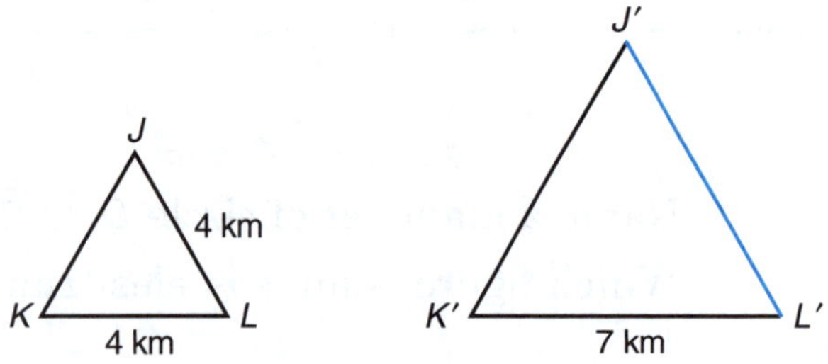
   - JL = 4 km, KL = 4 km
   - J'L' = 7 km
   - __7__ km

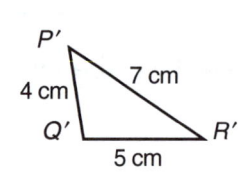

3. 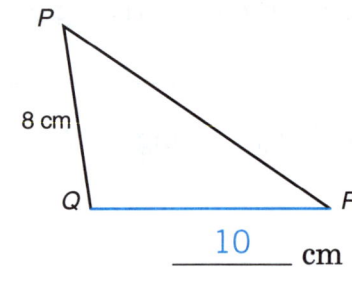
   - PQ = 8 cm, QR = 10 cm
   - P'Q' = 4 cm, P'R' = 7 cm, Q'R' = 5 cm
   - __10__ cm

Use the Pythagorean Theorem and the table on page **149** to help you find the length of each side shown in colour below.

4.
   - legs 6 cm and 8 cm, hypotenuse a 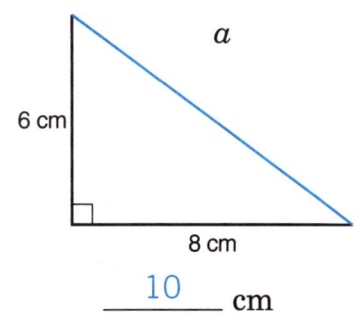
   - __10__ cm
   - legs 15 m and b, hypotenuse 17 m
   - __8__ m

5. 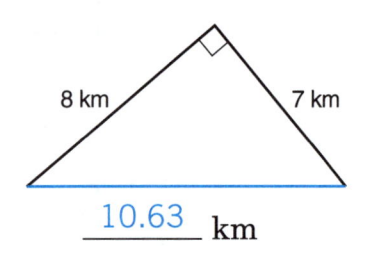
   - legs 8 km and 7 km, hypotenuse
   - __10.63__ km 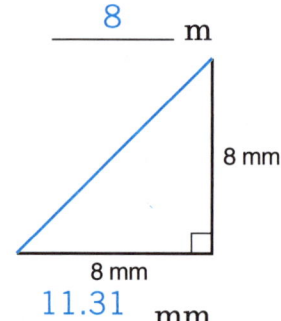
   - legs 8 mm and 8 mm, hypotenuse
   - __11.31__ mm

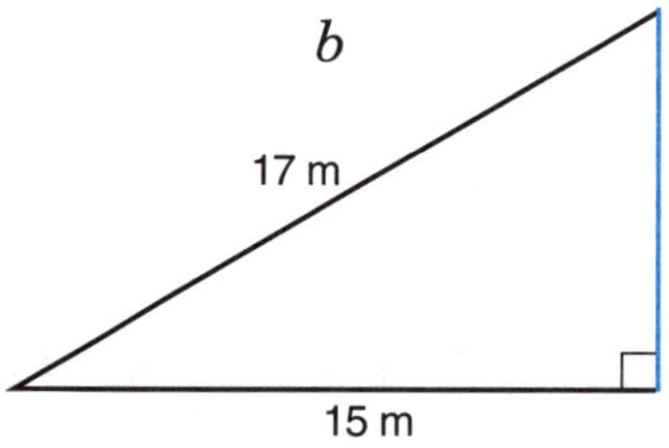

CHAPTER 10
Similar Triangles

Prerequisite Skills: naming angles in a triangle

Lesson Focus: identifying similar triangles by comparing angles
Possible Score: 12
Time Frame: 5–10 minutes

Lesson 1 Similar Triangles

When this photo was enlarged, the size of the angles did not change.

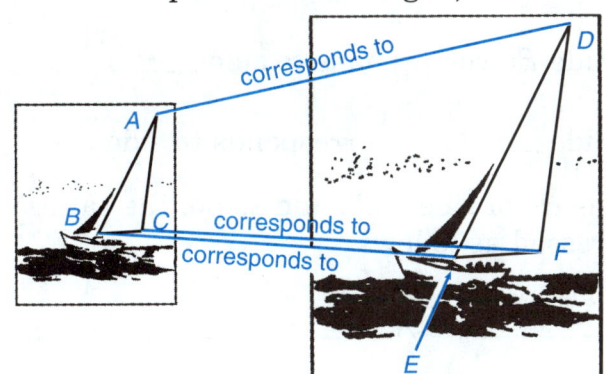

$\angle C \cong \angle F$
$\angle A \cong \angle D$
$\angle B \cong \angle E$

Did the lengths of the sides change? __yes__

Are the two triangles the same size? __no__

△ABC is similar to △DEF.

or

△ABC ~ △DEF

Two triangles are similar if their corresponding angles are congruent.

One triangle in each pair below is labelled with symbols like D′ (read D prime) and J′ (read J prime). This makes it easy to tell that D corresponds to D′ and J corresponds to J′, and so on. Complete the following.

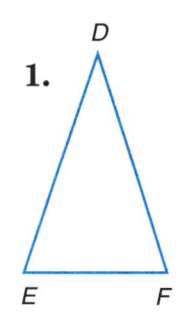

a   *b*

1. $\angle D \cong \angle D'$ $\angle J' \cong \angle J$
   $\angle E \cong \angle E'$ $\angle K' \cong \angle K$
   $\angle F \cong \angle F'$ $\angle L' \cong \angle L$
   △DEF ~ △__D′E′F′__   △J′K′L′ ~ △__JKL__

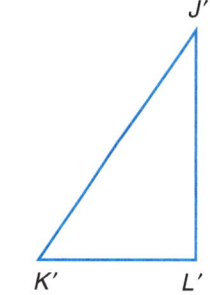

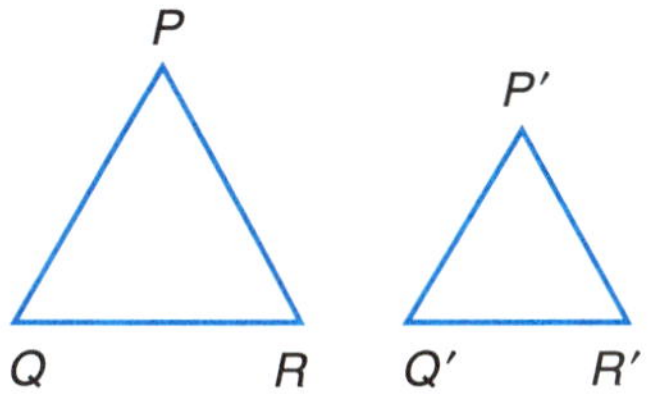

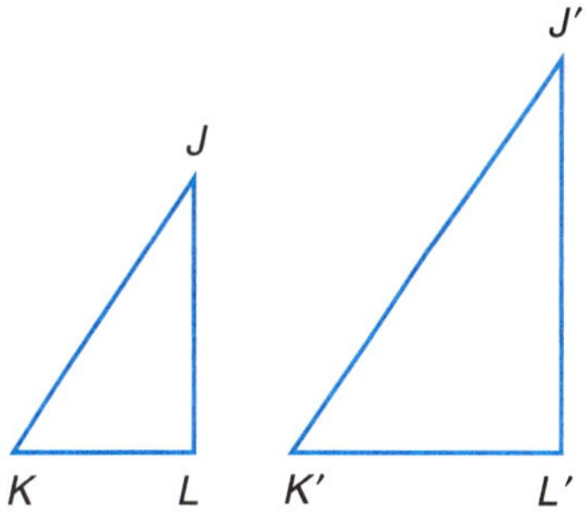

2. △PQR ~ △P′Q′R′   △MNO ~ △M′N′O′
   $\angle P \cong \angle$ __P′__ $\angle M \cong \angle$ __M′__
   $\angle Q \cong \angle$ __Q′__ $\angle N \cong \angle$ __N′__
   $\angle R \cong \angle$ __R′__ $\angle O \cong \angle$ __O′__

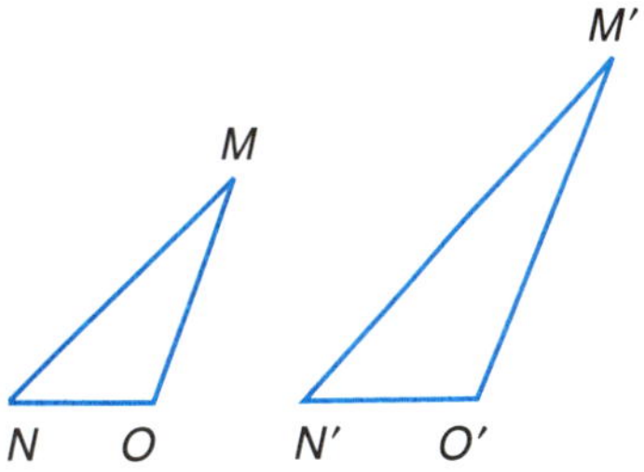

Write *True* or *False* after each statement below.

3. Two triangles that are similar must be the same size.   __False__

4. Two triangles that are similar have the same shape.   __True__

5. Two triangles that are similar have corresponding
   angles that are congruent.   __True__

6. All right triangles are similar.   __False__

LESSON FOLLOW-UP AND ERROR ANALYSIS

Number Correct
10–12: Have students choose one pair of triangles and draw a third triangle similar to those two.
7–9: Have students explain the difference between the symbol for congruent (≅) and the symbol for similar (~).
Less than 7: Ask students to explain why the corresponding angles must be congruent for triangles to be similar.

Prerequisite Skills: naming sides of a triangle

Lesson Focus: identifying similar triangles by comparing sides
Possible Score: 18
Time Frame: 10–15 minutes

Lesson 2 Similar Triangles

$\triangle ABC \sim \triangle A'B'C'$

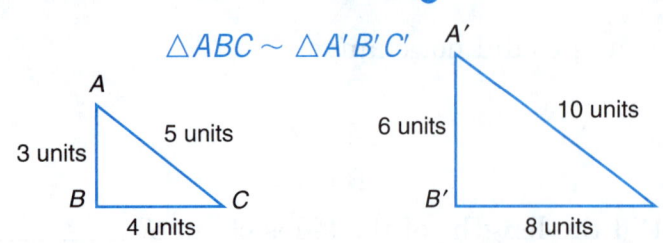

Side AB corresponds to side $A'B'$.

Side BC corresponds to side __$B'C'$__.

Side __CA__ corresponds to side $C'A'$.

If AB denotes the measure of side AB, $A'B'$ the measure of side $A'B'$, and so on, the ratios of the measures of corresponding sides can be expressed as follows.

$\dfrac{AB}{A'B'} = \dfrac{3}{6} = \dfrac{1}{2} \qquad \dfrac{BC}{B'C'} = \dfrac{4}{8} = \dfrac{1}{2} \qquad \dfrac{CA}{C'A'} = \dfrac{5}{10} = \dfrac{1}{2}$

If two triangles are similar, the ratios of the measures of their corresponding sides are equal.

For each pair of similar triangles, complete the following to show that the ratios of the measures of corresponding sides are equal.

1.

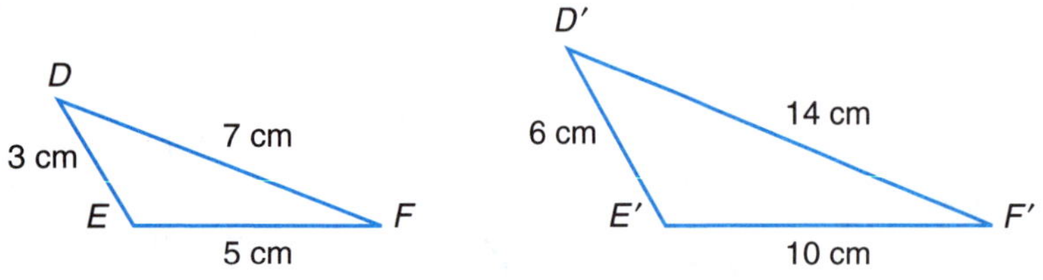

$\dfrac{DE}{D'E'} = \dfrac{3}{6} = \dfrac{1}{2}$

$\dfrac{EF}{E'F'} = \dfrac{5}{10} = \dfrac{1}{2}$

$\dfrac{FD}{F'D'} = \dfrac{7}{14} = \dfrac{1}{2}$

2.

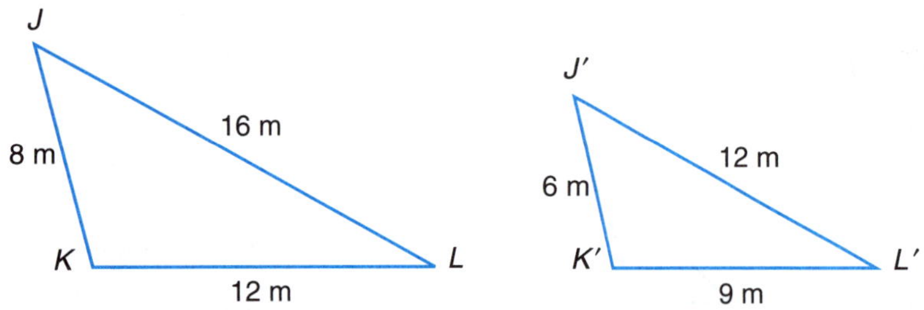

$\dfrac{JK}{J'K'} = \dfrac{8}{6} = \dfrac{4}{3}$

$\dfrac{KL}{K'L'} = \dfrac{12}{9} = \dfrac{4}{3}$

$\dfrac{LJ}{L'J'} = \dfrac{16}{12} = \dfrac{4}{3}$

3.

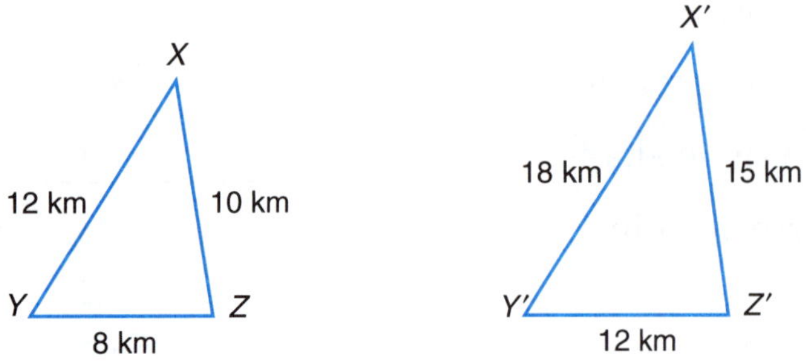

$\dfrac{XY}{X'Y'} = \dfrac{12}{18} = \dfrac{2}{3}$

$\dfrac{YZ}{Y'Z'} = \dfrac{8}{12} = \dfrac{2}{3}$

$\dfrac{ZX}{Z'X'} = \dfrac{10}{5} = \dfrac{2}{3}$

Number Correct 144

LESSON FOLLOW-UP AND ERROR ANALYSIS

15–18: Have students choose one pair of triangles and draw a third triangle similar to those two.
12–14: Before students redo the page, review with them that a ratio is a comparison of one number to another.
Less than 12: Have students explain how to substitute the measures of each side in the given ratios.

Prerequisite Skills: ratio and proportion; recognizing corresponding sides of triangles

Lesson Focus: finding a missing side in similar triangles
Possible Score: 10
Time Frame: 10–15 minutes

Lesson 3   Similar Triangles   PRE-ALGEBRA

A 5-m post casts a shadow 8 m long while a nearby flagpole casts a shadow 48 m long. What is the height of the flagpole?

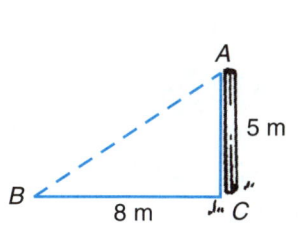

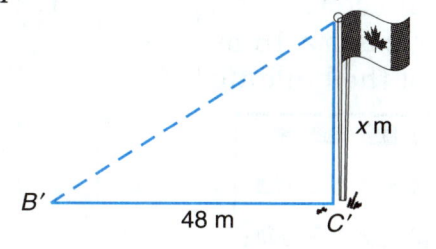

$$\frac{BC}{B'C'} = \frac{CA}{C'A'}$$

$$\frac{8}{48} = \frac{5}{x}$$

$$8 \times x = 48 \times 5$$

$$30 = x$$

The height of the flagpole is __30__ m.

Find the length of the side shown in colour in each pair of similar triangles below.

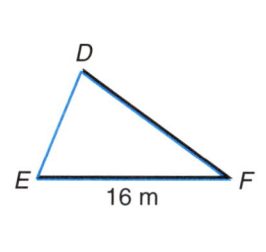

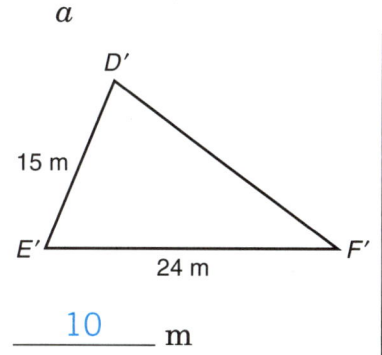

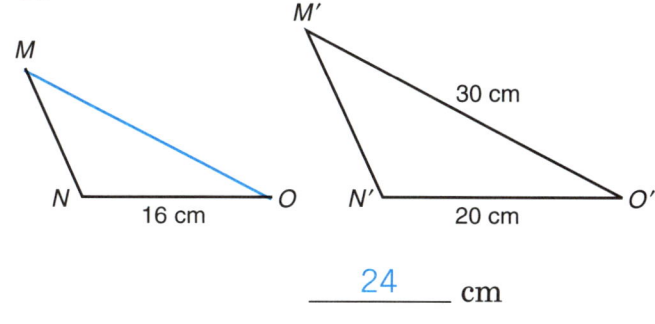

a

1. __10__ m

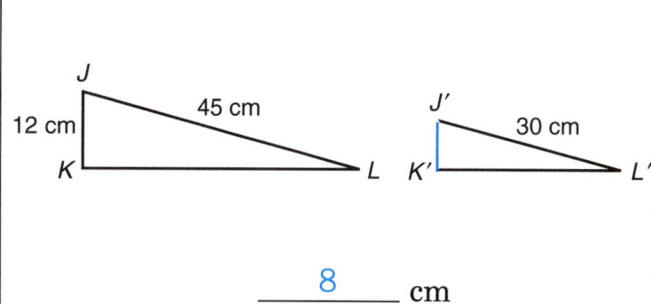

b

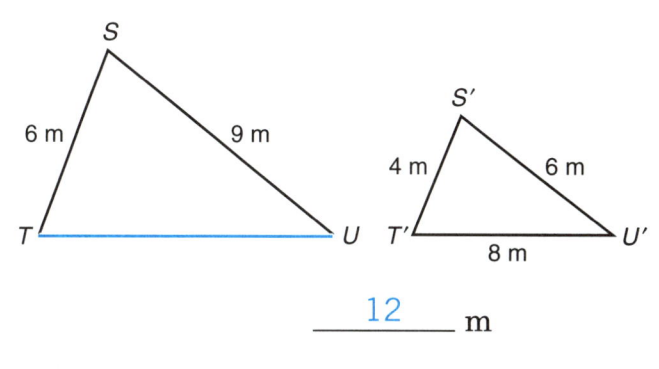

__8__ cm

2. __24__ cm

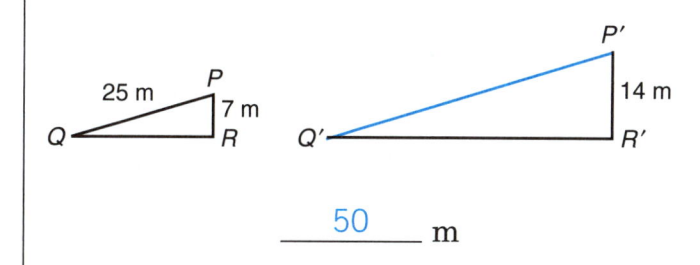

__50__ m

3. __12__ m

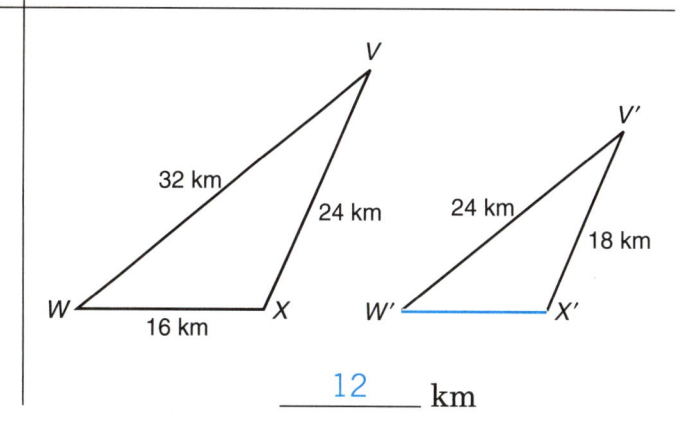

__12__ km

Number Correct   LESSON FOLLOW-UP AND ERROR ANALYSIS

8–10: Have students choose one pair of similar triangles and write a story problem using those triangles.
6–7: Suggest that students first set up a proportion using the names of the sides, substitute the lengths, and then solve.
Less than 6: Ask students to explain the importance of using corresponding sides when setting up each proportion.

Lesson 3 Problem Solving   PRE-ALGEBRA

Solve each problem.

1. A tree 8 m high casts a 4-m shadow. At the same time, a nearby building casts a 16-m shadow. What is the height of the building?

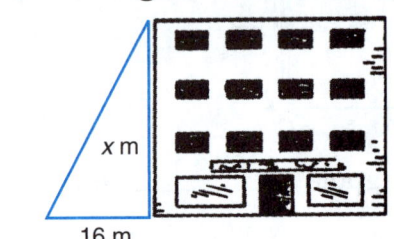

   The height of the building is ___32___ m.

2. If △CAB ~ △EDC, what is the length of the pond shown below?

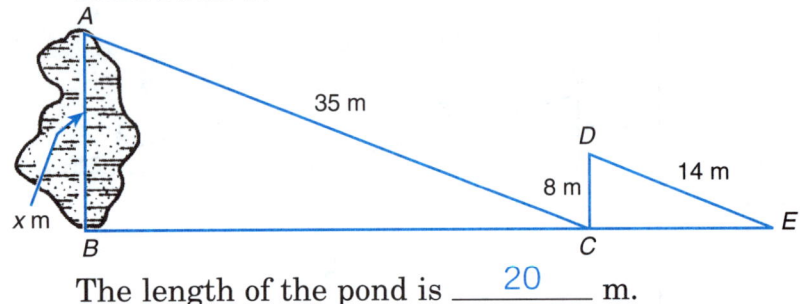

   The length of the pond is ___20___ m.

3. △JKL ~ △PQL, what is the height of the building shown below?

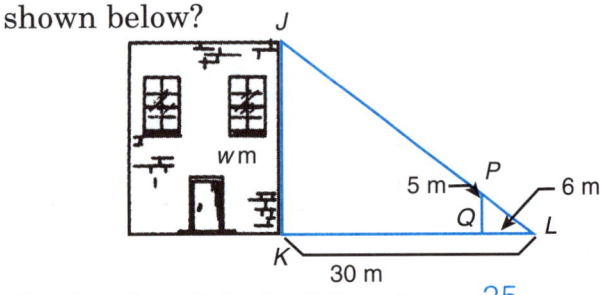

   The height of the building is ___25___ m.

4. A pole 8 m high casts a shadow 6 m long. At the same time, a TV tower casts a shadow 30 m long. How high is the TV tower?

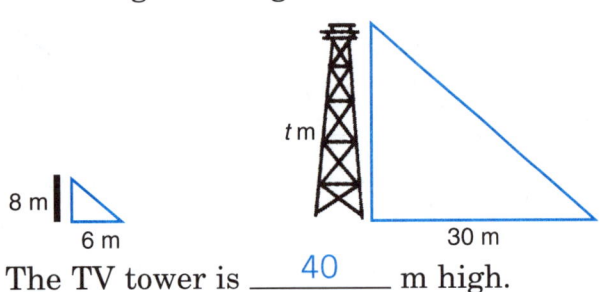

   The TV tower is ___40___ m high.

CHAPTER 10
Similar Triangles

146

Prerequisite Skills: multiplication, and division facts

Lesson Focus: finding squares and square roots
Possible Score: 26
Time Frame: 10–15 minutes

Lesson 4 Squares and Square Roots

6^2 is read 6 *squared*.
6^2 means 6×6.

$\sqrt{36}$ is read *the square root of 36*.
$\sqrt{36}$ is some positive number a so that $a \times a = 36$.

$6^2 = \underline{\ 6\ } \times \underline{\ 6\ } = \underline{\ 36\ }$
$9^2 = \underline{\ 9\ } \times \underline{\ 9\ } = \underline{\ 81\ }$
$3^2 = \underline{\ 3\ } \times \underline{\ 3\ } = \underline{\ 9\ }$

$\sqrt{36} = \sqrt{\underline{\ 6\ } \times \underline{\ 6\ }} = \underline{\ 6\ }$
$\sqrt{81} = \sqrt{\underline{\ 9\ } \times \underline{\ 9\ }} = \underline{\ 9\ }$
$\sqrt{9} = \sqrt{\underline{\ 3\ } \times \underline{\ 3\ }} = \underline{\ 3\ }$

Complete the following.

a

1. $5^2 = \underline{\ 5\ } \times \underline{\ 5\ } = \underline{\ 25\ }$
2. $8^2 = \underline{\ 8\ } \times \underline{\ 8\ } = \underline{\ 64\ }$
3. $2^2 = \underline{\ 2\ } \times \underline{\ 2\ } = \underline{\ 4\ }$
4. $10^2 = \underline{\ 10\ } \times \underline{\ 10\ } = \underline{\ 100\ }$
5. $4^2 = \underline{\ 4\ } \times \underline{\ 4\ } = \underline{\ 16\ }$
6. $12^2 = \underline{\ 12\ } \times \underline{\ 12\ } = \underline{\ 144\ }$
7. $20^2 = \underline{\ 20\ } \times \underline{\ 20\ } = \underline{\ 400\ }$
8. $11^2 = \underline{\ 11\ } \times \underline{\ 11\ } = \underline{\ 121\ }$
9. $19^2 = \underline{\ 19\ } \times \underline{\ 19\ } = \underline{\ 361\ }$
10. $25^2 = \underline{\ 25\ } \times \underline{\ 25\ } = \underline{\ 625\ }$
11. $31^2 = \underline{\ 31\ } \times \underline{\ 31\ } = \underline{\ 961\ }$
12. $43^2 = \underline{\ 43\ } \times \underline{\ 43\ } = \underline{\ 1849\ }$
13. $50^2 = \underline{\ 50\ } \times \underline{\ 50\ } = \underline{\ 2500\ }$

b

$\sqrt{25} = \sqrt{\underline{\ 5\ } \times \underline{\ 5\ }} = \underline{\ 5\ }$
$\sqrt{64} = \sqrt{\underline{\ 8\ } \times \underline{\ 8\ }} = \underline{\ 8\ }$
$\sqrt{4} = \sqrt{\underline{\ 2\ } \times \underline{\ 2\ }} = \underline{\ 2\ }$
$\sqrt{100} = \sqrt{\underline{\ 10\ } \times \underline{\ 10\ }} = \underline{\ 10\ }$
$\sqrt{16} = \sqrt{\underline{\ 4\ } \times \underline{\ 4\ }} = \underline{\ 4\ }$
$\sqrt{144} = \sqrt{\underline{\ 12\ } \times \underline{\ 12\ }} = \underline{\ 12\ }$
$\sqrt{400} = \sqrt{\underline{\ 20\ } \times \underline{\ 20\ }} = \underline{\ 20\ }$
$\sqrt{121} = \sqrt{\underline{\ 11\ } \times \underline{\ 11\ }} = \underline{\ 11\ }$
$\sqrt{361} = \sqrt{\underline{\ 19\ } \times \underline{\ 19\ }} = \underline{\ 19\ }$
$\sqrt{625} = \sqrt{\underline{\ 25\ } \times \underline{\ 25\ }} = \underline{\ 25\ }$
$\sqrt{961} = \sqrt{\underline{\ 31\ } \times \underline{\ 31\ }} = \underline{\ 31\ }$
$\sqrt{1849} = \sqrt{\underline{\ 43\ } \times \underline{\ 43\ }} = \underline{\ 43\ }$
$\sqrt{2500} = \sqrt{\underline{\ 50\ } \times \underline{\ 50\ }} = \underline{\ 50\ }$

CHAPTER 10

Number Correct

LESSON FOLLOW-UP AND ERROR ANALYSIS

22–26: Have students complete columns *a* and *b* for two numbers less than 10 not already shown.
17–21: Have students explain why in column *b* they write the two identical numbers that are factors of the first number.
Less than 17: Ask students to explain the difference between *squares* and *square roots* before they redo the page.

Prerequisite Skills: reading a table

Lesson Focus: finding squares and square roots using a table
Possible Score: 22
Time Frame: 10–15 minutes

Lesson 5 Squares and Square Roots (table)

Study how the table is used to find the square and the square root of a number n. ($\doteq$ is read *is approximately equal to*.)

n	n^2	$\sqrt{n}$
1	1	1.00
2	4	1.41
3	9	1.73
4	16	2.00
5	25	2.24
6	36	2.45
7	49	2.65
8	64	2.83
9	81	3.00

If $n = 2$, then $2^2 =$ __4__ and $\sqrt{2} \doteq$ __1.41__.

If $n = 4$, then $4^2 =$ __16__ and $\sqrt{4} =$ __2.00 or 2__.

If $n = 7$, then $7^2 =$ __49__ and $\sqrt{7} \doteq$ __2.65__.

If $n = 9$, then $9^2 =$ __81__ and $\sqrt{9} \doteq$ __3.00 or 3__.

Use the table on page **149** to help you complete the following.

1. If $n = 18$, then $18^2 =$ __324__ and $\sqrt{18} \doteq$ __4.24__.

2. If $n = 25$, then $25^2 =$ __625__ and $\sqrt{25} =$ __5__.

3. If $n = 45$, then $45^2 =$ __2025__ and $\sqrt{45} \doteq$ __6.71__.

4. If $n = 64$, then $64^2 =$ __4096__ and $\sqrt{64} =$ __8__.

5. If $n = 83$, then $83^2 =$ __6889__ and $\sqrt{83} \doteq$ __9.11__.

6. If $n = 75$, then $75^2 =$ __5625__ and $\sqrt{75} \doteq$ __8.66__.

7. If $n = 90$, then $90^2 =$ __8100__ and $\sqrt{90} \doteq$ __9.49__.

8. If $n = 104$, then $104^2 =$ __10 816__ and $\sqrt{104} \doteq$ __10.2__.

9. If $n = 135$, then $135^2 =$ __18 225__ and $\sqrt{135} \doteq$ __11.62__.

10. If $n = 147$, then $147^2 =$ __21 609__ and $\sqrt{147} \doteq$ __12.12__.

11. If $n = 150$, then $150^2 =$ __22 500__ and $\sqrt{150} \doteq$ __12.25__.

LESSON FOLLOW-UP AND ERROR ANALYSIS

Number Correct
- **19–22:** Have students explain why some numbers have exact square roots and others are only approximate.
- **14–18:** Check to see if students can find the appropriate column when finding squares and square roots.
- **Less than 14:** Before students redo incorrect answers, have them explain how to use the *Table of Squares and Square Roots*.

Prerequisite Skills: reading a table of squares and square roots

Lesson Focus: finding square roots using a table
Possible Score: 36
Time Frame: 10–15 minutes

Lesson 6 Squares and Square Roots (table)

Table of Squares and Square Roots

n	n^2	$\sqrt{n}$	n	n^2	$\sqrt{n}$	n	n^2	$\sqrt{n}$
1	1	1.00	51	2 601	7.14	101	10 201	10.05
2	4	1.41	52	2 704	7.21	102	10 404	10.10
3	9	1.73	53	2 809	7.28	103	10 609	10.15
4	16	2.00	54	2 916	7.35	104	10 816	10.20
5	25	2.24	55	3 025	7.42	105	11 025	10.25
6	36	2.45	56	3 136	7.48	106	11 236	10.30
7	49	2.65	57	3 249	7.55	107	11 449	10.34
8	64	2.83	58	3 364	7.62	108	11 664	10.39
9	81	3.00	59	3 481	7.68	109	11 881	10.44
10	100	3.16	60	3 600	7.75	110	12 100	10.49
11	121	3.32	61	3 721	7.81	111	12 321	10.54
12	144	3.46	62	3 844	7.87	112	12 544	10.58
13	169	3.61	63	3 969	7.94	113	12 769	10.63
14	196	3.74	64	4 096	8.00	114	12 996	10.68
15	225	3.87	65	4 225	8.06	115	13 225	10.72
16	256	4.00	66	4 356	8.12	116	13 456	10.77
17	289	4.12	67	4 489	8.19	117	13 689	10.82
18	324	4.24	68	4 624	8.25	118	13 924	10.86
19	361	4.36	69	4 761	8.31	119	14 161	10.91
20	400	4.47	70	4 900	8.37	120	14 400	10.95
21	441	4.58	71	5 041	8.43	121	14 641	11.00
22	484	4.69	72	5 184	8.49	122	14 884	11.05
23	529	4.80	73	5 329	8.54	123	15 129	11.09
24	576	4.90	74	5 476	8.60	124	15 376	11.14
25	625	5.00	75	5 625	8.66	125	15 625	11.18
26	676	5.10	76	5 776	8.72	126	15 876	11.22
27	729	5.20	77	5 929	8.77	127	16 129	11.27
28	784	5.29	78	6 084	8.83	128	16 384	11.31
29	841	5.39	79	6 241	8.89	129	16 641	11.36
30	900	5.48	80	6 400	8.94	130	16 900	11.40
31	961	5.57	81	6 561	9.00	131	17 161	11.45
32	1 024	5.66	82	6 724	9.06	132	17 424	11.49
33	1 089	5.74	83	6 889	9.11	133	17 689	11.53
34	1 156	5.83	84	7 056	9.17	134	17 956	11.58
35	1 225	5.92	85	7 225	9.22	135	18 225	11.62
36	1 296	6.00	86	7 396	9.27	136	18 496	11.66
37	1 369	6.08	87	7 569	9.33	137	18 769	11.70
38	1 444	6.16	88	7 744	9.38	138	19 044	11.75
39	1 521	6.24	89	7 921	9.43	139	19 321	11.79
40	1 600	6.32	90	8 100	9.49	140	19 600	11.83
41	1 681	6.40	91	8 281	9.54	141	19 881	11.87
42	1 764	6.48	92	8 464	9.59	142	20 164	11.92
43	1 849	6.56	93	8 649	9.64	143	20 449	11.96
44	1 936	6.63	94	8 836	9.70	144	20 736	12.00
45	2 025	6.71	95	9 025	9.75	145	21 025	12.04
46	2 116	6.78	96	9 216	9.80	146	21 316	12.08
47	2 209	6.86	97	9 409	9.85	147	21 609	12.12
48	2 304	6.93	98	9 604	9.90	148	21 904	12.17
49	2 401	7.00	99	9 801	9.95	149	22 201	12.21
50	2 500	7.07	100	10 000	10.00	150	22 500	12.25

CHAPTER 10

Number Correct

LESSON FOLLOW-UP AND ERROR ANALYSIS

30–36: Have students choose one value for *n* from the table, multiply it by itself, and compare the answer to the table.
23–29: Because of the square root sign over the number, students may have been tempted to look in the third rather than the second column for answers.
Less than 23: Ask students to explain how to use the table on page 149 to redo these pages.

Lesson 6 Square Roots

Study how the table on page **149** can be used to find the square root of a number greater than 150.

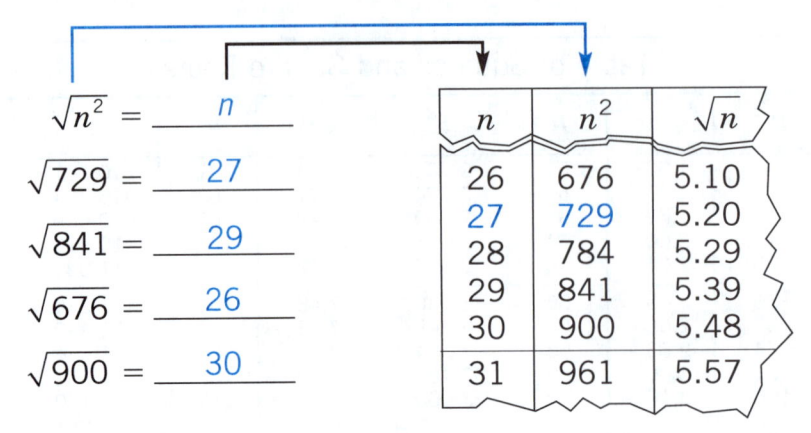

$\sqrt{n^2}$ = __n__

$\sqrt{729}$ = __27__

$\sqrt{841}$ = __29__

$\sqrt{676}$ = __26__

$\sqrt{900}$ = __30__

n	n^2	$\sqrt{n}$
26	676	5.10
27	729	5.20
28	784	5.29
29	841	5.39
30	900	5.48
31	961	5.57

Use the table on page **149** to help you complete each of the following.

	a	b	c
1.	$\sqrt{169}$ = __13__	$\sqrt{529}$ = __23__	$\sqrt{784}$ = __28__
2.	$\sqrt{256}$ = __16__	$\sqrt{361}$ = __19__	$\sqrt{961}$ = __31__
3.	$\sqrt{1225}$ = __35__	$\sqrt{2209}$ = __47__	$\sqrt{3969}$ = __63__
4.	$\sqrt{1681}$ = __41__	$\sqrt{3136}$ = __56__	$\sqrt{4761}$ = __69__
5.	$\sqrt{5329}$ = __73__	$\sqrt{6084}$ = __78__	$\sqrt{6889}$ = __83__
6.	$\sqrt{7921}$ = __89__	$\sqrt{8649}$ = __93__	$\sqrt{9604}$ = __98__
7.	$\sqrt{10\,201}$ = __101__	$\sqrt{11\,025}$ = __105__	$\sqrt{11\,449}$ = __107__
8.	$\sqrt{12\,544}$ = __112__	$\sqrt{17\,424}$ = __132__	$\sqrt{22\,201}$ = __149__
9.	$\sqrt{4900}$ = __70__	$\sqrt{1849}$ = __43__	$\sqrt{16\,900}$ = __130__
10.	$\sqrt{10\,000}$ = __100__	$\sqrt{12\,100}$ = __110__	$\sqrt{2500}$ = __50__
11.	$\sqrt{8464}$ = __92__	$\sqrt{19\,321}$ = __139__	$\sqrt{13\,924}$ = __118__
12.	$\sqrt{8281}$ = __91__	$\sqrt{18\,225}$ = __135__	$\sqrt{21\,316}$ = __146__

CHAPTER 10
Similar Triangles

Prerequisite Skills: reading a table of squares and square roots

Lesson Focus: finding the hypotenuse using the Pythagorean Theorem
Possible Score: 16
Time Frame: 20–25 minutes

Lesson 7   The Pythagorean Theorem   PRE-ALGEBRA

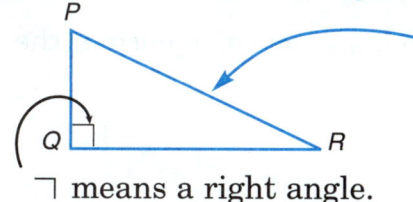

In a right triangle the side opposite the right angle is called the **hypotenuse**.

⌐ means a right angle.

Which side is the hypotenuse in △PQR? __PR__

Pythagorean Theorem: The square of the measure of the hypotenuse of a right triangle is equal to the sum of the squares of the measures of the other two sides.

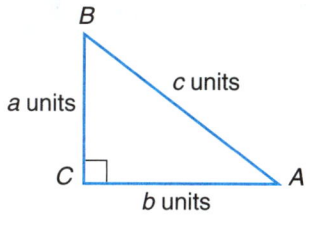

$c^2 = a^2 + b^2$ or $c^2 = b^2 + a^2$

Find c if $a = 3$ and $b = 4$.
$c^2 = a^2 + b^2$
$c^2 = 3^2 + 4^2$
$c^2 = 9 + 16$ or 25
$\sqrt{c^2} = \sqrt{25}$
$c =$ __5__

Use △ABC above and the table on page **149** to help you complete the following.

1. If $a = 6$ and $b = 8$, then $c =$ __10__.

2. If $a = 7$ and $b = 24$, then $c =$ __25__.

3. If $a = 5$ and $b = 7$, then $c \doteq$ __8.6__.

4. If $a = 7$ and $b = 9$, then $c \doteq$ __11.4__.

5. If $a = 5$ and $b = 12$, then $c =$ __13__.

6. If $a = 8$ and $b = 8$, then $c \doteq$ __11.31__.

7. If $a = 8$ and $b = 15$, then $c =$ __17__.

8. If $a = 3$ and $b = 8$, then $c \doteq$ __8.54__.

9. If $a = 20$ and $b = 21$, then $c =$ __29__.

10. If $a = 12$ and $b = 2$, then $c \doteq$ __12.17__.

11. If $a = 45$ and $b = 28$, then $c =$ __53__.

CHAPTER 10

Number Correct   **LESSON FOLLOW-UP AND ERROR ANALYSIS**
13–16: Have students draw several right triangles and use the Pythagorean Theorem to find a missing side of each.
10–12: Check to see if students added the two given sides without squaring them first.
Less than 10: Ask students to explain how to use the table on page 149 to redo these pages.

Lesson 7 Problem Solving   PRE-ALGEBRA

Use the Pythagorean Theorem and the table on page **149** to help you solve each of the following.

1. The foot of a ladder is placed 5 m from a building. The top of the ladder rests 12 m up on the building. How long is the ladder?

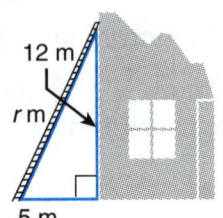

   The ladder is ____13____ m long.

2. The roof of a house is to be built as shown. How long should each rafter be?

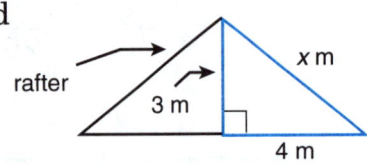

   Each rafter should be ____10____ m long.

3. A ship left port and sailed 5 km east and then 7 km north. How far was the ship from the port then?

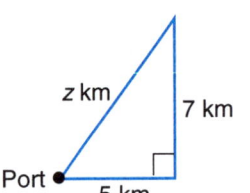

   The ship was about ____8.6____ km from the port.

4. What is the length of the lake shown below?

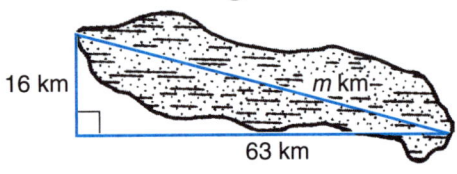

   The length is ____65____ km.

5. An inclined ramp rises 5 m over a horizontal distance of 11 m. What is the length of the ramp?

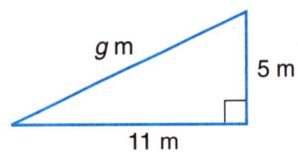

   The length is ____12.08____ m.

CHAPTER 10
Similar Triangles

Lesson 7
The Pythagorean Theorem

Prerequisite Skills: reading a table of squares and square roots

Lesson Focus: finding a side of a triangle using the Pythagorean Theorem
Possible Score: 18
Time Frame: 20–25 minutes

Lesson 8   Using the Pythagorean Theorem   PRE-ALGEBRA

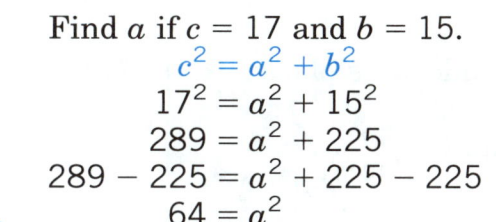

Find a if $c = 17$ and $b = 15$.
$$c^2 = a^2 + b^2$$
$$17^2 = a^2 + 15^2$$
$$289 = a^2 + 225$$
$$289 - 225 = a^2 + 225 - 225$$
$$64 = a^2$$
$$\sqrt{64} = \sqrt{a^2}$$
$$\underline{\;8\;} = a$$

Find b if $c = 13$ and $a = 12$.
$$c^2 = b^2 + a^2$$
$$13^2 = b^2 + 12^2$$
$$169 = b^2 + 144$$
$$169 - 144 = b^2 + 144 - 144$$
$$25 = b^2$$
$$\sqrt{25} = \sqrt{b^2}$$
$$\underline{\;5\;} = b$$

Use the triangle above and the table on page **149** to help you complete the following.

1. If $c = 25$ and $a = 24$, then $b = \underline{\;7\;}$.

2. If $c = 41$ and $b = 40$, then $a = \underline{\;9\;}$.

3. If $c = 61$ and $a = 60$, then $b = \underline{\;11\;}$.

4. If $c = 25$ and $b = 22$, then $a \doteq \underline{\;11.87\;}$.

5. If $c = 26$ and $a = 24$, then $b = \underline{\;10\;}$.

6. If $c = 20$ and $b = 17$, then $a \doteq \underline{\;10.54\;}$.

7. If $c = 89$ and $a = 80$, then $b = \underline{\;39\;}$.

8. If $c = 65$ and $b = 63$, then $a = \underline{\;16\;}$.

9. If $c = 72$ and $a = 71$, then $b \doteq \underline{\;11.96\;}$.

10. If $c = 73$ and $b = 48$, then $a = \underline{\;55\;}$.

11. If $c = 38$ and $a = 36$, then $b \doteq \underline{\;12.17\;}$.

12. If $c = 85$ and $b = 36$, then $a = \underline{\;77\;}$.

Number Correct — LESSON FOLLOW-UP AND ERROR ANALYSIS

15–18: Have students give the hypotenuse and one side of a right triangle, and then find the other side.
12–14: Before students redo these pages, have them explain the example at the top of page 153.
Less than 12: Have students explain $c^2 = a^2 + b^2$. (c is the measure of the hypotenuse and a or b is one of the other sides)

Lesson 8   Problem Solving   PRE-ALGEBRA

Use the Pythagorean Theorem and the table on page **149** to help you solve each of the following.

1. Suppose the foot of a 12-m ladder was placed 4 m from the building. How high up on the building would the top of the ladder reach?

   The ladder will reach about __11.31__ m.

2. A ship is 24 km east of port. How far north must the ship sail to reach a point that is 25 km from the port?

   The ship must sail __7__ km north.

3. A telephone pole is braced by a guy wire as shown. How high up on the pole is the wire fastened?

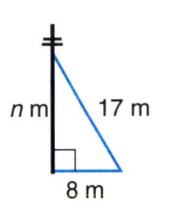

   The wire is fastened __15__ m above the ground.

4. How far is it across the pond shown below?

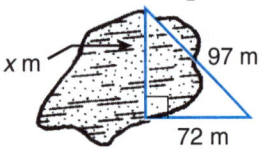

   It is __65__ m across the pond.

5. Alstown, Donville, and Maxburg are located as shown below. How many kilometres is it from Alstown to Maxburg?

   It is about __11.53__ km from Alstown to Maxburg.

6. A sail is shaped as shown. How high is the sail?

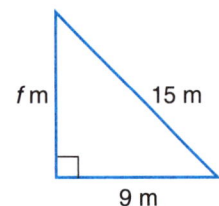

   The sail is __12__ m high.

CHAPTER 10
Similar Triangles

Lesson 8
Using the Pythagorean Theorem

Prerequisite Skills: reading a table of square roots; Pythagorean Theorem

Lesson Focus: finding the measure of missing sides of similar triangles
Possible Score: 18
Time Frame: 25–30 minutes

Lesson 9 Similar Right Triangles   PRE-ALGEBRA

Study how the Pythagorean Theorem and the ratios of similar triangles are used to find the measure of $\overline{AB}$, the measure of $\overline{A'C'}$, and the measure of $\overline{B'C'}$.

right $\triangle ABC \sim$ right $\triangle A'B'C'$

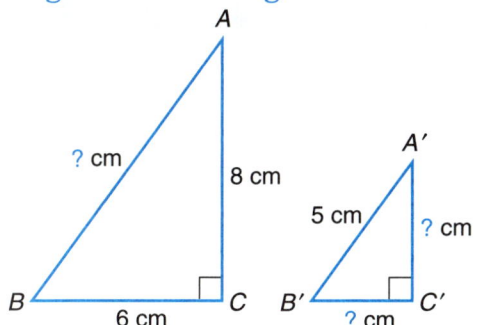

Step 1
Use $c^2 = a^2 + b^2$ to find the measure of $\overline{AB}$.

$c^2 = a^2 + b^2$
$c^2 = 6^2 + 8^2$
$c^2 = 100$
$c = 10$
$AB = \underline{\ 10\ }$

Step 2
Use the measure of $\overline{AB}$ from Step 1 and find the measure of $\overline{A'C'}$ and $\overline{B'C'}$.

$\dfrac{AB}{A'B'} = \dfrac{AC}{A'C'}$ $\dfrac{AB}{A'B'} = \dfrac{BC}{B'C'}$

$\dfrac{10}{5} = \dfrac{8}{A'C'}$ $\dfrac{10}{5} = \dfrac{6}{B'C'}$

$A'C' = \underline{\ 4\ }$ $B'C' = \underline{\ 3\ }$

$\overline{AB}$ is __10__ cm long. $\overline{A'C'}$ is __4__ cm long. $\overline{B'C'}$ is __3__ cm long.

Find the length of each side shown in colour in each pair of similar right triangles below. You may use the table on page **149** if necessary.

1.

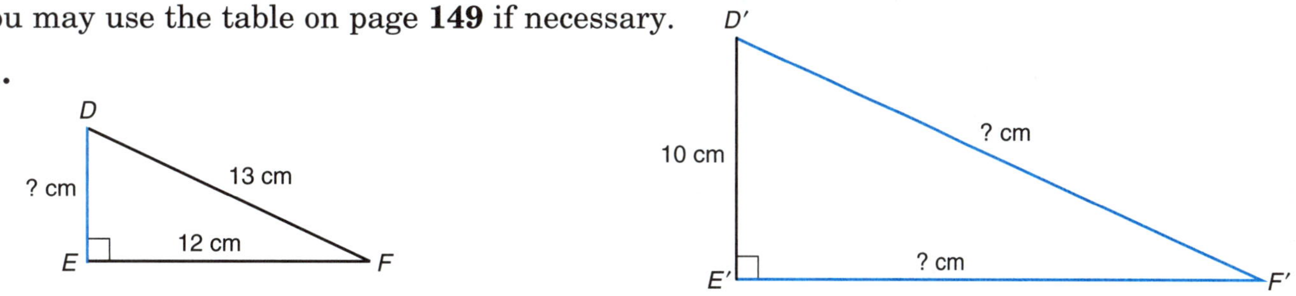

$\overline{DE}$ is __5__ cm long. $\overline{E'F'}$ is __24__ cm long. $\overline{D'F'}$ is __26__ cm long.

2.

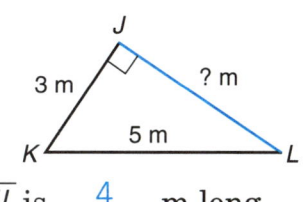

$\overline{JL}$ is __4__ m long. $\overline{J'K'}$ is __4.5__ m long. $\overline{K'L'}$ is __7.5__ m long.

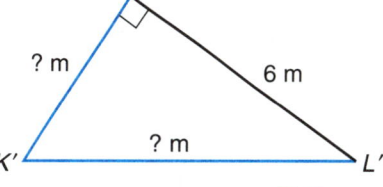

3.

$\overline{PR}$ is __8__ km long. $\overline{Q'R'}$ is __$18\tfrac{3}{4}$__ km long. $\overline{Q'P'}$ is __$21\tfrac{1}{4}$__ km long.

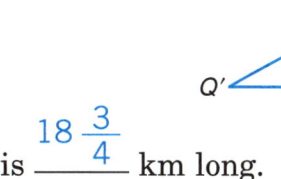

Number Correct   LESSON FOLLOW-UP AND ERROR ANALYSIS

15–18: Have students choose one pair of triangles and draw a third triangle similar to those two.
12–14: Have students explain how to find the missing side of the first triangle, and then use it to find the sides of the second triangle.
Less than 12: Ask students to explain the example at the top of page 155 before they redo the pages.

155

Lesson 9 Problem Solving   PRE-ALGEBRA

Solve each problem. If necessary, use the table on page 149.

1. If △ABD ~ △ECD, how far is it from the pier to the island? From the boathouse to the campsite? From the lodge to the campsite?

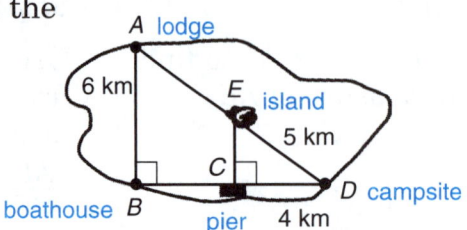

   It is __3__ km from the pier to the island.

   It is __8__ km from the boathouse to the campsite.

   It is __10__ km from the lodge to the campsite.

2. A cellphone tower is steadied by guy wires as shown. If △JKL ~ △MKN, what are the lengths of the guy wires? How high is the upper guy wire fastened above the ground?

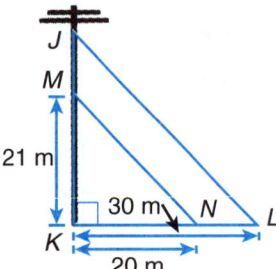

   The shorter wire is __29__ m long.

   The longer wire is __$43\frac{1}{2}$__ m long.

   The upper guy wire is fastened __$31\frac{1}{2}$__ m above the ground.

3. Two inclined ramps are shaped as shown below and △PQR ~ △XYZ. What is the length of $\overline{QR}$? What is the height of each ramp?

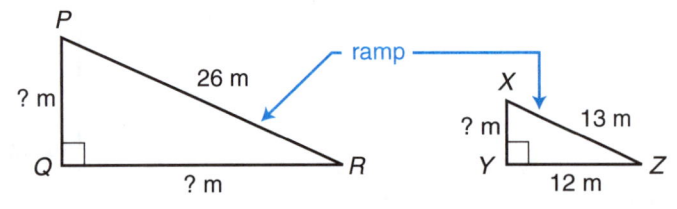

   The length of the taller ramp is __24__ m.

   The height of the shorter ramp is __5__ m.

   The height of the taller ramp is __10__ m.

CHAPTER 10
Similar Triangles

Lesson 9
Similar Right Triangles

For further evaluation, copy the Chapter Test on page 274.
For maintaining skills, use the Cumulative Review on pages 231–232.

Possible Score: 10
Time Frame: 25–30 minutes

CHAPTER 10 PRACTICE TEST
Similar Triangles

Use the similar triangles below to help you complete the following.

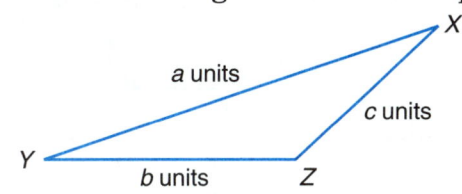

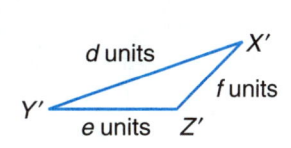

$$\frac{a}{d} = \frac{b}{e} = \frac{c}{f}$$

1. If $a = 6$, $d = 9$, and $b = 8$, then $e =$ __12__.

2. If $b = 24$, $e = 12$, and $c = 16$, then $f =$ __8__.

3. If $c = 8$, $f = 4$, and $e = 6$, then $b =$ __12__.

4. If $e = 9$, $b = 3$, and $d = 6$, then $a =$ __2__.

Use the triangle below and the table on page 149 to help you complete the following.

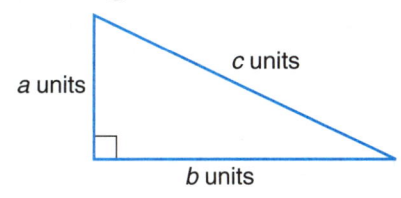

$$c^2 = a^2 + b^2 \text{ or } c^2 = b^2 + a^2$$

5. If $a = 8$ and $b = 6$, then $c =$ __10__.

6. If $a = 16$ and $c = 65$, then $b =$ __63__.

7. If $b = 20$ and $c = 23$, then $a \doteq$ __11.36__.

8. If $a = 9$ and $b = 8$, then $c \doteq$ __12.04__.

Solve each of the following.

9. A post and a flagpole cast shadows as shown below. What is the height of the flagpole?

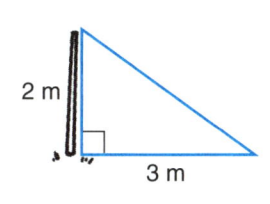

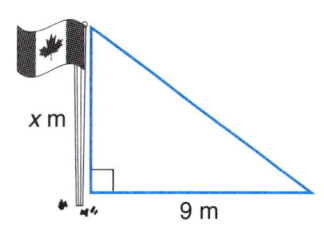

   2 m, 3 m, x m, 9 m

   The height of the flagpole is __6__ m.

10. A windowpane is 18 cm by 18 cm. What is the distance between opposite corners of the windowpane?

    The distance is about __25.5__ cm.

Number Correct
LESSON FOLLOW-UP AND ERROR ANALYSIS
9–10: Have students choose a problem from rows 5–8 and write a story problem it would solve.
7–8: Ask students to explain how to use proportions to solve problems such as 9 and 10.
Less than 7: Have students explain how to substitute numbers into formulas (problems 1–8) before they redo incorrect answers.

CHAPTER 11 PRETEST
Perimeter, Area, and Volume

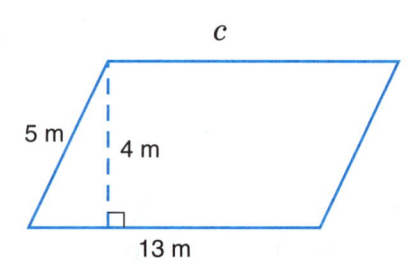

Find the perimeter and area of each figure.

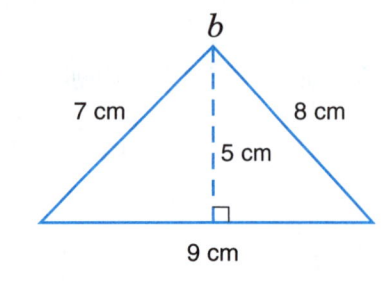

1.   a   b   c

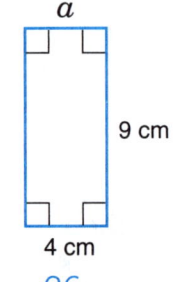

perimeter: __26__ cm   __24__ cm   __36__ m
area: __36__ cm²   __22.5__ cm²   __52__ m²

Complete the table below. Use 3.14 for π. Find the approximate circumference and area.

	diameter	radius	circumference	area
2.	12 cm	__6__ cm	about $37\frac{5}{7}$ cm	about $113\frac{1}{7}$ cm²
3.	__6__ mm	3 mm	about $18\frac{6}{7}$ mm	about $28\frac{2}{7}$ mm²

Find the surface area of each figure. Use 3.14 for π.

4.   a   b   c

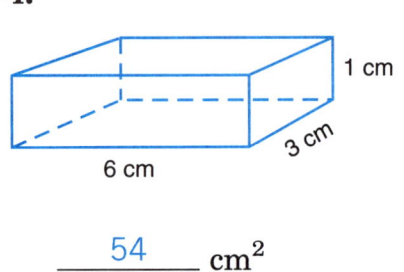

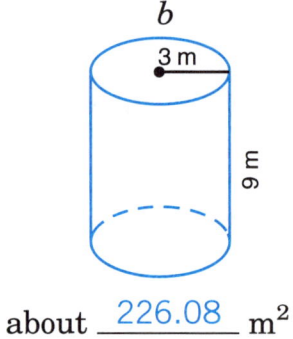

__54__ cm²   about __226.08__ m²   __272__ cm²

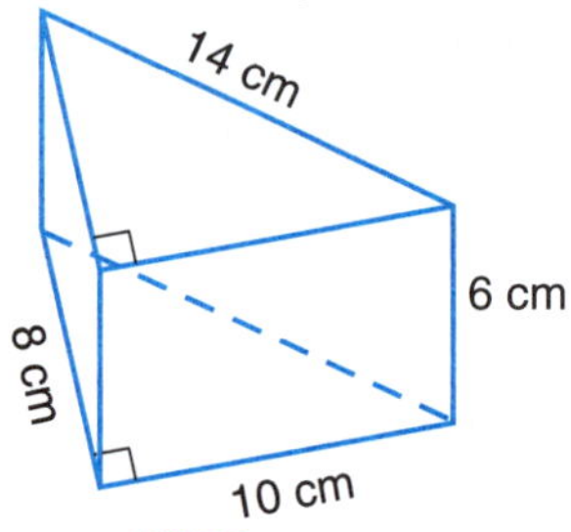

Find the volume of each figure. Use 3.14 for π.

5.

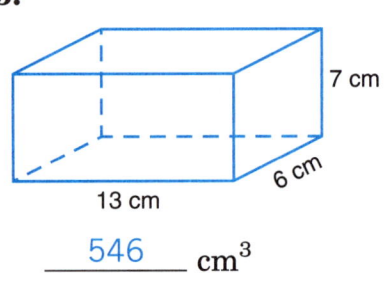

__546__ cm³   about __62.8__ m³   __500__ mm³

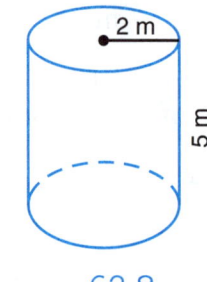

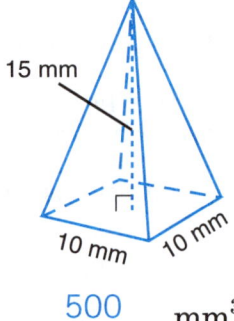

CHAPTER 11
Perimeter, Area, and Volume

Prerequisite Skills: addition and multiplication

Lesson Focus: finding perimeter of polygons
Possible Score: 9
Time Frame: 5–10 minutes

Lesson 1 Perimeter   PRE-ALGEBRA

The perimeter measure (P) of a figure is equal to the sum of the measures of its sides.

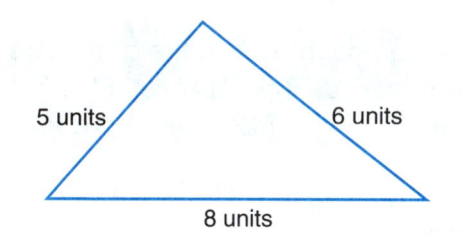

Find P if $a = 5$, $b = 6$, and $c = 8$.

$P = a + b + c$
$= 5 + 8 + 6$
$= \underline{19}$

The perimeter is $\underline{19}$ units.

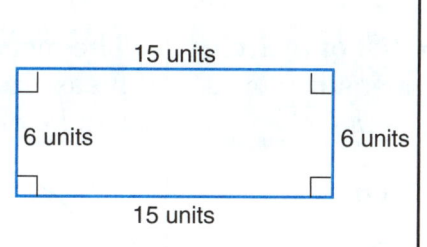

Find P if $l = 15$ and $w = 6$.

$P = l + w + l + w$
$= 2(l + w)$
$= 2(15 + 6)$
$= 2 \times 21$ or $\underline{42}$

The perimeter is $\underline{42}$ units.

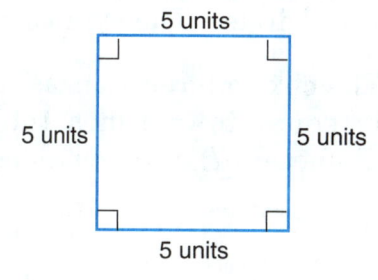

Find P if $s = 5$.

$P = s + s + s + s$
$= 4s$
$= 4 \times 5$
$= \underline{20}$

The perimeter is $\underline{20}$ units.

Find the perimeter of each figure below.

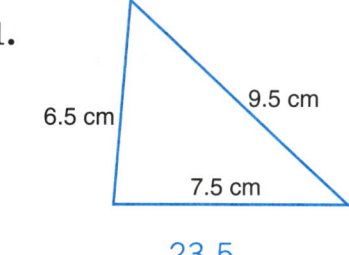

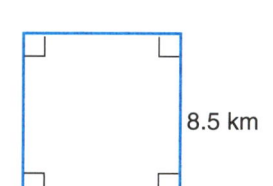

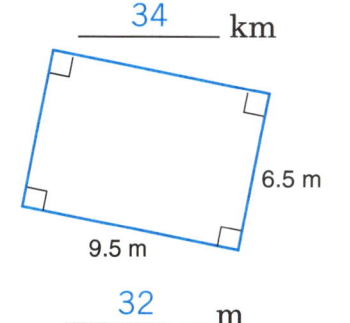

  a   b   c

1. triangle: 6.5 cm, 9.5 cm, 7.5 cm   rectangle: 13 m, 7 m   square: 7.4 cm
   $\underline{23.5}$ cm $\underline{40}$ m $\underline{29.6}$ cm

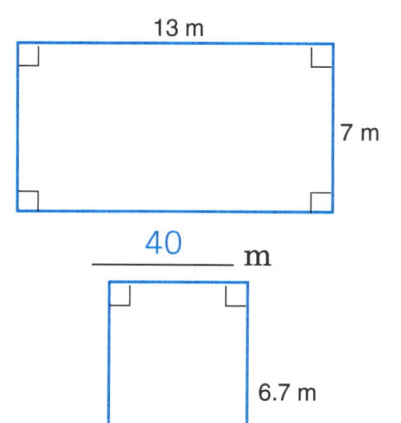

2. square: 8.5 km   rectangle: 6.7 m, 4.2 m   pentagon: 7 mm, 8 mm, 9 mm, 8 mm, 10 mm
   $\underline{34}$ km $\underline{21.8}$ m 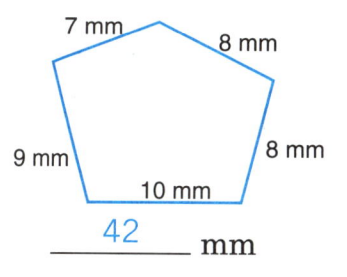 $\underline{42}$ mm

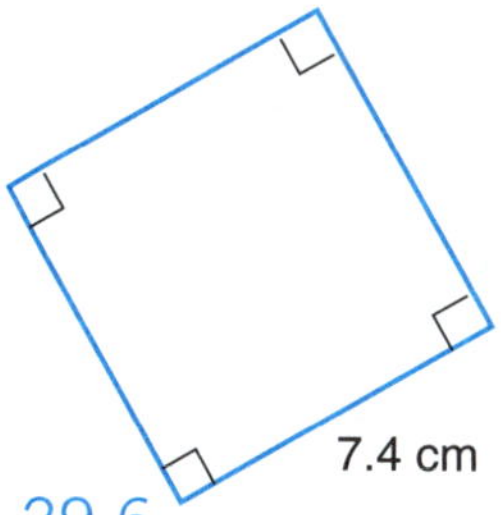

3. quadrilateral: 6.5 m, 9.5 m   parallelogram: 7.6 cm, 13.2 cm   hexagon: 7 m
   $\underline{32}$ m 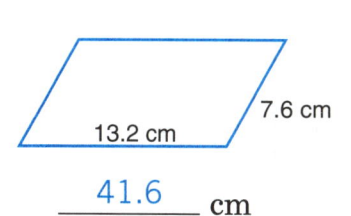 $\underline{41.6}$ cm 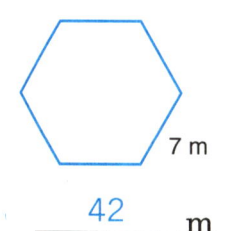 $\underline{42}$ m

---

Number Correct   LESSON FOLLOW-UP AND ERROR ANALYSIS

8–9: Have students find the perimeter of their classroom to the nearest metre.
6–7: Ask students to explain what to do if only one measure is given for a figure. (Assume that all the other sides shown have the same measure.)
Less than 6: Have students explain how to find the perimeter of a figure using the different formulas at the top of the page.

Prerequisite Skills: multiplication of mixed numerals and decimals

Lesson Focus: finding circumference of circles
Possible Score: 15
Time Frame: 15–20 minutes

Lesson 2   Circumference   PRE-ALGEBRA

The ratio of the measure of the circumference to the measure of a diameter is the same for all circles. The symbol π stands for this ratio. π is approximately equal to 3.14.

| The circumference measure (C) of a circle is equal to π times the measure of a diameter (d) of the circle. $C = \pi d$ | The measure of a diameter (d) is twice the measure of a radius (r). Hence, $C = \pi d$ can be changed to $C = \pi(2r)$ or $C = 2\pi r$. |

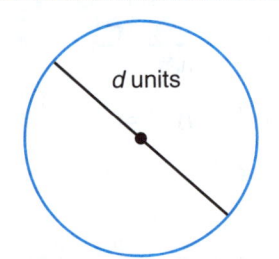

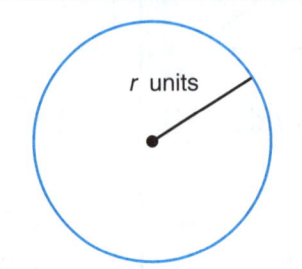

Find C if $d = 7$.

$C = \pi d$
$\doteq 3.14 \times 7$
$\doteq \underline{21.98}$

The circumference is about __21.98__ units.

Find C if $r = 6$.

$C = 2\pi r$
$\doteq 2 \times 3.14 \times 6$
$\doteq \underline{37\frac{5}{7}}$

The circumference is about __$37\frac{5}{7}$__ units.

Find the approximate circumference of each circle below. Use 3.14 for π.

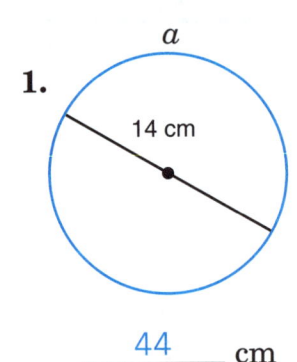

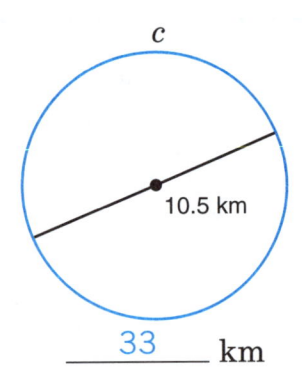

1. a) 14 cm → __44__ cm
    b) 2.8 m → __17.6__ m
    c) 10.5 km → __33__ km

Find the approximate circumference of each circle described below. Use 3.14 for π.

a
	diameter	approximate circumference
2.	6 m	__18.84__ m
3.	15 cm	__47.1__ cm
4.	6.8 km	__21.352__ km
5.	81 mm	__254.34__ mm
6.	27 mm	__84.78__ mm
7.	4.2 m	__13.188__ m

b
radius	approximate circumference
21 mm	__131.88__ mm
6.7 cm	__42.076__ cm
48 cm	__301.44__ cm
37 mm	__232.36__ mm
9.6 m	__60.288__ m
4 km	__25.12__ km

LESSON FOLLOW-UP AND ERROR ANALYSIS

Number Correct 160

13–15: Have students measure several circles and determine their circumferences.
10–12: Before students redo the page, discuss that π (read *pi*) can be expressed either as a mixed numeral ($3\frac{1}{7}$) or as a decimal (3.14).
Less than 10: Ask students to explain how to find the circumference given the diameter or the radius.

Prerequisite Skills: multiplication of whole numbers, mixed numerals, and decimals

Lesson Focus: finding area of rectangles
Possible Score: 11
Time Frame: 10–15 minutes

Lesson 3   Area of a Rectangle   PRE-ALGEBRA

The area measure (A) of a rectangle is equal to the product of the measure of its length (l) and the measure of its width (w). $A = l \times w$ or $A = lw$

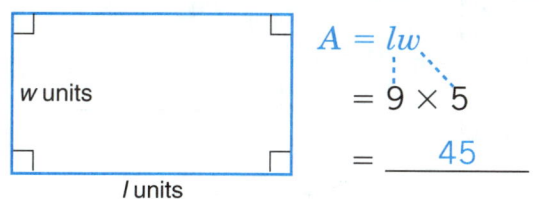

Find A if $l = 9$ and $w = 5$.
$A = lw$
$= 9 \times 5$
$= \underline{45}$

The area is $\underline{45}$ square units.

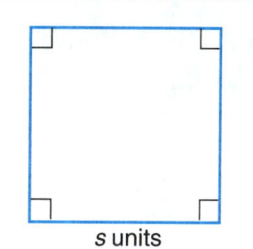

Find A if $s = 3$.
$A = s \times s$ or s^2
$= 3 \times 3$
$= \underline{9}$

The area is $\underline{9}$ square units.

Find the area of each rectangle below.

1.
a. 13 cm × 7 cm → $\underline{91}$ cm²
b. 8.5 km (square) → $\underline{72.25}$ km²
c. 6.5 m × 9.5 m → $\underline{61.75}$ m²

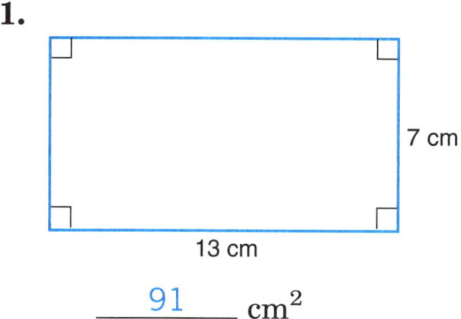

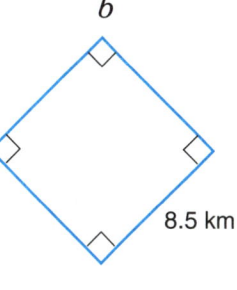

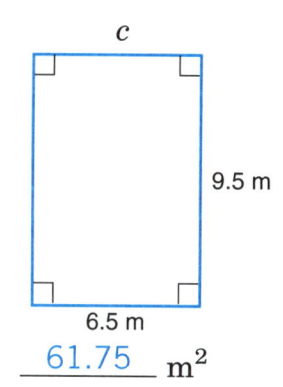

Find the area of each rectangle described below.

	length	width	area
2.	33 cm	27 cm	$\underline{891}$ cm²
3.	5.3 m	3.5 m	$\underline{18.55}$ m²
4.	3.8 km	2 km	$\underline{7.6}$ km²
5.	6.7 m	6.7 m	$\underline{44.89}$ m²
6.	9.2 cm	7.7 cm	$\underline{70.84}$ cm²
7.	18 m	4.6 m	$\underline{82.8}$ m²
8.	3.6 km	3.6 km	$\underline{12.96}$ km²
9.	9.5 cm	6.6 cm	$\underline{62.7}$ cm²

Number Correct

LESSON FOLLOW-UP AND ERROR ANALYSIS

9–11: Have students find the area of their classroom in metres.
7–8: Before students redo the page, ask them to explain area and perimeter and how each is calculated.
Less than 7: Discuss that when dealing with area, the answer is always given in *square units*.

Prerequisite Skills: multiplication of whole numbers, mixed numerals, and decimals

Lesson Focus: finding area of triangles
Possible Score: 11
Time Frame: 10–15 minutes

Lesson 4 Area of a Triangle   PRE-ALGEBRA

The area measure (A) of a triangle is equal to $\frac{1}{2}$ the product of the measure of its base (b) and the measure of its height (h). $A = \frac{1}{2}bh$ or $A = 0.5bh$

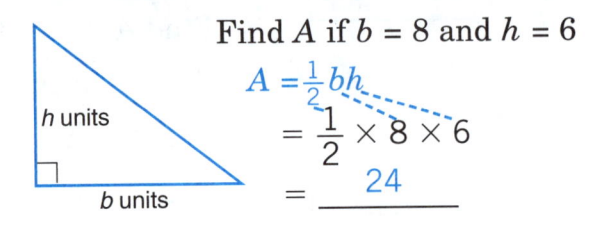

Find A if $b = 8$ and $h = 6$

$A = \frac{1}{2}bh$
$= \frac{1}{2} \times 8 \times 6$
$= \underline{24}$

The area is __24__ square units.

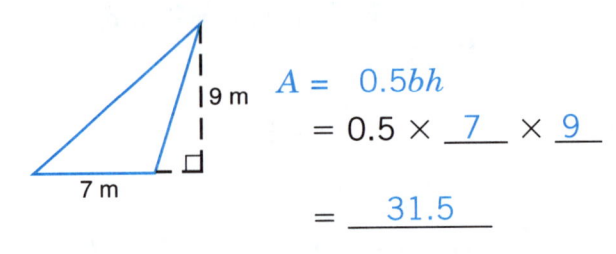

$A = 0.5bh$
$= 0.5 \times \underline{7} \times \underline{9}$
$= \underline{31.5}$

The area is __31.5__ m².

Find the area of each triangle below.

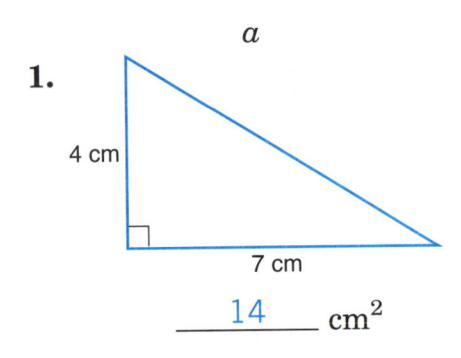

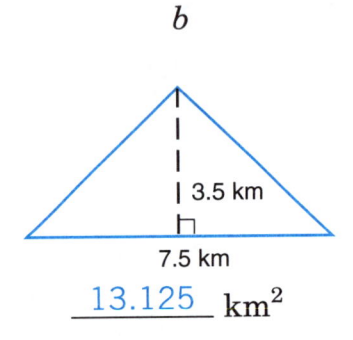

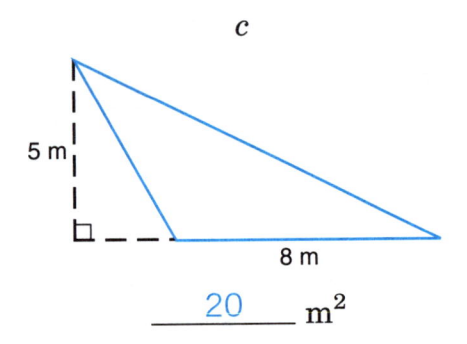

1.
a. 4 cm, 7 cm → __14__ cm²
b. 3.5 km, 7.5 km → __13.125__ km²
c. 5 m, 8 m → __20__ m²

Find the area of each triangle described below.

	base	height	area
2.	15 m	9 m	__67.5__ m²
3.	$3\frac{1}{2}$ mm	$6\frac{1}{2}$ mm	__11.375__ mm²
4.	7.4 cm	6.5 cm	__24.05__ cm²
5.	$11\frac{1}{2}$ m	7 m	__40.25__ m²
6.	154 mm	37 mm	__2849__ mm²
7.	85 cm	35 cm	__1487.5__ cm²
8.	18.8 m	7.5 m	__70.5__ m²
9.	9.5 km	6.6 km	__31.35__ km²

Number Correct

9–11: Have students draw several triangles and determine the area of each one.
7–8: Ask students to explain the two formulas for finding the area of a triangle—what each letter represents, that 0.5 and $\frac{1}{2}$ are equal, and so on.
Less than 7: Before students redo incorrect answers, discuss that only in a *right triangle* is the height the same as one side.

LESSON FOLLOW-UP AND ERROR ANALYSIS

Prerequisite Skills: multiplication of whole numbers, mixed numerals, and decimals

Lesson Focus: finding area of circles
Possible Score: 23
Time Frame: 15–20 minutes

Lesson 5   Area of a Circle   PRE-ALGEBRA

The area measure (A) of a circle is equal to the product of π and the square of the measure of a radius (r^2) of the circle. $A = \pi r^2$

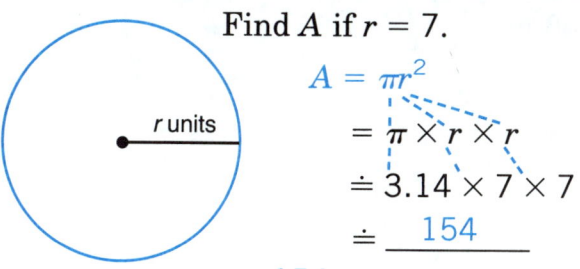

Find A if $r = 7$.

$A = \pi r^2$
$= \pi \times r \times r$
$\doteq 3.14 \times 7 \times 7$
$\doteq \underline{154}$

The area is about __154__ square units.

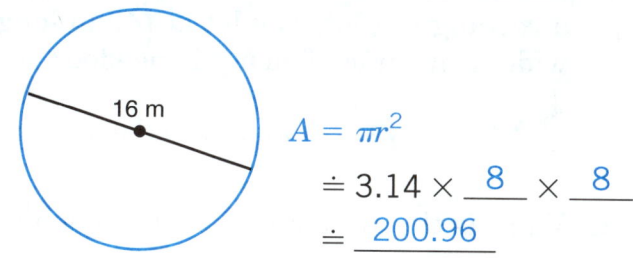

$A = \pi r^2$
$\doteq 3.14 \times \underline{8} \times \underline{8}$
$\doteq \underline{200.96}$

The area is about __200.96__ square units.

Find the approximate area of each circle below. Use 3.14 for π.

1.
a) 5 cm — __78.5__ cm²
b) 4.6 m — __16.6106__ m²
c) 3 m — __28.26__ m²

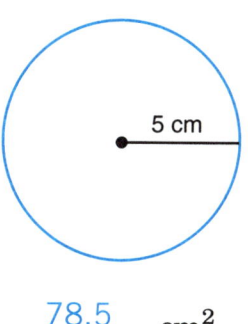

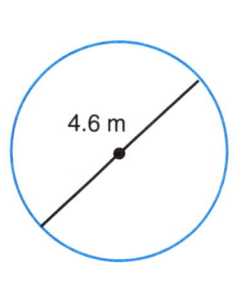

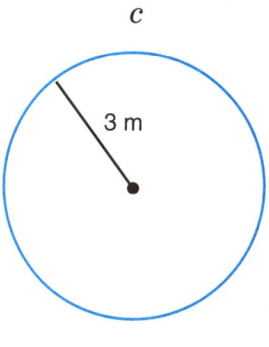

Find the approximate area of each circle described below. Use 3.14 for π.

a)

	radius	approximate area
2.	9 cm	__254.57__ cm²
3.	14 mm	__616__ mm²
4.	$3\frac{1}{2}$ m	__38.5__ m²
5.	56 cm	__9856__ cm²
6.	5.3 mm	__88.25__ mm²
7.	45 km	__6364.29__ km²

b)

diameter	approximate area
28 mm	__616__ mm²
42 cm	__1386__ cm²
72 m	__4073.14__ m²
126 mm	__12 474__ mm²
84 cm	__5544__ cm²
1.8 km	__2.54__ km²

Number Correct   LESSON FOLLOW-UP AND ERROR ANALYSIS
20–23: Have students give practical applications for finding area of a circle. (painting a round table, making a tablecloth)
15–19: In problems where the diameter is given, check to see if students forgot to square the radius after dividing the diameter by 2.
Less than 15: Have students identify whether it is the radius or the diameter that is given in each problem that they redo.

Lesson 5 Problem Solving   PRE-ALGEBRA

Solve each problem. Use 3.14 for π.

1. The Redfords would like to build a fence around a rectangular lot. The lot is 140 m long and 50 m wide. How much fencing is needed?

   __380__ m of fencing are needed.

2. What is the area of the lot in problem **1**?

   The area is __7000__ m^2.

3. Mr. McDaniel wants to put carpeting in a room that is 4 m long and 3 m wide. How many square metres of carpeting does he need?

   He needs __12__ m^2 of carpeting.

4. The lengths of the sides of a triangular-shaped garden are 17 m, 26 m, and 35 m. What is the perimeter of the garden?

   The perimeter is __78__ m.

5. The diameter of a circular pond is 28 m. What is the circumference of the pond?

   The circumference is about __88__ m.

6. What is the area of the pond in problem **5**?

   The area is about __616__ m^2.

7. Mrs. Witt is refinishing a circular table with a radius of 60 cm. Find the area of the tabletop.

   The area is about __11 304__ cm^2.

8. Find the circumference of the tabletop in problem **7**.

   The circumference is about __376.8__ cm.

CHAPTER 11
Perimeter, Area, and Volume

Lesson 5
Area of a Circle

Prerequisite Skills: multiplication of whole numbers, mixed numerals, and decimals

Lesson Focus: finding area of parallelograms
Possible Score: 10
Time Frame: 10–15 minutes

Lesson 6 Area of a Parallelogram   PRE-ALGEBRA

The area measure (A) of a parallelogram is equal to the product of the measure of its base (b) and the measure of its height (h). $A = bh$

Find A if $b = 14$ and $h = 9$.

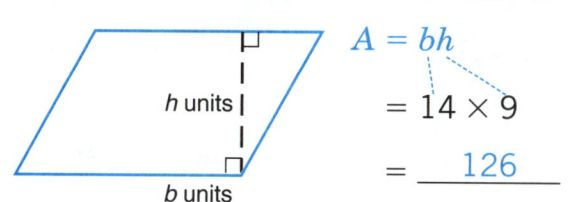

$A = bh$
$= 14 \times 9$
$= \underline{126}$

The area is __126__ square units.

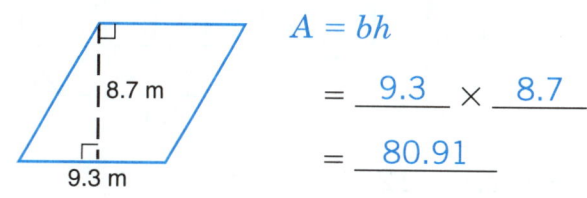

$A = bh$
$= \underline{9.3} \times \underline{8.7}$
$= \underline{80.91}$

The area is __80.91__ m².

Find the area of each parallelogram below.

1.

a) 4.5 cm, 5 cm → __22.5__ cm²

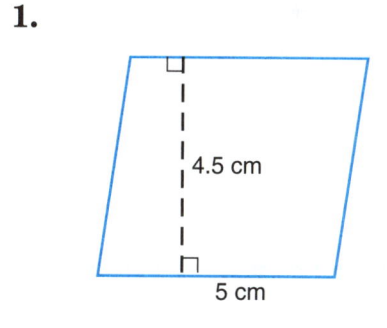

b) 7.3 cm, 13.6 cm → __99.28__ cm²

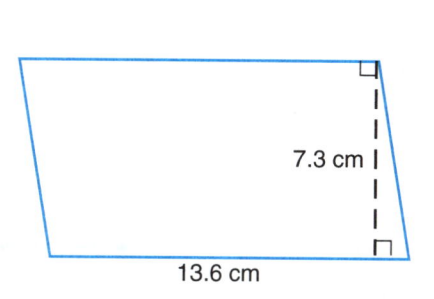

c) 12.5 m, 8.5 m → __106.25__ m²

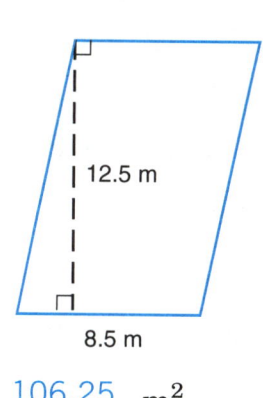

Find the area of each parallelogram described below.

	base	height	area
2.	72 mm	24 mm	__1728__ mm²
3.	7.5 cm	5 cm	__37.5__ cm²
4.	4.8 km	3.8 km	__18.24__ km²
5.	7.2 m	6 m	__43.2__ m²
6.	9.4 cm	6.7 cm	__62.98__ cm²
7.	9 m	7.3 m	__65.7__ m²
8.	16 km	12.4 km	__198.4__ km²

CHAPTER 11

Number Correct — LESSON FOLLOW-UP AND ERROR ANALYSIS

9–10: Have students give practical applications for finding the area of parallelograms.
7–8: Have students explain how to find the area of a parallelogram. Emphasize that it is the height, not a side, that is multiplied by the base.
Less than 7: Before students redo incorrect answers, discuss that when finding area, the answer is always given in *square units*.

165

Prerequisite Skills: addition and multiplication of whole numbers, mixed numerals, and fractions

Lesson Focus: find surface area of rectangular solids
Possible Score: 7
Time Frame: 15–20 minutes

Lesson 7 Surface Area of a Rectangular Prism

The surface area (SA) of a rectangular prism is the sum of the areas of all its faces.

Find the surface area of the figure shown.

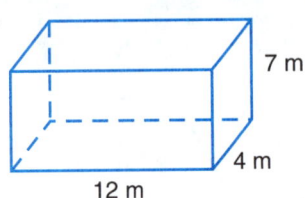

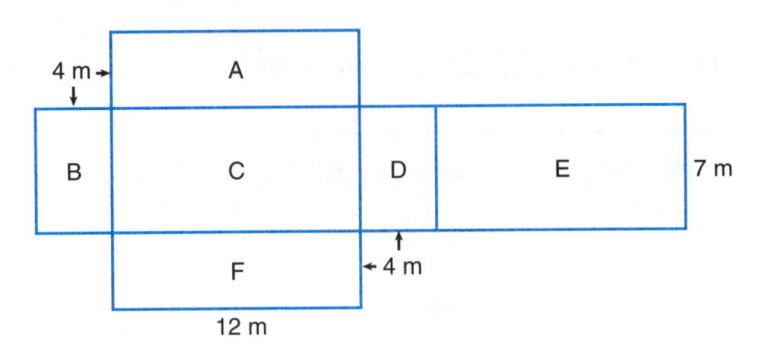

area $A = 12 \times 4 = 48$

area $B = 4 \times 7 = 28$

area $C = 12 \times 7 = 84$

area $D = 4 \times 7 = 28$

area $E = 12 \times 7 = 84$

area $F = 12 \times 4 = 48$

$SA = A + B + C + D + E + F$

$SA = 48 + 28 + 84 + 28 + 84 + 48 = 320$

The surface area is __320__ m².

Imagine the rectangular prism as a flat surface.

Find the surface area of each rectangular prism below.

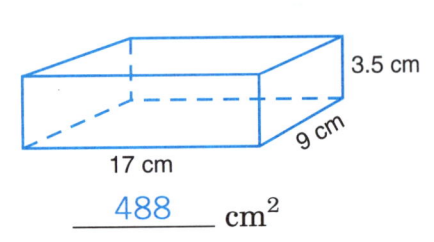

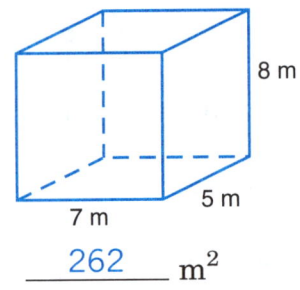

    a   b   c

1.   17 cm, 9 cm, 3.5 cm   7 m, 5 m, 8 m   18 m, 8 m, 11 m

       __488__ cm²   __262__ m²   __968__ m²

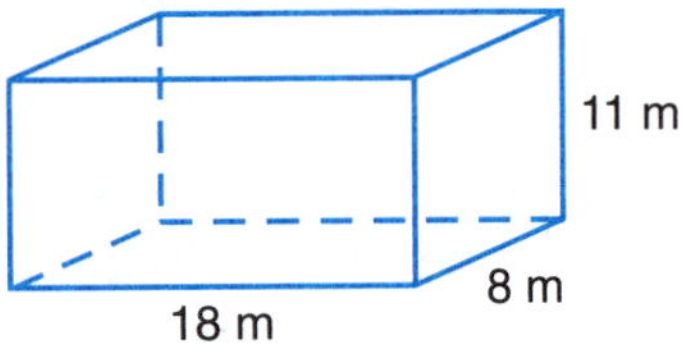

Find the surface area of each rectangular prism described below.

	length	width	height	surface area
2.	8 mm	11 mm	13 mm	__670__ mm²
3.	24 cm	20 cm	37 cm	__4216__ cm²
4.	6.5 m	14.2 m	9.7 m	__586.18__ m²
5.	4.5 cm	7.8 cm	12.3 cm	__372.78__ cm²

Number Correct

166

LESSON FOLLOW-UP AND ERROR ANALYSIS

6–7: Have students give practical applications for finding the surface area of a rectangular solid.
4–5: Review how to multiply decimals and mixed numerals before students redo the page.
Less than 4: Before students redo incorrect answers, review how to find the area of a rectangle.

Prerequisite Skills: addition and multiplication of whole numbers, mixed numerals, and fractions

Lesson Focus: find surface area of triangular prisms
Possible Score: 6
Time Frame: 15–20 minutes

Lesson 8 Surface Area of a Triangular Prism

The surface area (SA) of a triangular prism is the sum of the areas of all its faces.

Find the surface area of the figure shown.

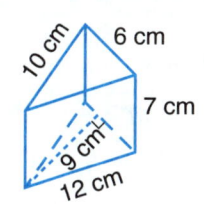

Imagine the triangular prism as a flat surface.

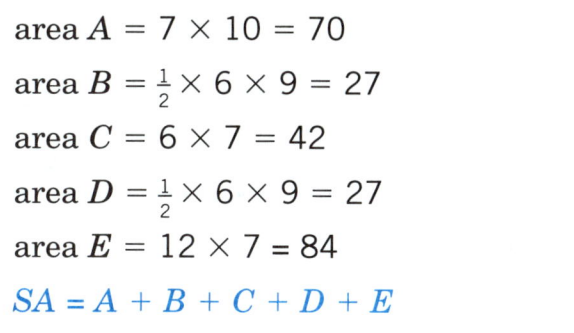

area $A = 7 \times 10 = 70$

area $B = \frac{1}{2} \times 6 \times 9 = 27$

area $C = 6 \times 7 = 42$

area $D = \frac{1}{2} \times 6 \times 9 = 27$

area $E = 12 \times 7 = 84$

$SA = A + B + C + D + E$

$SA = 70 + 27 + 42 + 27 + 84 = 250$

The surface area is __250__ cm².

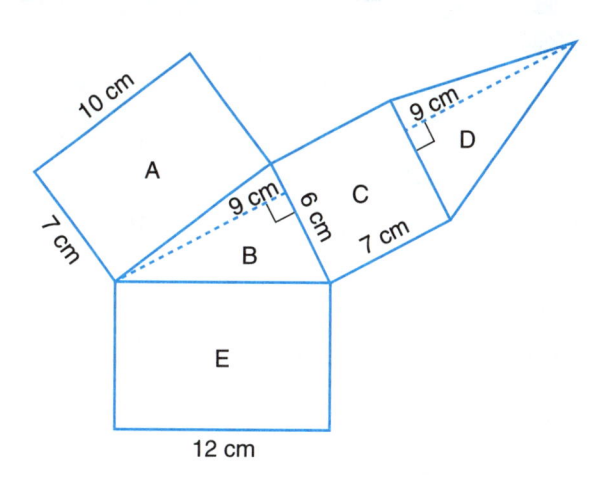

Find the surface area of each triangular prism below.

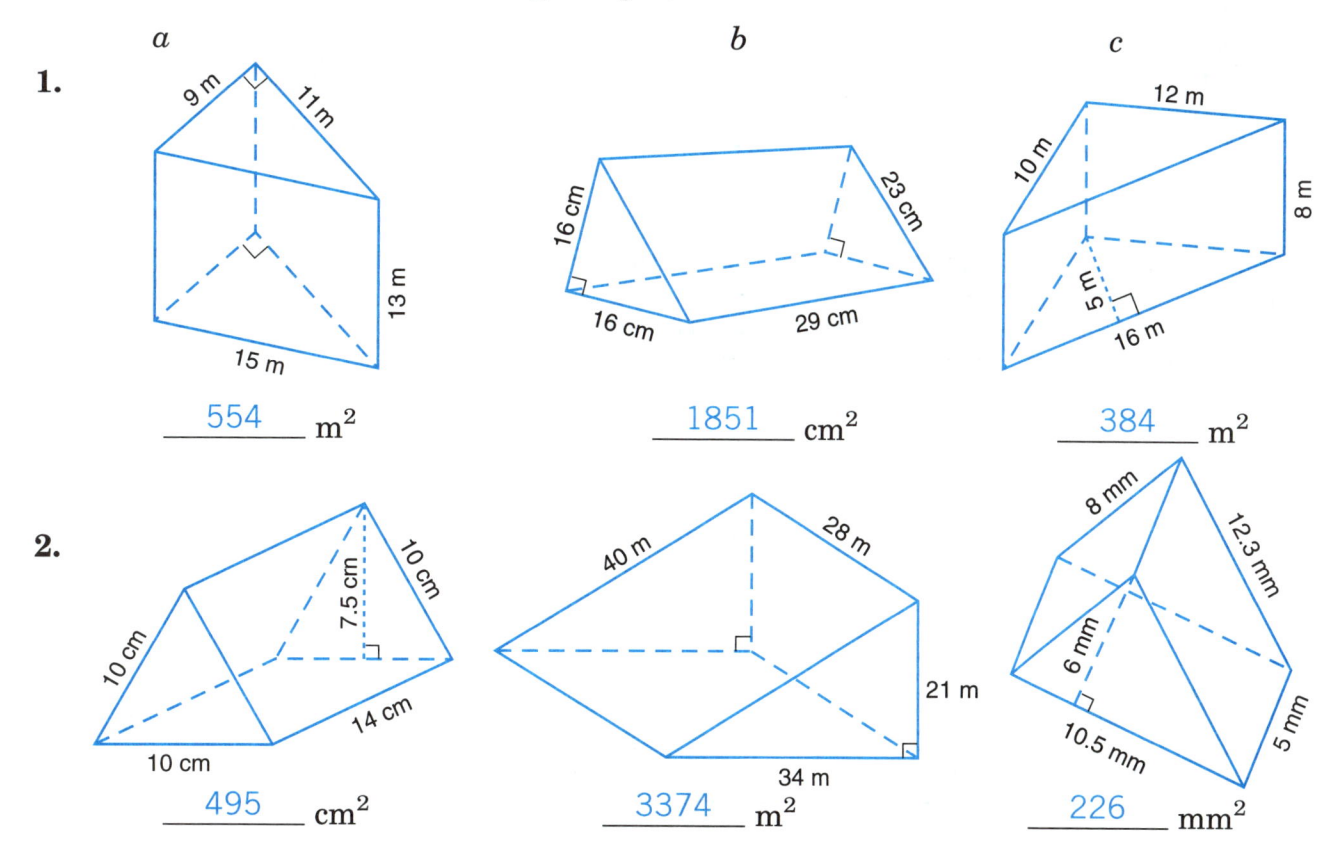

1.
 a. __554__ m²
 b. __1851__ cm²
 c. __384__ m²

2.
 a. __495__ cm²
 b. __3374__ m²
 c. __226__ mm²

Number Correct — **LESSON FOLLOW-UP AND ERROR ANALYSIS**
 5–6: Have students give practical applications for finding the surface area of a triangular prism.
 3–4: Before students redo incorrect answers, review the formula for finding the area of a triangle.
 Less than 3: Remind students that they need to find the area of all 5 faces in order to find the surface area.

Prerequisite Skills: addition and multiplication of whole numbers and fractions

Lesson Focus: find surface area of cylinders
Possible Score: 9
Time Frame: 15–20 minutes

Lesson 9   Surface Area of a Cylinder

The surface area (SA) of a cylinder is the sum of the lateral area and twice the area of the circular base. $SA = 2\pi rh + 2\pi r^2$

Find the surface area of the figure shown. Use 3.14 for π.

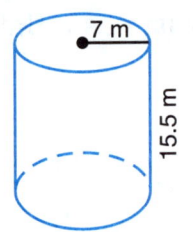

$SA = 2\pi rh + 2\pi r^2$
$\doteq (2 \times 3.14 \times 7 \times 15.5) + (2 \times 3.14 \times 7^2)$
$\doteq 681.38 + 307.72$
$\doteq 989.1$

The surface area is about __989.1__ m².

Find the approximate surface area of each cylinder below. Use 3.14 for π.

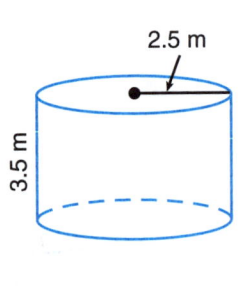

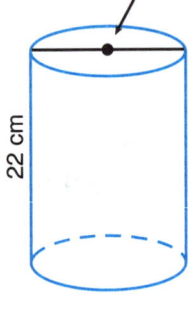

1.
a. __94.2__ m²
b. __1055.04__ cm²
c. __596.6__ mm²

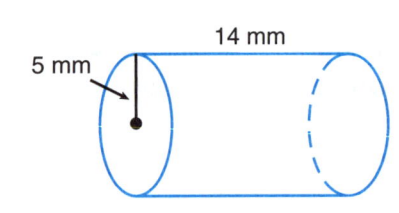

Find the approximate surface area of each cylinder described below. Use 3.14 for π.

	radius	height	approximate surface area
2.	5 cm	12 cm	__533.8__ cm²
3.	18 m	12.5 m	__3447.72__ m²
4.	13.5 cm	5 cm	__1568.43__ cm²
5.	0.75 m	1.25 m	__9.42__ m²
6.	53 cm	71 cm	__41 272.16__ cm²
7.	8.15 mm	16.75 mm	__1274.4318__ mm²

Number Correct

168

LESSON FOLLOW-UP AND ERROR ANALYSIS

8–9: Have students draw a cylinder, give its dimensions, and find its surface area.
6–7: Before they redo the page, have students explain how to multiply decimals.
Less than 6: Have students explain how to find the area of a circle, which is the area of the base of a cylinder.

Prerequisite Skills: multiplication of whole numbers, mixed numerals, and decimals

Lesson Focus: finding volume of rectangular prisms
Possible Score: 11
Time Frame: 10–15 minutes

Lesson 10 Volume of a Rectangular Prism   PRE-ALGEBRA

The volume measure (V) of a rectangular prism is equal to the product of the area measure of its base (B) and the measure of its height (h). $V = Bh$

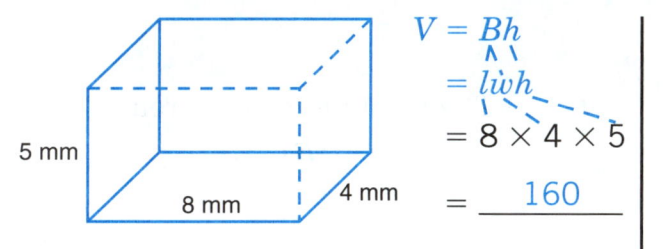

$V = Bh$
$= lwh$
$= 8 \times 4 \times 5$
$= \underline{160}$

5 mm, 8 mm, 4 mm

The volume is __160__ mm³.

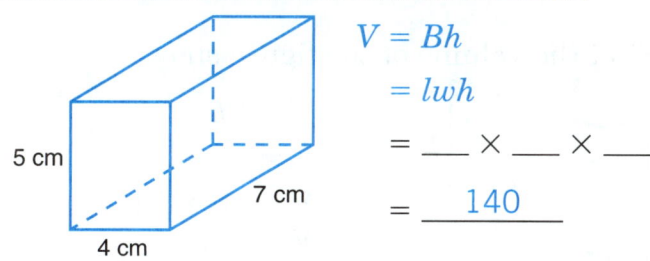

$V = Bh$
$= lwh$
$= \underline{} \times \underline{} \times \underline{}$
$= \underline{140}$

5 cm, 7 cm, 4 cm

The volume is __140__ cm³.

Find the volume of each rectangular prism below.

1.
a: 3 m, 4 m, 7 m — __84__ m³
b: 5 cm, 4 cm, 13.5 cm — __270__ cm³
c: 4.1 m, 4.1 m, 4.1 m — __68.921__ m³

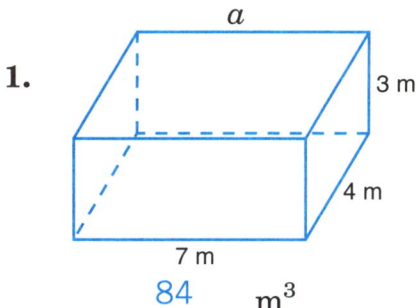

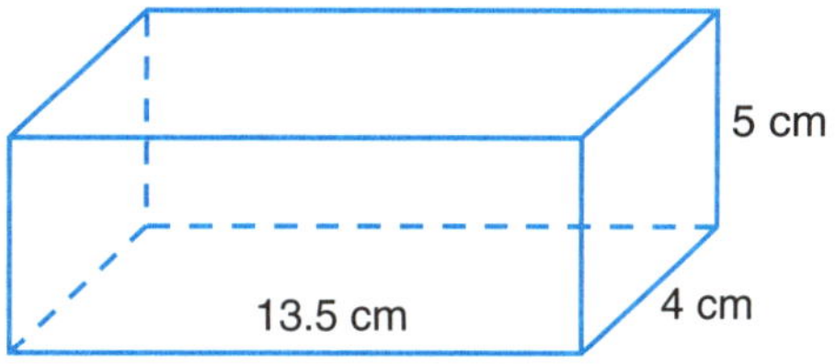

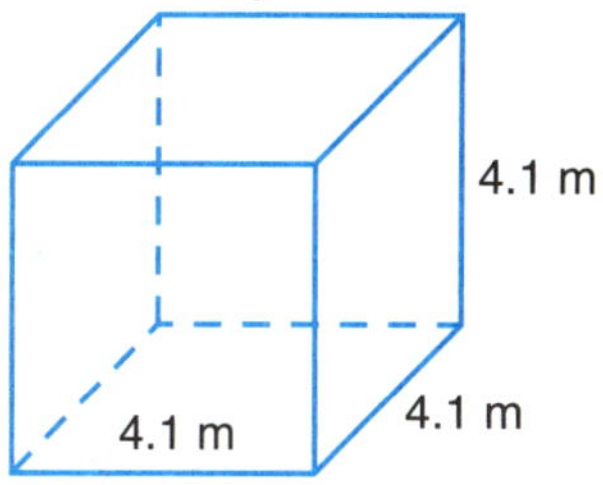

2. 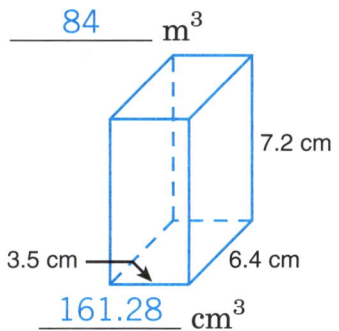
a: 7.2 cm, 3.5 cm, 6.4 cm — __161.28__ cm³ 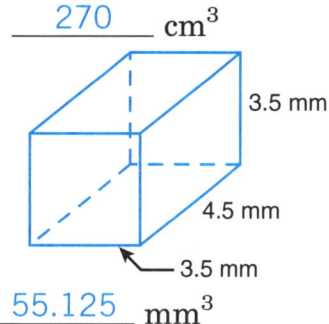
b: 3.5 mm, 4.5 mm, 3.5 mm — __55.125__ mm³ 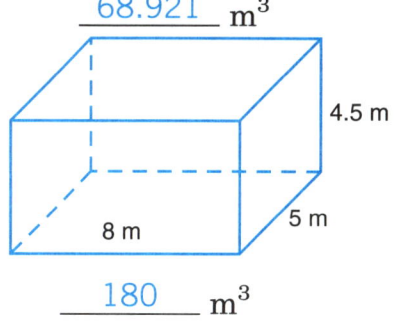
c: 4.5 m, 8 m, 5 m — __180__ m³

Find the volume of each rectangular prism described below.

	length	width	height	volume
3.	6 cm	7 cm	8 cm	__336__ cm³
4.	4.1 m	3.7 m	2.6 m	__39.442__ m³
5.	3.5 cm	3.5 cm	3.5 cm	__42.875__ cm³
6.	28 mm	36 mm	14 mm	__14 112__ mm³
7.	7.3 m	2.5 m	5.7 m	__104.025__ m³

Number Correct LESSON FOLLOW-UP AND ERROR ANALYSIS
9–11: Have students draw a figure, give dimensions, and determine its volume.
7–8: Check to see if students made multiplication errors. If so, you may want to have them use a calculator to redo the page.
Less than 7: Before students redo incorrect answers, discuss that when dealing with volume, the answer is always given in *cubic units*.

Prerequisite Skills: multiplication of whole numbers, mixed numerals, and fractions

Lesson Focus: find volume of triangular prisms
Possible Score: 9
Time Frame: 15–20 minutes

Lesson 11 Volume of a Triangular Prism   PRE-ALGEBRA

The volume (V) of a triangular prism is equal to the product of the area measure of its base (B) and the measure of its height. $V = Bh$

Find the volume of the figure shown.

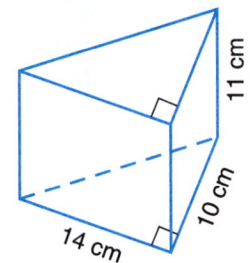

$V = Bh$
$V = (\frac{1}{2} \times 14 \times 10) \times 11$
$V = 70 \times 11$
$V = 770$

The formula for the area of the base (B) is $A = \frac{1}{2}bh$.

The volume is __770__ cm³.

Find the volume of each triangular prism below.

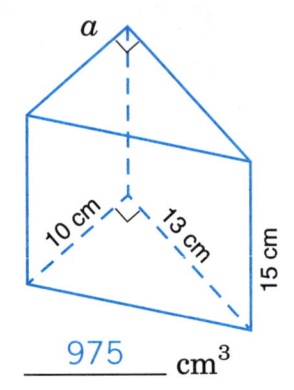

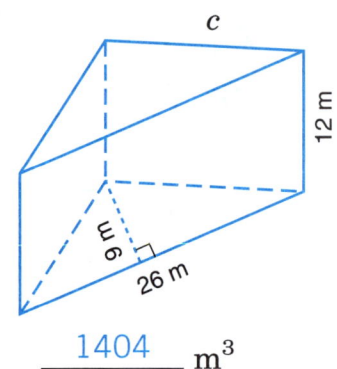

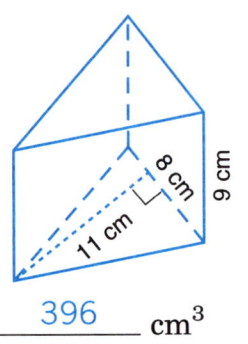

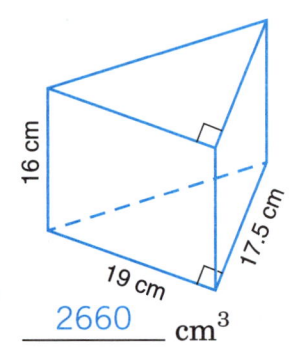

1.
a. __975__ cm³
b. __82.5__ mm³
c. __1404__ m³

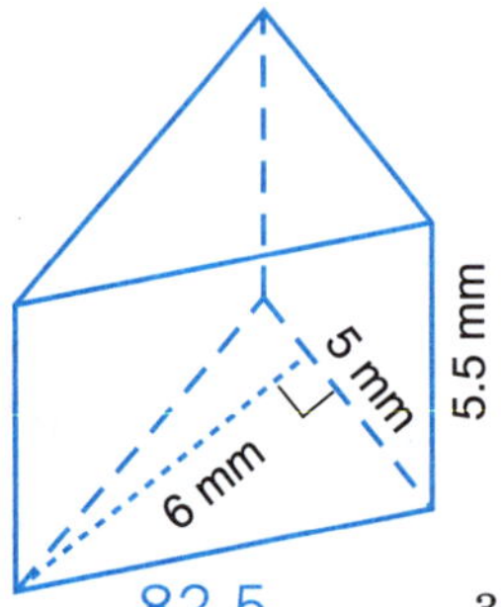

2.
a. __396__ cm³
b. __9135__ m³
c. __2660__ cm³

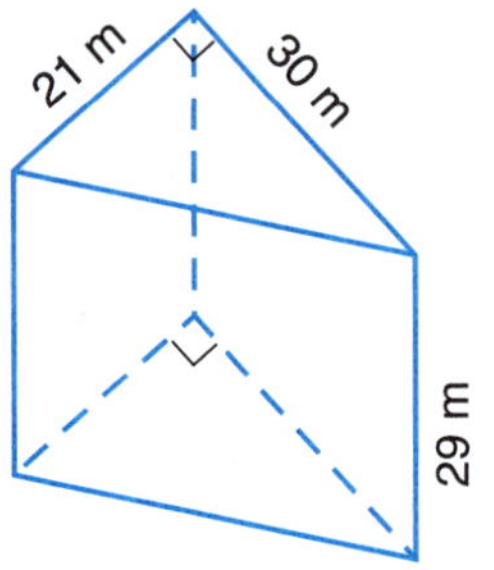

3.
a. __4.375__ m³
b. __5780__ mm³
c. __253.575__ m³

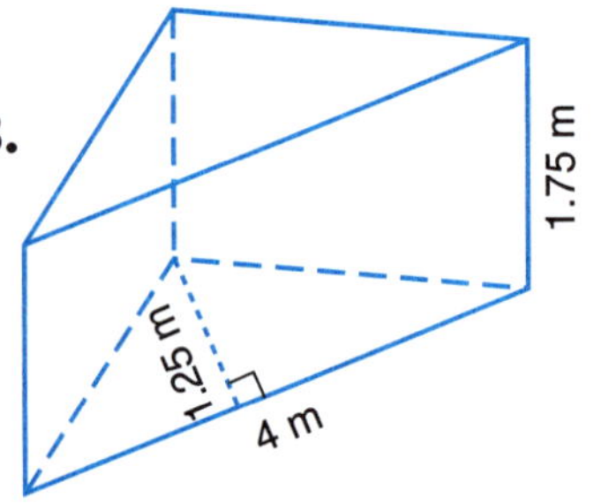

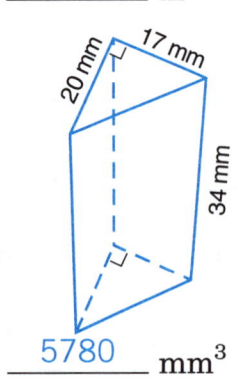

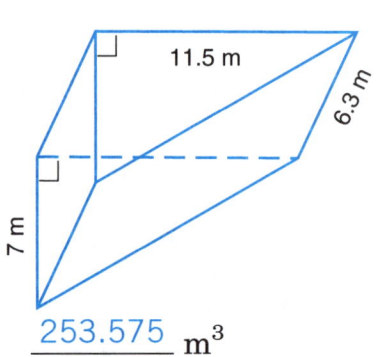

Number Correct

LESSON FOLLOW-UP AND ERROR ANALYSIS

170

- 8–9: Have students draw a triangular prism, give its dimensions, and find its volume.
- 6–7: Check to see if students made multiplication errors. If so, you may want to have them use a calculator to redo the page.
- Less than 6: Have students explain how to find the area of a triangle.

Prerequisite Skills: multiplication of whole numbers, mixed numerals, and decimals

Lesson Focus: finding volume of cylinders
Possible Score: 17
Time Frame: 15–20 minutes

Lesson 12 Volume of a Cylinder PRE-ALGEBRA

The volume measure (V) of a cylinder is equal to the product of the area measure of its base (B) and the measure of its height (h). $V = Bh$

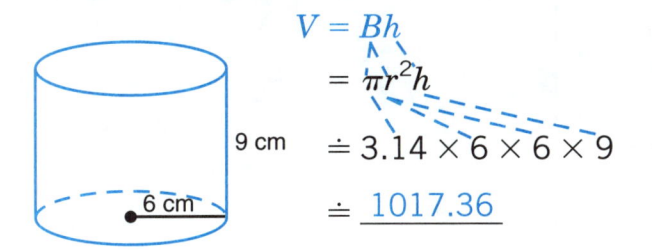

$V = Bh$
$= \pi r^2 h$
$\doteq 3.14 \times 6 \times 6 \times 9$
$\doteq 1017.36$

The volume is about __1017.36__ cm³.

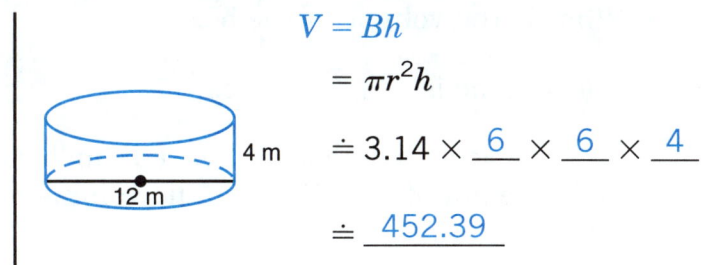

$V = Bh$
$= \pi r^2 h$
$\doteq 3.14 \times \underline{6} \times \underline{6} \times \underline{4}$
$\doteq 452.39$

The volume is about __452.39__ m³.

Find the approximate volume of each cylinder. Use 3.14 for π.

1.

a) 12 m, 7 m → __1848__ m³

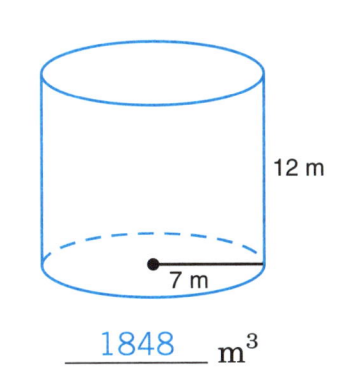

b) 14 cm, 6 cm → __396__ cm³

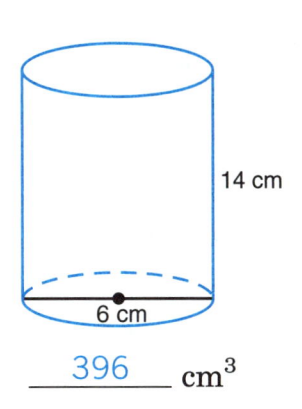

c) 4 m, 3.5 m → __154__ m³

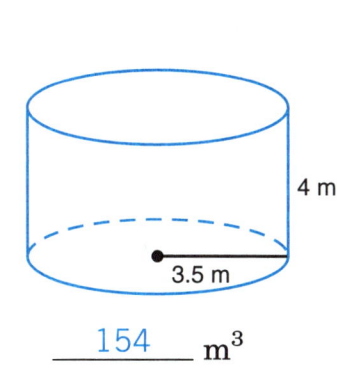

Find the approximate volume of each cylinder described below. Use 3.14 for π.

	radius	height	approximate volume
2.	8 cm	6 cm	__1205.76__ cm³
3.	18 mm	9 mm	__9156.24__ mm³
4.	1.7 m	3.4 m	__30.85364__ m³
5.	14 mm	6.5 mm	__4000.36__ mm³
6.	9 cm	14 cm	__3560.76__ cm³
7.	7 m	3.8 m	__584.668__ m³

Number Correct

LESSON FOLLOW-UP AND ERROR ANALYSIS

14–17: Have students give practical applications for finding volume of cylinders. (finding containers to hold a drink)
11–13: Before they redo the pages, have students explain how to multiply decimals and mixed numerals.
Less than 11: Have students explain how to find the area of a circle, which is the area of the base of a cylinder.

171

Lesson 12   Problem Solving   PRE-ALGEBRA

Solve each problem. Use 3.14 for π.

1. A box is 6 cm long, 4 cm wide, and 3 cm high. What is the volume of the box?

   The volume is ___72___ cm³.

2. A cylindrical storage tank has a diameter of 7 m and a height of 5 m. What is the volume of the storage tank?

   The volume is about ___192.42___ m³.

3. Cereal A comes in a rectangular box 20 cm wide, 6 cm deep, and 25 cm high. Find the volume of that box.

   The volume is about ___3000___ cm³.

4. Cereal B comes in a cylindrical box that has a diameter of 13 cm and a height of 25 cm. What is the volume of that box?

   The volume is about ___3316.625___ cm³.

5. Which cereal comes in the box with the larger volume? How much larger?

   Cereal ___B___ comes in a box that has a volume about ___316.625___ cm³ larger.

6. A classroom is 11 m long, 8 m wide, and 3 m high. What is the volume of the classroom?

   The volume is ___264___ m³.

7. Courtney has a cylindrical juice container with a diameter of 10 cm. Its height is 20 cm. How many cubic centimetres of juice will the container hold?

   The container will hold about ___1570.8___ cm³.

CHAPTER 11
Perimeter, Area, and Volume

Lesson 12
Volume of a Cylinder

Prerequisite Skills: multiplication of whole numbers and fractions

Lesson Focus: find volume of cones
Possible Score: 9
Time Frame: 15–20 minutes

Lesson 13 Volume of a Cone   PRE-ALGEBRA

The volume (V) of a cone is equal to $\frac{1}{3}$ the volume of a cylinder with the same base. $V = \frac{1}{3}Bh$

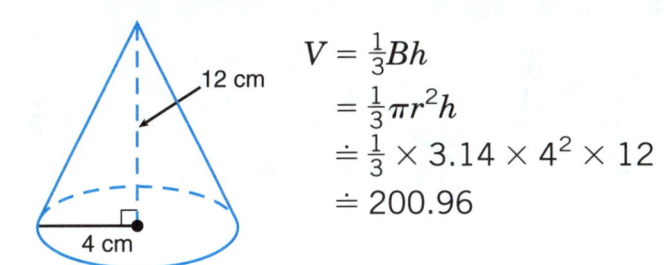

$V = \frac{1}{3}Bh$
$= \frac{1}{3}\pi r^2 h$
$\doteq \frac{1}{3} \times 3.14 \times 4^2 \times 12$
$\doteq 200.96$

12 cm, 4 cm

The volume is about __200.96__ cm³.

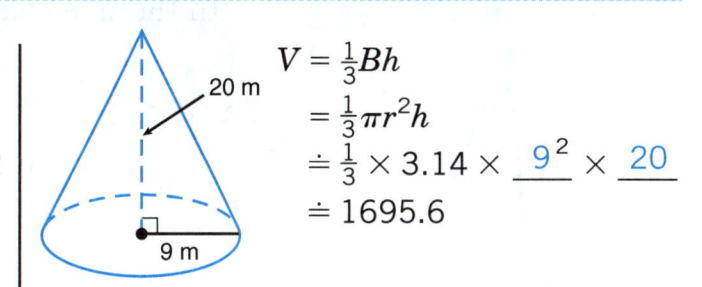

$V = \frac{1}{3}Bh$
$= \frac{1}{3}\pi r^2 h$
$\doteq \frac{1}{3} \times 3.14 \times \underline{9^2} \times \underline{20}$
$\doteq 1695.6$

20 m, 9 m

The volume is about __1695.6__ m³.

Find the approximate volume of each cone. Use 3.14 for π.

1.
 a. (16 cm, 7 cm) about __820.59__ cm³
 b. (34 cm, 37 cm) about __11 192__ cm³
 c. (15 m, 19.5 m) about __1148.06__ m³

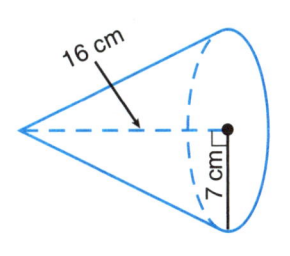

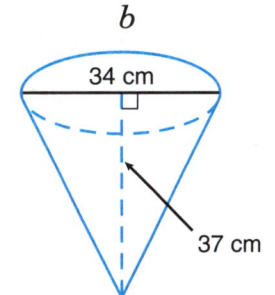

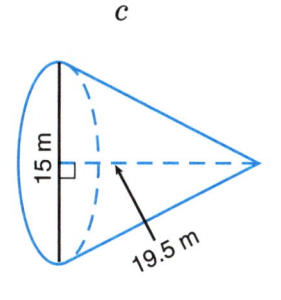

Find the approximate volume of each cone described below. Use 3.14 for π.

	radius	height	approximate volume
2.	10 cm	15 cm	__1570__ cm³
3.	4 m	5.25 m	__87.92__ m³
4.	34 mm	57 mm	__68 966.96__ mm³
5.	11 m	1.5 m	__189.97__ m³
6.	19 cm	24.75 cm	__9351.705__ cm³
7.	0.58 m	1.35 m	__0.475 333 2__ m³

Number Correct — LESSON FOLLOW-UP AND ERROR ANALYSIS
8–9: Have students draw a cone, give its dimensions, and find its approximate volume.
6–7: Before they redo the page, have students explain how to multiply decimals.
Less than 6: Remind students to square the radius when finding the volume of a cone.

Prerequisite Skills: multiplication of whole numbers, mixed numerals, and fractions

Lesson Focus: find volume of pyramids
Possible Score: 9
Time Frame: 10–15 minutes

Lesson 14 Volume of a Pyramid   PRE-ALGEBRA

The volume (V) of a pyramid is equal to $\frac{1}{3}$ the volume of a rectangular prism with the same base. $V = \frac{1}{3}Bh$

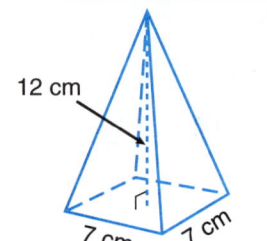

$V = \frac{1}{3}Bh$
$= \frac{1}{3} \times 7 \times 7 \times 12$
$= 196$

The volume is __196__ cm³.

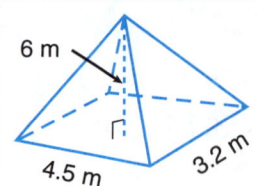

$V = \frac{1}{3}Bh$
$= \frac{1}{3} \times 3.2 \times \underline{4.5} \times \underline{6}$
$= 28.8$

The volume is __28.8__ m³.

Find the volume of each pyramid.

1.

a: __115.5__ mm³

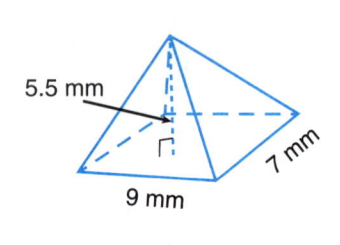

b: __2.625__ m³

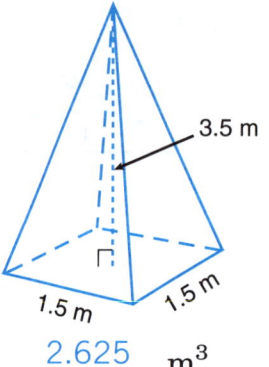

c: __452.83__ cm³

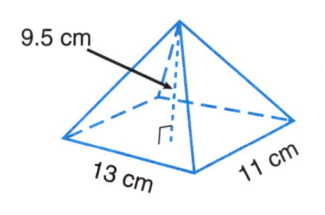

Find the volume of each pyramid described below.

	length of base	width of base	height	volume
2.	9 cm	9 cm	15 cm	__405__ cm³
3.	12 mm	8 mm	10 mm	__320__ mm³
4.	9 cm	15 cm	9.9 cm	__445.5__ cm³
5.	0.6 m	0.4 m	0.8 m	__0.064__ m³
6.	8.25 cm	10.5 cm	6 cm	__173.25__ cm³
7.	12.75 mm	12.75 mm	5 mm	__270.9375__ mm³

Number Correct

174

LESSON FOLLOW-UP AND ERROR ANALYSIS

8–9: Have students choose the most difficult problem on the page and explain their choices.
6–7: Remind students to multiply by $\frac{1}{3}$ when finding the volume of a pyramid.
Less than 6: Before students redo incorrect answers, discuss that when finding volume the answer is always given in cubic units.

Prerequisite Skills: finding perimeter, circumference, area, and volume

Lesson Focus: finding perimeter, circumference, area, and volume
Possible Score: 20
Time Frame: 25–30 minutes

Lesson 15 Perimeter, Area, and Volume   PRE-ALGEBRA

Find the perimeter or circumference of each figure below. Use 3.14 for π.

1.
   a. Triangle with sides 8 m, 16.5 m, 22.3 m — __46.8__ m
   b. Circle with diameter 21 cm — about __66__ cm

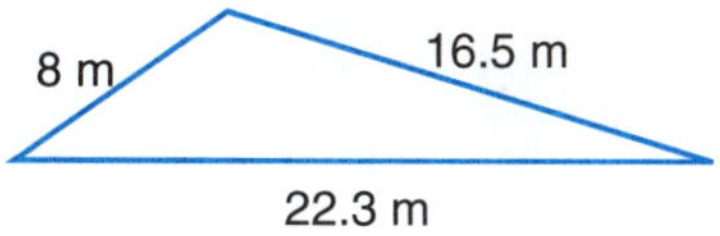

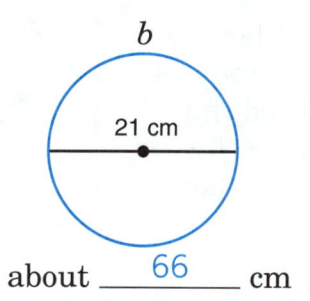

Find the area of each figure below. Use 3.14 for π.

2.
   a. Square 8 km — __64__ km²
   b. Rectangle 7 m by 13 m — __91__ m²
   c. Triangle height 6 cm, base 18 cm — __54__ cm²

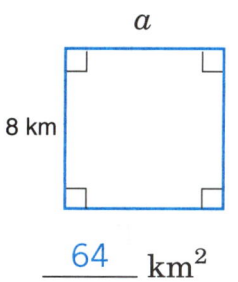

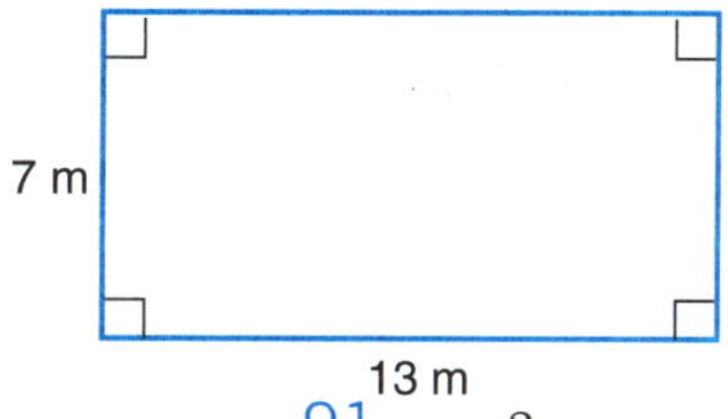

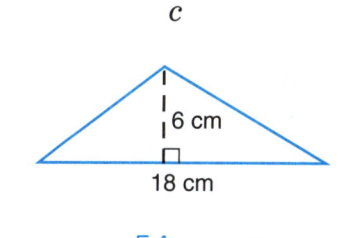

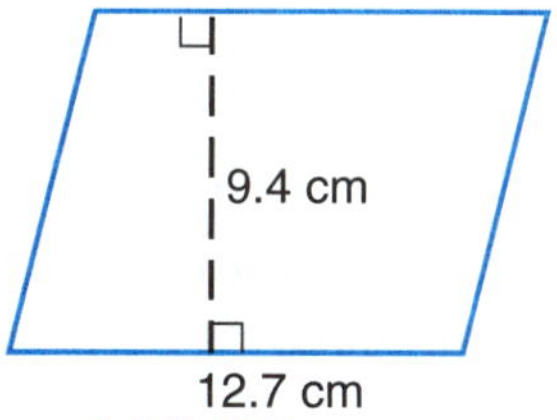

3. 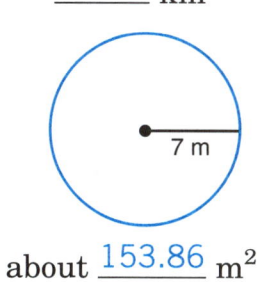
   a. Circle radius 7 m — about __153.86__ m²
   b. Parallelogram height 9.4 cm, base 12.7 cm — __119.38__ cm²
   c. Parallelogram height 9 mm, base 14 mm — __126__ mm²

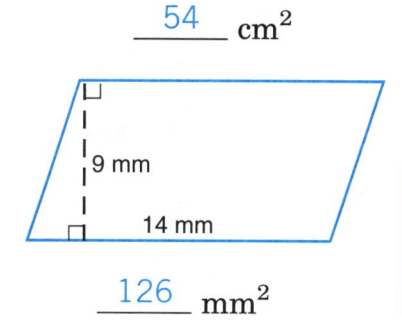

Find the volume of each figure below. Use 3.14 for π.

4.
   a. Cube 7 m × 7 m × 7 m — __343__ m³
   b. Rectangular prism 5 cm × 7 cm × 3 cm — __105__ cm³

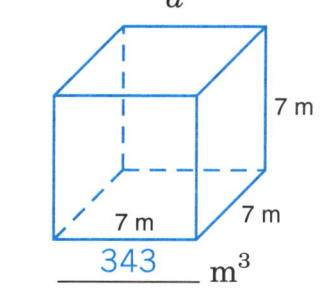

5.
   a. Rectangular prism 8.4 cm × 3.6 cm × 2.8 cm — __84.672__ cm³
   b. Cylinder radius 7 mm, height 9 mm — about __1386__ mm³

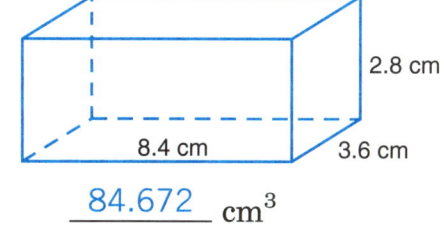

Number Correct

LESSON FOLLOW-UP AND ERROR ANALYSIS

17–20: Have students draw and find the perimeter, area, and volume of three figures of their own.
13–16: Check to see if students confused the formulas for perimeter, area, and volume.
Less than 13: Have students write the appropriate formulas to help them find the perimeter, area, or volume needed to redo the pages.

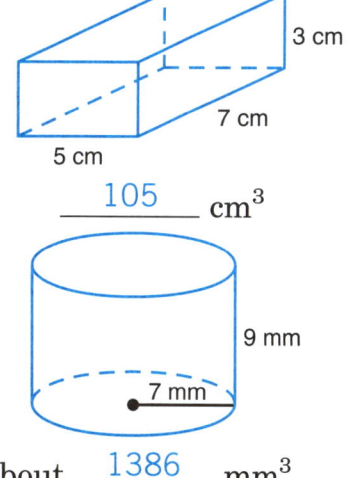

Lesson 15 Problem Solving   PRE-ALGEBRA

Solve each problem. Use 3.14 for π.

1. A carpenter cut a circular shelf from a square piece of wood as shown at the right. Find the area of the square piece of wood. Find the area of the circular piece of wood.

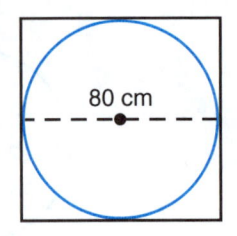

   80 cm

   The area of the square piece is __6400__ cm².
   The area of the circular piece is about __5024__ cm².

2. The carpenter threw away the wood left over after cutting out the circular piece. How much wood was thrown away?

   __1376__ cm² were thrown away.

3. Find the circumference of the circular piece of wood in problem **1**.

   The circumference is about __251.2__ cm.

4. A farmer has a field shaped like a parallelogram. The base is 1500 m. The height is 1200 m. Find the area of the field.

   The area is __1 800 000__ m².

5. If the farmer puts a fence around the field in problem **4**, how much fencing will be needed?

   __5400__ m of fencing will be needed.

6. How many cubic metres of earth will be removed to dig a well 2 m in diameter and 28 m deep?

   About __87.92__ m³ of earth will be removed.

7. A tank is 150 cm long, 120 cm wide, and 185 cm deep. Find its volume.

   The volume is __3 330 000__ cm³.

CHAPTER 11
Perimeter, Area, and Volume

176

Lesson 15
Perimeter, Area, and Volume

For further evaluation, copy the Chapter Test on page 275.
For maintaining skills, use the Cumulative Review on pages 233–234.

Possible Score: 18
Time Frame: 25–30 minutes

CHAPTER 11 PRACTICE TEST
Perimeter, Area, and Volume

Find the perimeter and area of each figure.

1.

a — 6 m square
perimeter: __24__ m
area: __36__ m²

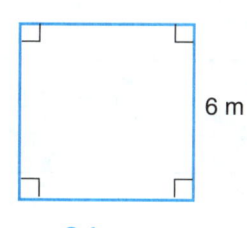

b — triangle with sides 8 m, 11 m, base 13 m, height 7 m
perimeter: __32__ m
area: __45.5__ m²

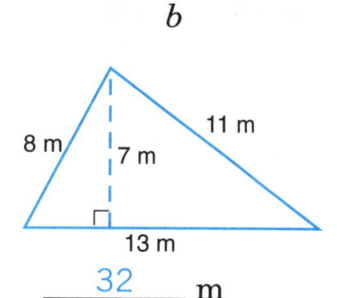

c — parallelogram with side 13 cm, height 12 cm, base 18.5 cm
perimeter: __63__ cm
area: __222__ cm²

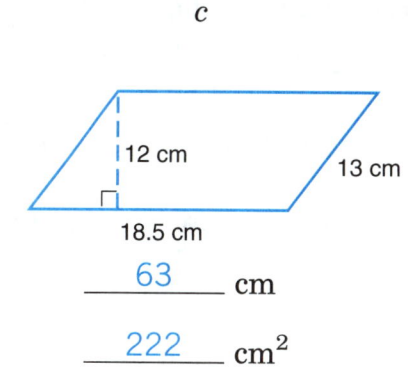

Complete the table below. Use 3.14 for π. Find the approximate circumference and area.

	diameter	radius	approximate circumference	approximate area
2.	8 cm	__4__ cm	about __25.13__ cm	about __50.27__ cm²
3.	__10__ m	5 m	about __31.42__ m	about __78.54__ m²

Find the surface area of each figure. Use 3.14 for π.

4.

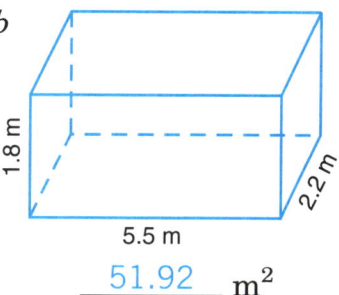

a — triangular prism (8 cm, 12 cm, 9 cm, 14 cm, 11 cm)
__500__ cm²

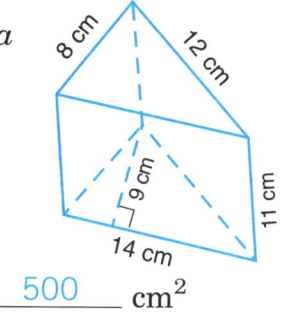

b — rectangular prism (1.8 m, 5.5 m, 2.2 m)
__51.92__ m²

c — cylinder (radius 10 mm, height 23 mm)
about __2072.4__ mm²

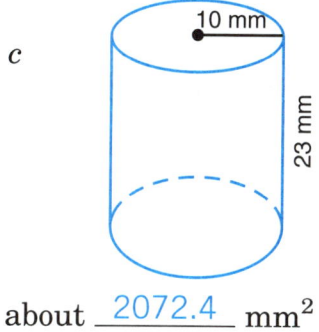

Find the volume of each figure. Use 3.14 for π.

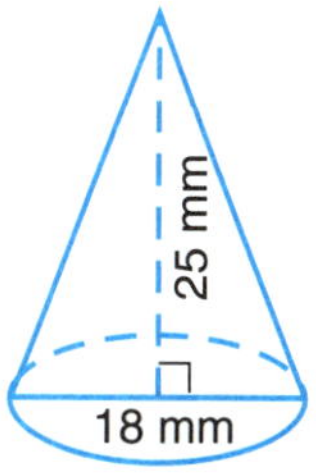

5.

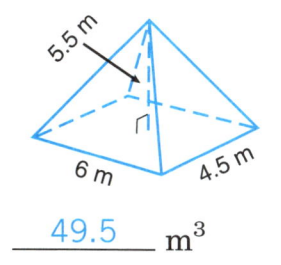

pyramid (5.5 m, 6 m, 4.5 m)
__49.5__ m³

cone (25 mm height, 18 mm diameter)
about __2119.5__ mm³

rectangular prism (10.7 cm, 14 cm, 8.5 cm)
__1273.3__ cm³

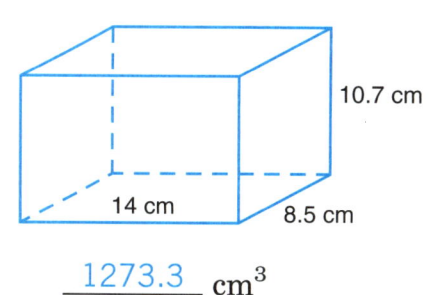

Number Correct	LESSON FOLLOW-UP AND ERROR ANALYSIS
15–18:	Have students draw and find the perimeter, area, surface area, and volume of four figures of their own.
11–14:	Check to see if students confused the formulas for perimeter, area, surface area, and volume.
Less than 11:	Have students write the appropriate formula for each problem as they redo the page.

CHAPTER 12 PRETEST
Graphs

Use the bar graph to answer each question.

1. Who read the most books over the summer?
   Amy

2. How many books did Zoe read in June? 0

3. Who read twice as many books in August as in July? Zoe

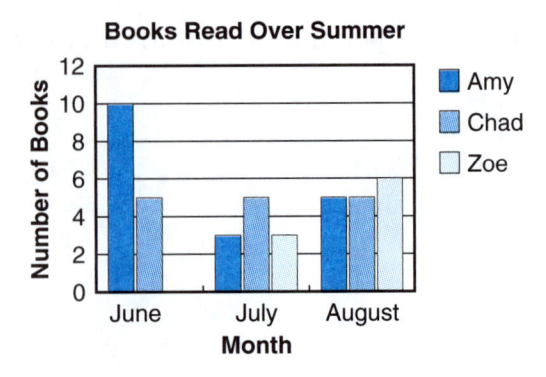

Use the line graph to answer each question.

4. What was the general trend of temperature in the P.M. hours?
   the temperature fell or decreased

5. What was the temperature difference from 7 A.M. to 7 P.M.? 8°

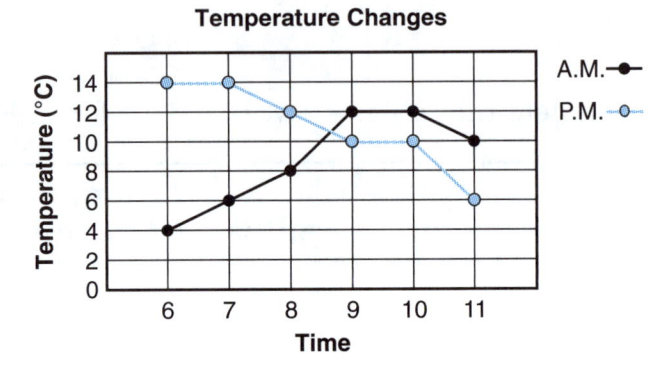

6. In a scatter plot, if the one set of data increases as the other set increases, the scatter plot has what type of correlation?
   positive

7. In a scatter plot, if the one set of data decreases as the other set increases, the scatter plot has what type of correlation? negative

8. The stronger the correlation, the more the points on the scatter plot resemble a line.

9. A plot with a positive correlation slants upward to the right.

10. A plot with a negative correlation slants downward to the right.

11. Describe the type of correlation that the graph at the right shows.

    The graph shows a negative correlation.

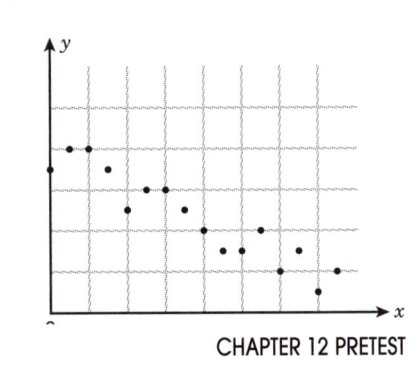

CHAPTER 12
Graphs

CHAPTER 12 PRETEST

178

Prerequisite Skills: interpreting simple bar graphs, reading a number line

Lesson Focus: interpret and make multiple bar graphs
Possible Score: 17
Time Frame: 15–20 minutes

Lesson 1   Multiple Bar Graphs

Multiple bar graphs are used to compare data from more than one set of data. The graph provides a way to make visual comparisons among the data sets.

A different colour or pattern is used to identify each set of data in the bar graph.

3 children ages 9–12 chose BBQ sauce.

1 child age 9–12 chose honey sauce.

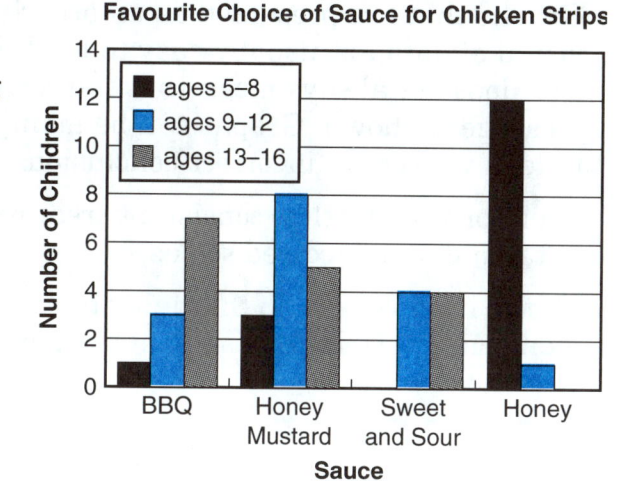

Favourite Choice of Sauce for Chicken Strips

Use the bar graph above to answer each question.

1. Which sauce did children ages 5–8 not choose? __sweet and sour__

2. How many children ages 9–12 chose honey mustard sauce? __8__ children

3. Which sauce was the overwhelming favourite of children ages 5–8? __honey__

4. How many children ages 13–16 chose BBQ sauce? __7__ children

5. Which age groups chose sweet and sour sauce equally? __9–12 and 13–16__

6. How many more children ages 5–8 than children ages 13–16 chose honey sauce? __12__ children

Use the information in the chart below to complete a multiple bar graph.

7. A hotel that provides breakfast kept track of the preferred breakfast juice for one Saturday morning.

	men	women	children
orange	30	16	28
apple	12	11	18
pineapple	5	12	4
grapefruit	3	10	0

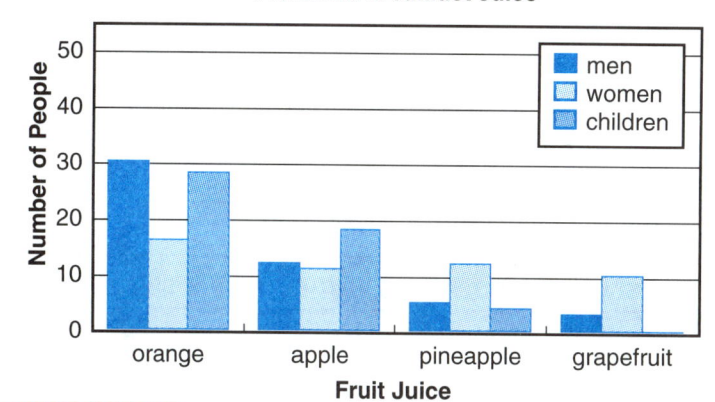

Preferred Breakfast Juice

Number Correct

LESSON FOLLOW-UP AND ERROR ANALYSIS

15–17: Have students describe various types of data that would be best displayed in a multiple bar graph.
10–14: Remind students to include a key when creating a multiple bar graph.
Less than 10: Have students practice counting up the number line on the graph.

179

Prerequisite Skills: bar graphs; multiple bar graphs; number lines

Lesson Focus: interpret misleading graphs
Possible Score: 8
Time Frame: 5–10 minutes

Lesson 2 Misleading Bar Graphs

Graphs that are presented as an orderly complete visual of data are usually easy to read. However, this does not always mean that the correct message is shown. Graphs can be manipulated to lead viewers to incorrect conclusions.

Look for use of titles, consistent treatment of data, and standardized scales.

Notice in the graph to the right that the scale along the horizontal axis is not standardized.

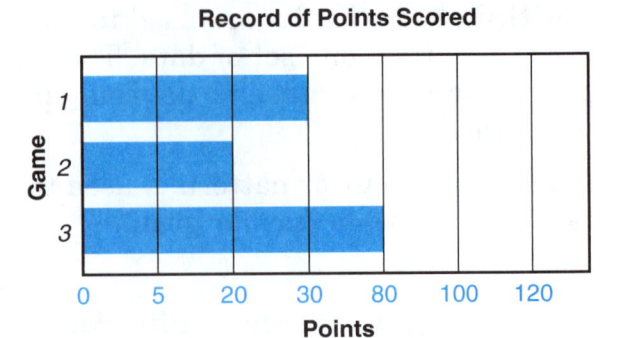

Record of Points Scored

Use the graph above to answer each question.

1. At first glance, what may you see as the points scored for game 1? Why?
   15; Since the first line represents 5 points, you expect each line to represent 5 points.

2. Is the difference in points scored from game 1 to game 2 the same as the difference in points scored from game 1 to game 3? Explain. Even though the distances on the scale are equal, the labelled units indicate that from game 1 to game 2 the difference was 10 points. The labelled units indicate that from game 1 to game 3 the difference was 50 points.

3. If the games were shown as game 2, then game 1 and game 3, what conclusion might you draw? You might conclude that the scores are increasing with each game because the length of the bars would increase from game to game.

Use the graph of Adam and Jamal's scores to answer each question.

4. How many increases of scores did Adam have? __3__

5. Is this graph constructed correctly? Explain. The bars for Adam are not always shown first for each test.

6. Does this presentation of the data make it easier or harder to compare one boy's progress? Why? Harder, you cannot look just at the bars on the left for Adam's scores or on the right for Jamal's scores.

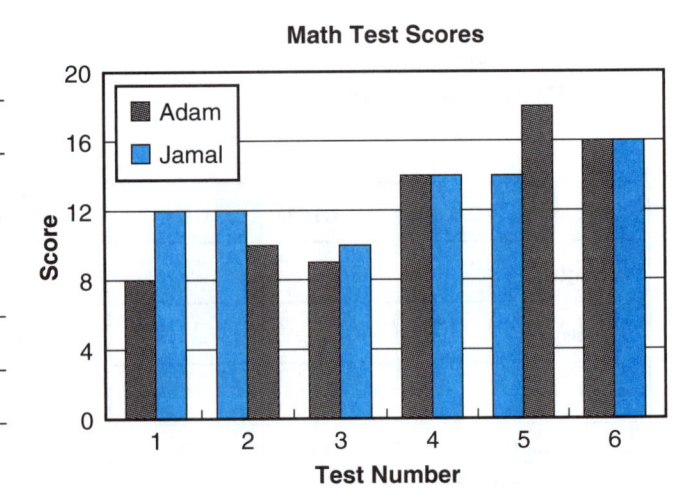

Number Correct

180

8: Ask students why a misleading graph might be created.
5–7: Check to see if students understand that data can be presented in a misleading manner.
Less than 5: Ask students to explain each question and what is being asked before they redo incorrect answers.

LESSON FOLLOW-UP AND ERROR ANALYSIS

Prerequisite Skills: interpreting simple line graphs, reading number lines

Lesson Focus: interpret and make multiple line graphs
Possible Score: 27
Time Frame: 15–20 minutes

Lesson 3 Multiple Line Graphs

Multiple line graphs are used to compare numbers from more than one set of data. The graph provides a way to make visual comparisons among the numbers.

A different style of line is used for each city's snowfall. The key shows what each line represents.

Complete the multiple line graph for the month of April showing that Riverview had no snow and Lakeville had 1 cm of snow.

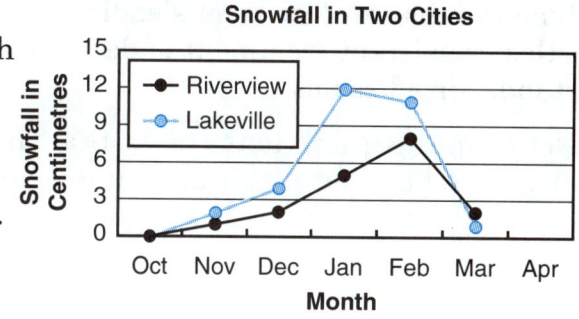

Use the line graph above to answer each question.

1. How many months did Riverview have more than 6 cm of snow? __1__ month(s)

2. How many months did Riverview have more snow than Lakeville? __1__ month(s)

3. In which two months did Lakeville have just 1 cm more of snow than Riverview?
   November and April

4. Which city had the highest snowfall and in what month? _New York, Jan_

5. In which month did both cities have the same amount of snow? _October_

6. In Lakeville, what month had twice as much snow as the previous month? _December_

7. Did Riverview have its highest snowfall in the same month as Lakeville? __no__

Use the information in the chart below to complete a double line graph.

8. Mrs. Smith charted the growth of her daughters' mass for two years.

	Jamie	Davida
birth	3	3.5
3 mo	3.5	4.25
6 mo	4	4.175
9 mo	5.75	6.5
1 year	7	7.5
15 mo	7.75	9
18 mo	8.75	9.5
21 mo	10.5	10.5
2 years	11.5	11.75

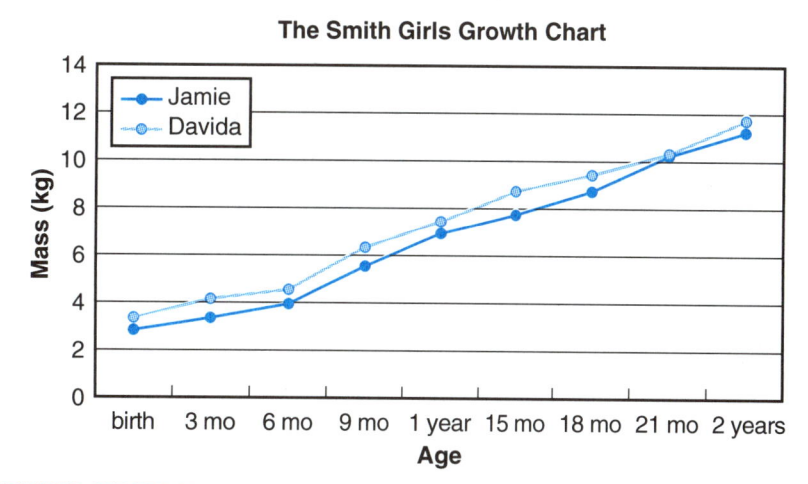

Number Correct — LESSON FOLLOW-UP AND ERROR ANALYSIS
- 24–27: Have students describe various types of data that would be best displayed in a multiple line graph.
- 20–23: Remind students to include a key when creating a multiple line graph.
- Less than 20: In problem 8, review how to graph a multiple line graph.

181

Prerequisite Skills: lines graphs; multiple line graphs; number lines

Lesson 4 Misleading Line Graphs

Lesson Focus: interpret misleading graphs
Possible Score: 4
Time Frame: 5–10 minutes

Line graphs can also be misleading. Look for use of titles, consistent treatment of data, and standardized scales.

Notice in the graph to the right that the months along the horizontal axis are out of sequence.

Furniture Sales

Use the graph above to answer each question.

1. If the months were listed in order, what trend would show?
   The number of sales is decreasing.

2. Name a reason why April may have been shown first. Answers will vary. Sample answer: Since April is the lowest, by showing it first, it appears as though the number of sales is increasing.

Use the graph of rainfall in British Columbia to answer each question below.

3. It appears that Kelowna gets a lot more rain than Vancouver. Is that a fair statement to make based on reading this graph?
   It does appear that Austin gets more rain, but once you look closely you notice that the rainfall is not recorded in the same units.

4. Name one way to revise this graph to show a fair comparison of the rainfall in Kelowna and Vancouver. In order to show both cities' data on the same graph, both cities' data must be in the same unit.

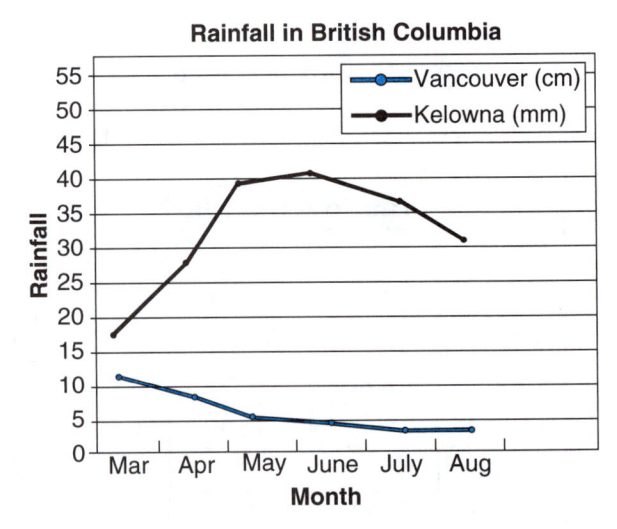

Rainfall in British Columbia

Number Correct

182

LESSON FOLLOW-UP AND ERROR ANALYSIS

4: Ask students why a misleading graph might be created.
3: Check to see if students understand that data can be presented in a misleading manner.
Less than 3: Ask students to explain each question and what is being asked before they redo incorrect answers.

Prerequisite Skills: reading number lines, plotting points on coordinate grid

Lesson Focus: interpret and make scatter plots
Possible Score: 23
Time Frame: 10–15 minutes

Lesson 5 Scatter Plots

Scatter plots are graphs that show the relationship of two variables in a set of data. Points are plotted on a coordinate grid to show how closely the variables are related.

Test Score	16	18	18	20	18	20	24	24	26	25	30	30	28	30	32	31	33	34	34	34
Number of Tests	1	2	3	4	5	6	7	8	9	10	11	12	13	14	15	16	17	18	19	20

To plot a pair of data, use the horizontal axis for one variable (in this case the number of practice tests) and the vertical axis for the other variable (score on timed test).

Complete the scatter plot for practice rounds greater than 14.

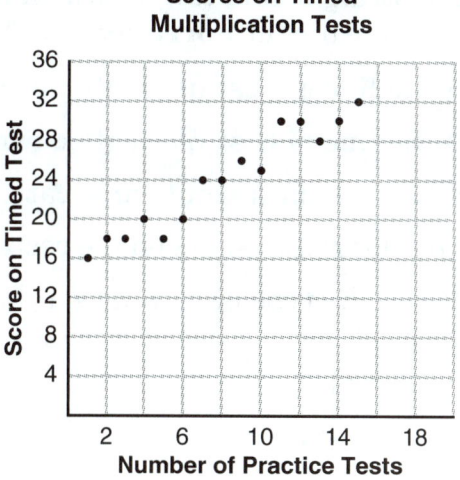

Use the scatter plot above to answer each question.

1. What score can a person who practised 18 or more rounds expect to get on the timed test? __34__

2. What test score did the person who practised 10 rounds get? __25__

3. Write a statement that explains the relationship between the practice rounds and the timed test score. _Answers will vary. Sample answer: The more practice rounds a person does, the higher the person will score on the timed test._

4. If the most points possible for the timed test is 34, what advice would you give to someone who wants a perfect score? _Answers will vary. Sample answer: Do at least 18 practice rounds before taking the timed test._

Use the data in the chart below to make a scatter plot.

5. Ben, Dr. Jackson's secretary, keeps track of visits to the dentist for cleanings and the number of cavities each patient has.

Cleanings	2	3	6	7	8	9	10	2	3	1	9	5	4	5
Cavities	7	6	4	3	3	1	0	9	5	8	0	4	5	5

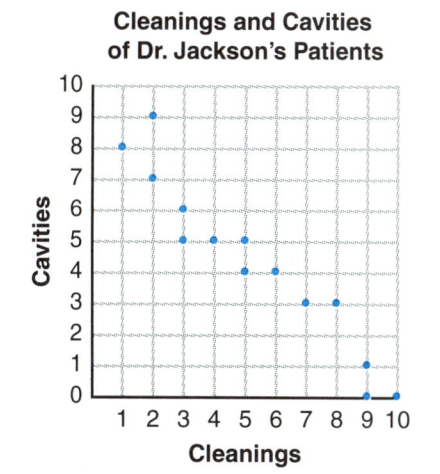

Number Correct
LESSON FOLLOW-UP AND ERROR ANALYSIS
21–23: Have students describe various types of data that would be best displayed in a scatter plot.
18–20: Remind students to use the first variable on the horizontal scale and the second variable on the vertical scale.
Less than 18: In problem 5, go through with the students how to graph at least two points on the scatter plot.

Prerequisite Skills: interpret scatter plots

Lesson Focus: determine the correlation of the variables in a data set
Possible Score: 7
Time Frame: 5–10 minutes

Lesson 6   Scatter Plots

If a straight line can be drawn following a pattern on a scatter plot, the data have a relationship, or **correlation.** This line is called a **line of best fit.**

If the line of best fit slants upward, the data have a **positive correlation.**

If the line of best fit slants downward, the data have a **negative correlation.**

When the data points are scattered throughout the graph, there is **no correlation.**

The closer the points on the scatter plot are to the line of best fit, the stronger the correlation.

Name the type of correlation that each graph shows.

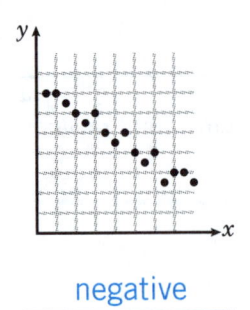

negative 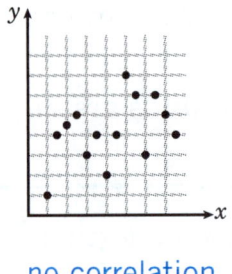   no correlation 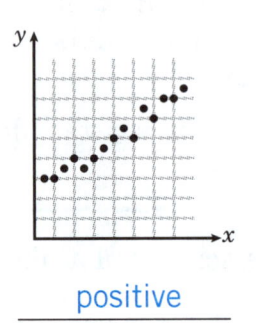   positive

---

Name the type of correlation that each graph shows.

a *b*

1. positive 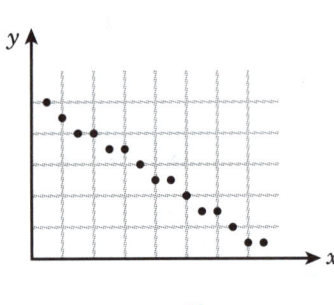   negative

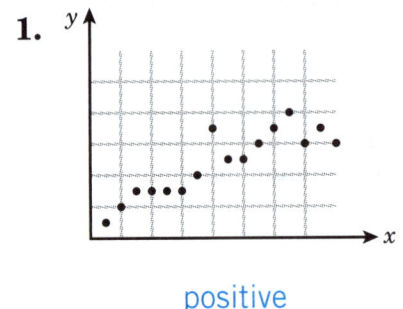

2. positive 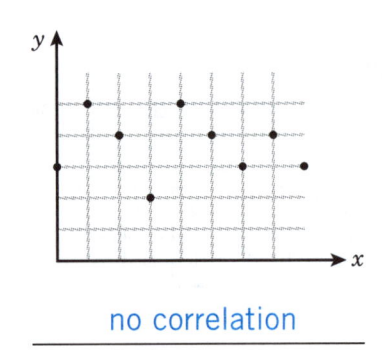   no correlation

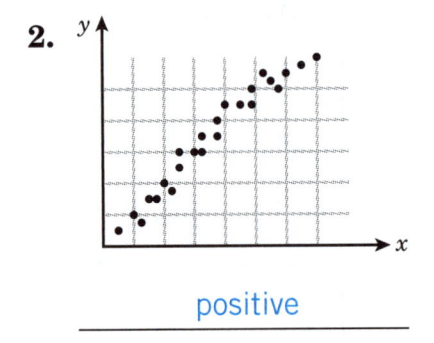

Number Correct

LESSON FOLLOW-UP AND ERROR ANALYSIS

6–7: Have students name some relationships that represent a positive correlation.
4–5: Remind students that an upward slant is positive and a downward slant is negative.
Less than 4: Have students use a straightedge to lay on top of the points to locate a line of best fit.

184

Prerequisite Skills: using a protractor to construct angles

Lesson Focus: separating a circle into circular regions
Possible Score: 18
Time Frame: 10–15 minutes

Lesson 7 Circles

A circle and its interior are called a **circular region.**

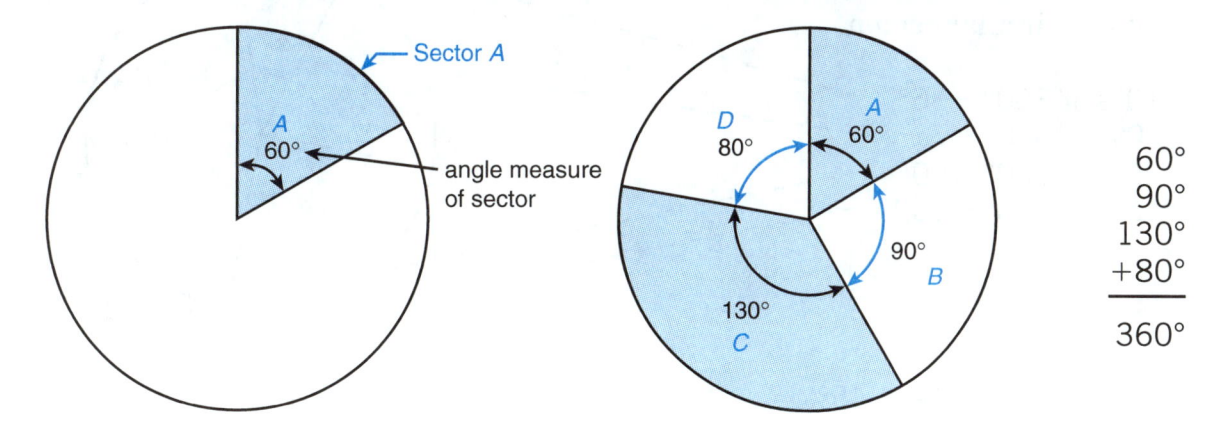

```
   60°
   90°
  130°
 + 80°
  ─────
  360°
```

What is the angle measure of sector B? __90°__ Sector C? __130°__ Sector D? __80°__

The sum of the angle measures of all the sectors in a circular region is __360°__.

Use a protractor to help you separate each circular region as directed. Label each sector with the proper letter and angle measurement.

1. 4 sectors with angle measures as follows:

Sector A	30°
Sector B	60°
Sector C	120°
Sector D	150°

2. 5 sectors with angle measures as follows:

Sector A	20°
Sector B	25°
Sector C	45°
Sector D	90°
Sector E	180°

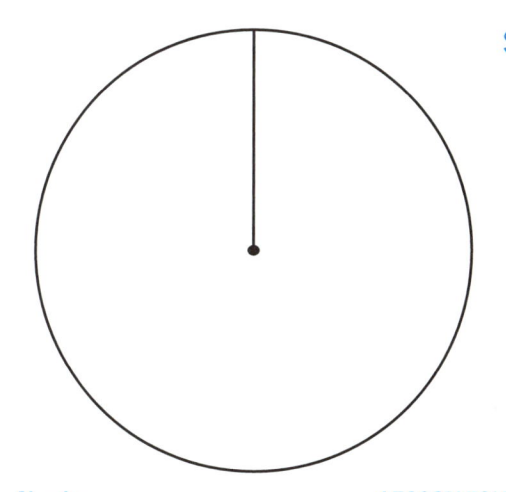

See students' work.

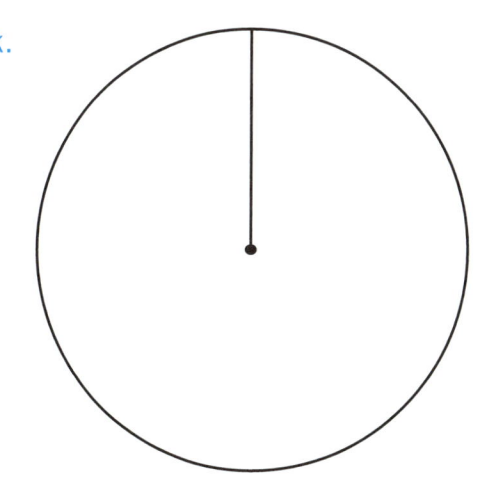

Number Correct
LESSON FOLLOW-UP AND ERROR ANALYSIS
- **15–18:** Have students total the number of degrees in problems 1 and then in 2, and finally compare those totals.
- **12–14:** Help students construct angles such as 120, 150, and 180 degrees before they redo the page.
- **Less than 12:** Ask students to explain how to use a protractor before they redo incorrect answers.

Prerequisite Skills: finding a percent of a number

Lesson Focus: finding the number of degrees in sectors of a circle
Possible Score: 20
Time Frame: 5–10 minutes

Lesson 8 Circles

Study how the circular region is separated into four sectors representing 10%, 20%, 25%, and 45% of the circular region.

10% of 360° = 36°
20% of 360° = 72°
25% of 360° = 90°
45% of 360° = 162°

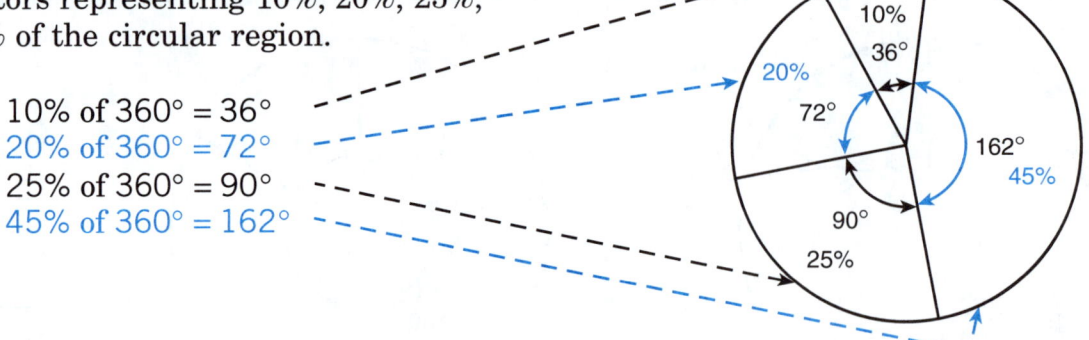

10% + 20% + 25% + 45% = __100__ %   36° + 72° + 90° + 162° = __360__ °

Complete each sentence. Then write the correct angle measurement in the appropriate sectors.

1. 10% of 360° = __36°__

20% of 360° = __72°__

30% of 360° = __108°__

40% of 360° = __144°__

10% + 20% + 30% + 40% = __100%__

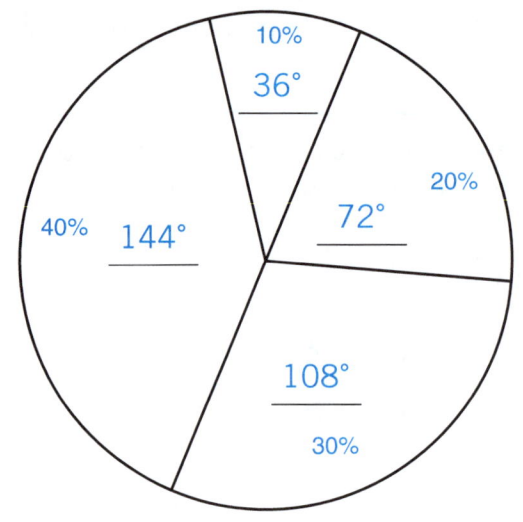

2. 5% of 360° = __18°__

15% of 360° = __54°__

35% of 360° = __126°__

45% of 360° = __162°__

5% + 15% + 35% + 45% = __100%__

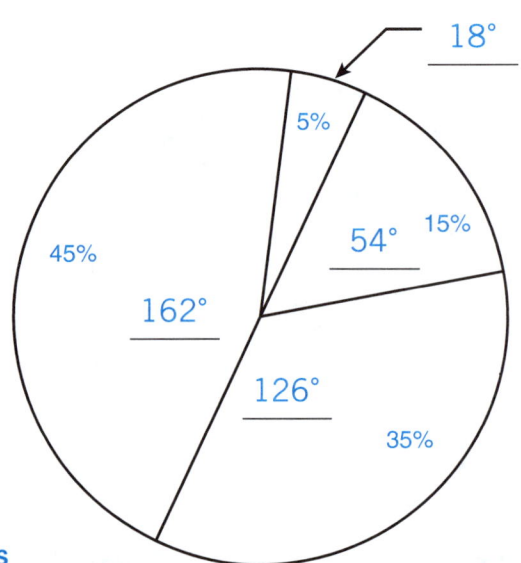

Number Correct

LESSON FOLLOW-UP AND ERROR ANALYSIS

17–20: Ask students to explain why each percent is multiplied by 360.
13–16: Have students explain when to insert a zero as they change a 1-digit percent to a decimal.
Less than 13: Ask students to explain how to change a percent to either a fraction or a decimal before they redo the page.

186

Prerequisite Skills: finding a percent of a number

Lesson Focus: reading circle graphs
Possible Score: 9
Time Frame: 10–15 minutes

Lesson 9   Circle Graphs

Study how a **circle graph** is used to present the following information in a clear and interesting way.

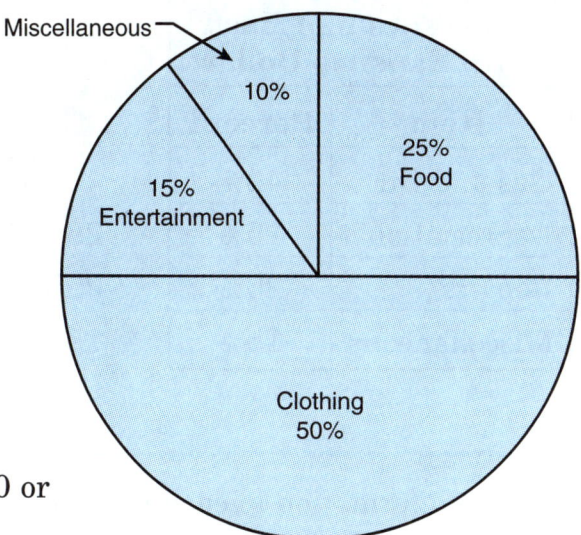

Ashlee spends her allowance as follows: 25% for food, 50% for clothing, 15% for entertainment, and 10% for miscellaneous expenses.

Assume Ashlee's allowance is $20.

On clothing she would spend 50% of $20 or $__10__.

On food she would spend 25% of $20 or $__5__.

On entertainment she would spend 15% of $20 or $__3__.

On miscellaneous expenses she would spend 10% of $20 or $__2__.

How Ashlee Spends Her Allowance
- Miscellaneous 10%
- 15% Entertainment
- 25% Food
- Clothing 50%

Complete each sentence.

How Lewis Spends His Allowance
- Miscellaneous 10%
- 40% Food
- 20% Entertainment
- 30% Clothing

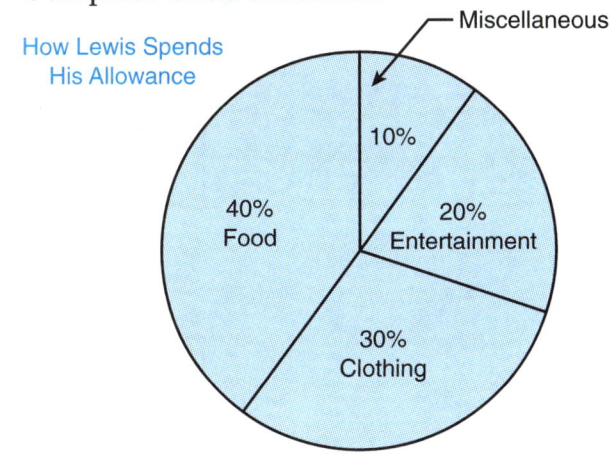

1. Assume Lewis' allowance is $25.

   He would spend $__7.50__ for clothing.

   He would spend $__10__ for food.

   He would spend $__5__ for entertainment.

   He would spend $__2.50__ for miscellaneous expenses.

2. Assume Ms. Adams' monthly income is $9000.

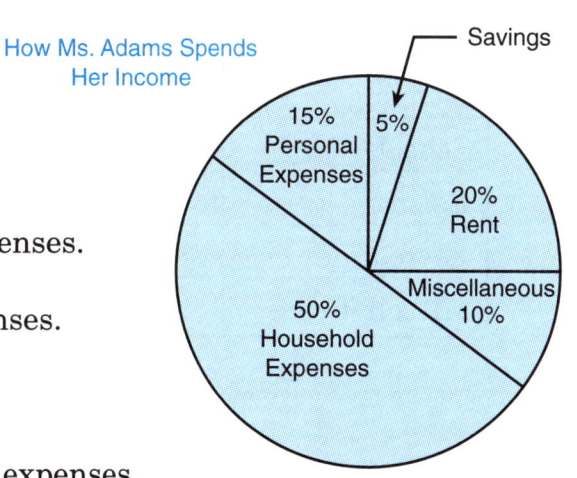

   She would spend $__1800__ for rent.

   She would spend $__4500__ for household expenses.

   She would spend $__1350__ for personal expenses.

   She would save $__450__.

   She would spend $__900__ for miscellaneous expenses.

How Ms. Adams Spends Her Income
- Savings 5%
- 15% Personal Expenses
- 20% Rent
- 50% Household Expenses
- Miscellaneous 10%

LESSON FOLLOW-UP AND ERROR ANALYSIS

Number Correct
- 8–9: Have students add the percents shown for each circle graph and then compare those sums.
- 6–7: Ask students to explain how to change a percent to either a fraction or a decimal.
- Less than 6: Have students explain how to multiply each percent times the amount given at the beginning of each problem.

187

Prerequisite Skills: reading circle graphs

Lesson Focus: making a circle graph
Possible Score: 18
Time Frame: 10–15 minutes

Lesson 10 Circle Graphs

Study how the information in the table can be presented on a circle graph.

Distribution of Each Auto Expense Dollar

Item	Percent
Gas and Oil	40%
Depreciation	25%
Repairs	20%
Miscellaneous	15%

40% of 360° = 144°
25% of 360° = 90°
20% of 360° = 72°
15% of 360° = 54°

Use the information in each table to help you complete each circle graph.

1. Distribution of Each Income Dollar

Expense	Percent
Rent	30%
Personal	20%
Household	35%
Miscellaneous	15%

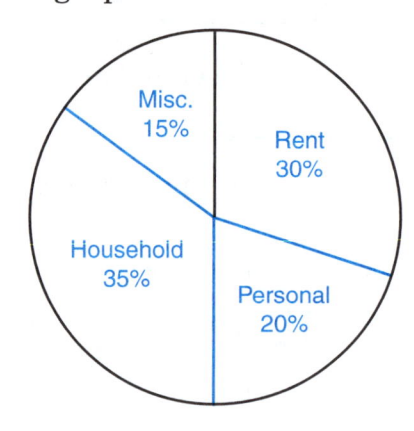

30% of 360° = 108°
20% of 360° = 72°
35% of 360° = 126°
15% of 360° = 54°

2. Distribution of Activities on an Average School Day

Activity	Percent
Sleeping	30%
School	25%
Eating	10%
Recreation	20%
Miscellaneous	15%

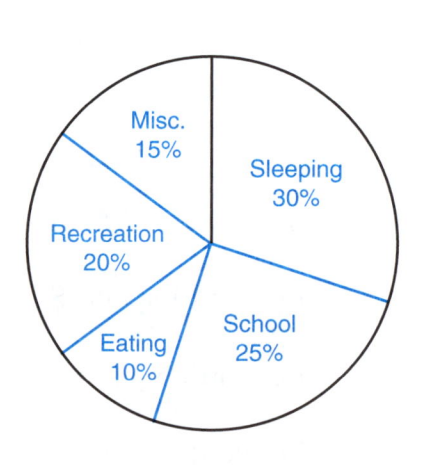

30% of 360° = 108°
25% of 360° = 90°
10% of 360° = 36°
20% of 360° = 72°
15% of 360° = 54°

LESSON FOLLOW-UP AND ERROR ANALYSIS

Number Correct
- **15–18:** Have students make and explain a circle graph of their own.
- **12–14:** Ask students to explain how to find the percents of 360 degrees and then how to divide each circle into sections.
- **Less than 12:** Have students explain how to find the percents of 360 degrees before they redo the page.

For further evaluation, copy the Chapter Test on page 276.
For maintaining skills, use the Cumulative Review on pages 235–236.

Possible Score: 15
Time Frame 10–15 minutes

CHAPTER 12   PRACTICE TEST
Graphs

Use the bar graph to answer each question.

1. On how many tests did Logan spell more words correctly than Sabrina? __4__ tests

2. How many words did Sabrina spell correctly on the second test? __5__ words

3. On which tests did Logan have the same score? __tests 1 and 3__

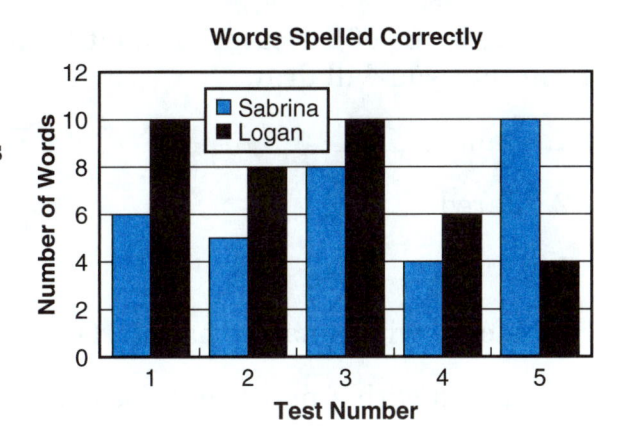

Use the line graph to determine how this graph is misleading.

4. By looking at the points and line graphed, what conclusion would you draw about the rainfall in Ashleyville? __The rainfall is decreasing.__

5. What is incorrect that causes you to reach the conclusion from question 4?
__The months are listed out of order.__

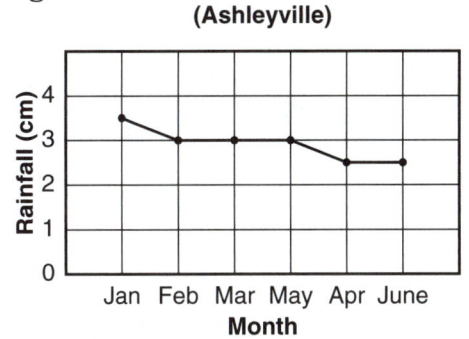

6. Use the chart to determine the number of degrees for each activity.

   work = __144°__

   travel = __18°__

   eating = __36°__

   recreation = __54°__

   miscellaneous = __108°__

Average Work Day	
Activity	Percent
Work	40%
Travel	5%
Eating	10%
Recreation	15%
Miscellaneous	30%

7. Create a circle graph using the information from question 6.

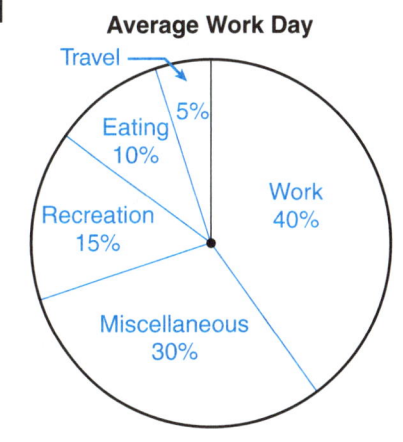

Number Correct	LESSON FOLLOW-UP AND ERROR ANALYSIS
13–15:	Have students make a circle graph that represents how they spend their time at school on an average day.
10–12:	Review how to find the degree represented by each percent of a circle graph.
Less than 10:	Review how to read and interpret data from multiple bar graphs and multiple line graphs.

189

CHAPTER 13 PRETEST
Probability

You draw one of the cards without looking. Write the probability as a fraction in simplest form that you will draw

1. Janet __$\frac{1}{3}$__
2. Jared __$\frac{1}{6}$__
3. Juan __$\frac{1}{2}$__

Jared	Janet	Janet
Juan	Juan	Juan

4. a card with a name that starts with J __1__

5. a card with a name that does not start with J __0__

6. Complete the sample space for tossing a penny and a nickel.

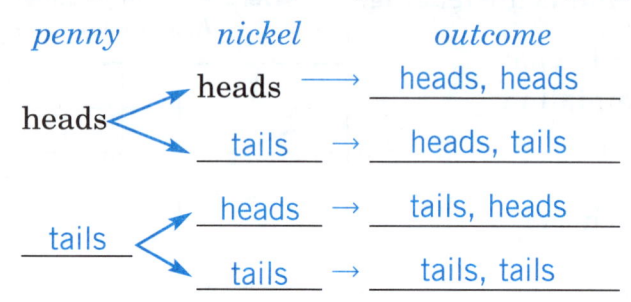

```
penny       nickel       outcome
         ┌ heads  →   heads, heads
heads ───┤
         └ tails  →   heads, tails

         ┌ heads  →   tails, heads
tails ───┤
         └ tails  →   tails, tails
```

Solve each problem. Write each probability as a percent.

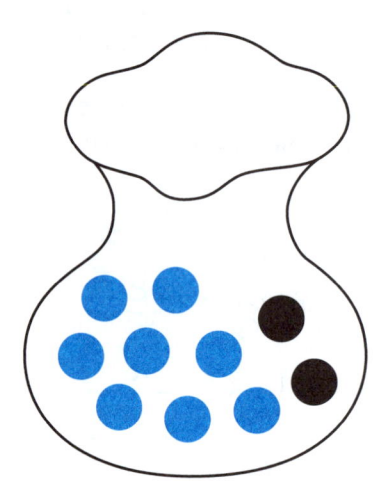

7. You draw one of the marbles without looking. What is the probability of drawing a blue marble?

    The probability is __80%__.

8. A company knows that 1% of the bolts they make are defective. If they produce 250 000 bolts, how many will be defective?

    __2500__ bolts will be defective.

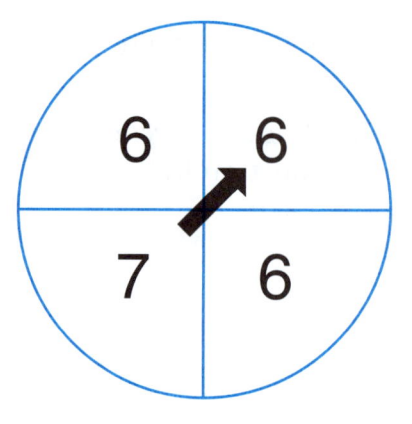

9. You spin the spinner at the right. What is the probability that the spinner will stop on 6?

    The probability is __75%__.

10. You spin the spinner at the right 20 times. Predict how many times the spinner will stop on 6.

    The spinner will stop on 6 __15__ times.

11. You spin the spinner at the right 200 times. Predict how many times the spinner will stop on 6.

    The spinner will stop on 6 __150__ times.

Prerequisite Skills: understanding and writing ratios

Lesson Focus: finding probability
Possible Score: 20
Time Frame: 10–15 minutes

Lesson 1 Probability

You draw one of the cards shown at the right without looking. You would like to know your *chance* or **probability** of getting a card that shows an **A**.

Each card (possible result) is called an **outcome**. There are 6 cards. There are 6 possible outcomes. Since you have the same chance to draw any of the cards, the outcomes are **equally likely**.

number of outcomes that show **A** $\longrightarrow$
number of possible outcomes $\longrightarrow$ $\frac{2}{6}$ or $\frac{1}{3}$ Write the probability in simplest form.

The probability of drawing a card that shows an **A** is $\frac{1}{3}$.

You pick a marble without looking. In simplest form, write the probability of picking

1. white $\frac{1}{5}$

2. black $\frac{3}{10}$

3. blue $\frac{1}{2}$

4. a marble that is **not** white $\frac{4}{5}$

You spin the spinner shown at the right. Find the probability of the spinner stopping on

5. a blue section $\frac{1}{2}$

6. a white section $\frac{1}{2}$

7. a 1 $\frac{1}{4}$

8. a 3 $\frac{1}{2}$

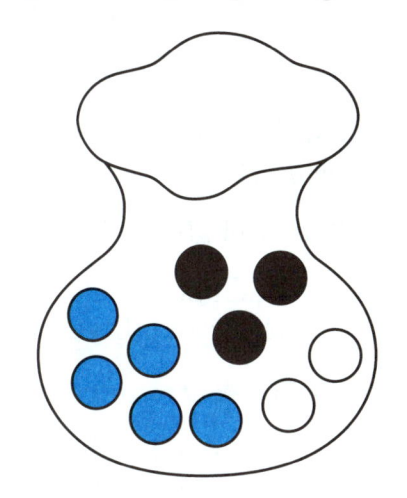

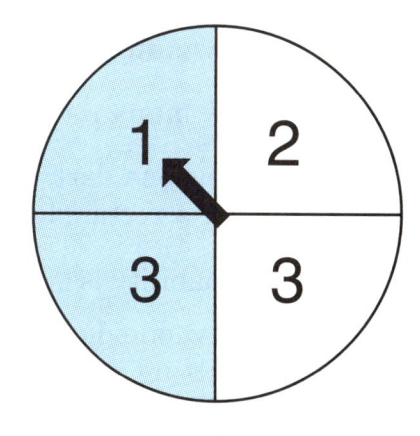

9. a 2 $\frac{1}{4}$

10. an odd number $\frac{3}{4}$

Number Correct

LESSON FOLLOW-UP AND ERROR ANALYSIS

17–20: Have students add a red marble to the bag and write the probability of picking each colour then.
13–16: Ask students to explain how to count the total number of outcomes when choosing a denominator for the probability.
Less than 13: Have students explain how to determine probability on a spinner.

Lesson 1 Problem Solving

Your company is having a box-lunch picnic. The company is furnishing free box lunches. The contents are on labels like those at the right. Solve each problem. Write each probability in simplest form.

Cheese Sandwich/ Apple	Roast Beef Sandwich/ Apple
Cheese Sandwich/ Pear	Roast Beef Sandwich/ Orange
Peanut Butter Sandwich/ Apple	Chicken Sandwich/ Pear
Taco/ Pear	Taco/ Apple

1. Suppose you do not care what kind of lunch you get, so you take one box without looking. What is the probability that you will get a cheese sandwich and an apple?

   The probability is $\frac{1}{8}$.

2. You take one box without looking. What is the probability that you will get an apple?

   The probability is $\frac{1}{2}$.

3. You take one box without looking. What is the probability that you will get a taco?

   The probability is $\frac{1}{4}$.

4. You take one box without looking. What is the probability that you will get a pear?

   The probability is $\frac{3}{8}$.

5. You take one box without looking. What is the probability that you will **not** get an orange?

   The probability is $\frac{7}{8}$.

A game has a board like the one shown below. Use the board to answer each question. Write each probability in simplest form.

6. Are the outcomes equally likely? __no__

Draw lines to make all of the rectangles the same size. Remember to label each section.

DART THROW

win	lose	win
lose	lose	lose
lose	win	lose
win	lose	win
lose	lose	lose
lose	win	lose

7. Now how many rectangles say *win*? __6__

8. Now how many rectangles say *lose*? __12__

9. You throw one dart. What is the probability of hitting a rectangle that says *win*?

   The probability is $\frac{1}{3}$.

10. You throw one dart. What is the probability of **not** hitting a rectangle that says *win*?

    The probability is $\frac{2}{3}$.

CHAPTER 13
Probability

Lesson 1
Probability

Prerequisite Skills: understanding and writing ratios

Lesson Focus: finding 0 and 1 probabilities
Possible Score: 15
Time Frame: 20–25 minutes

Lesson 2   0 and 1 Probabilities

You spin the spinner at the right.

Probability of the spinner stopping on 2	Probability of the spinner stopping on a number less than 5	Probability of the spinner stopping on 6
$\frac{1}{4}$	$\frac{4}{4}$ or 1 A probability of 1 means the outcome is **certain** to happen.	$\frac{0}{4}$ or 0 A probability of 0 means the outcome will **never** happen.

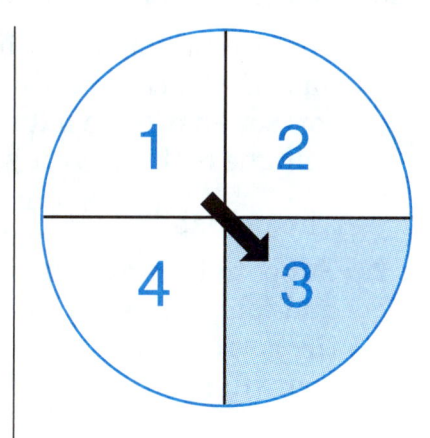

Solve each problem. Write each probability in simplest terms.

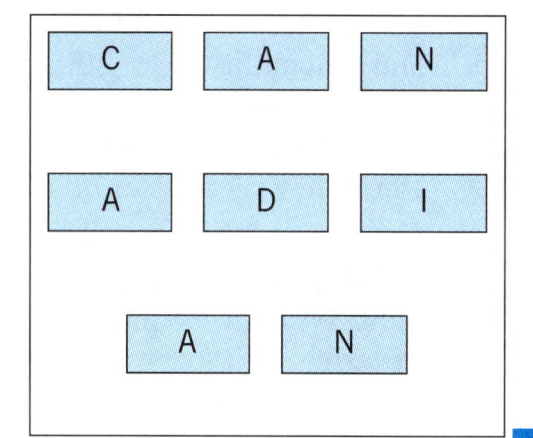

1. You pick one of the letter cards shown at the right without looking. What is the probability that you will pick a vowel (a, e, i, o, u)?

    The probability is $\frac{1}{2}$.

2. You pick one of the letter cards without looking. What is the probability that you will pick an A?

    The probability is $\frac{1}{4}$.

3. You pick one of the letter cards without looking. What is the probability that you will pick a B?

    The probability is 0.

4. You pick one of the letter cards without looking. What is the probability that the letter on the card is in the word CANADIAN?

    The probability is 1.

5. You pick one of the letter cards without looking. What is the probability that the letter on the card is **not** in the word CANADIAN?

    The probability is 0.

6. You pick one of the letter cards without looking. What is the probability that the letter on the card is in the word AID?

    The probability is $\frac{3}{8}$.

Number Correct	LESSON FOLLOW-UP AND ERROR ANALYSIS
13–15:	Have students give their own example of 0 and 1 probabilities.
10–12:	Have students explain why it is important to count all the possible outcomes when writing probabilities.
Less than 10:	Check to see if students wrote the correct probability but forgot to put it in simplest form.

CHAPTER 13

193

Lesson 2 Problem Solving

Solve each problem. Write each probability in simplest form.

1. You are taking a multiple-choice test. Each item has six choices. You have no idea which is the correct answer. What is the probability that you will guess the correct answer?

   The probability is $\frac{1}{6}$.

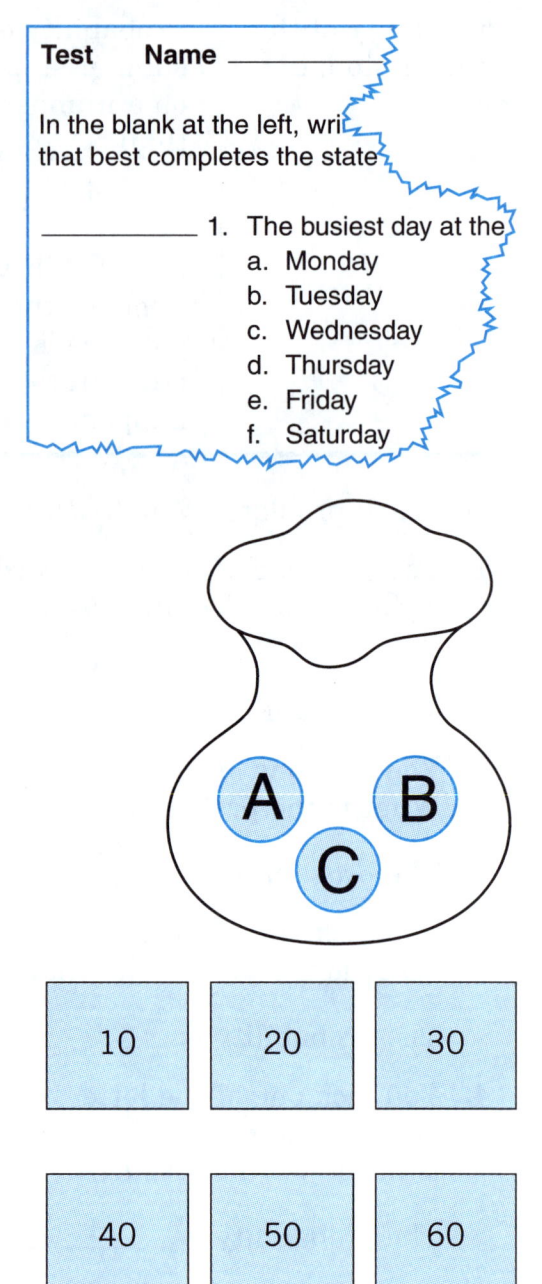

2. Suppose that each item on the test in problem **1** had four choices. You still have no idea which is the correct answer. What is the probability that you will guess the correct answer?

   The probability is $\frac{1}{4}$.

3. You draw one marble from the bag shown at the right. What is the probability that you will draw a marble with a number on it?

   The probability is 0.

4. You draw one marble from the bag shown at the right. What is the probability that you will draw a marble with a letter on it?

   The probability is 1.

5. You draw one marble from the bag shown at the right. What is the probability that you will draw a marble with a vowel on it?

   The probability is 0.

6. You pick one of the number cards shown at the right without looking. What is the probability that you will pick a number greater than 30?

   The probability is $\frac{1}{2}$.

7. You pick one of the number cards shown at the right without looking. What is the probability that you will pick a number less than 100?

   The probability is 1.

8. You pick one of the number cards shown without looking. What is the probability that you will pick a card with a 0 on it?

   The probability is 1.

9. You pick one of the number cards shown without looking. What is the probability that you will pick a card with a 7 on it?

   The probability is 0.

CHAPTER 13
Probability

Prerequisite Skills: finding probabilities

Lesson Focus: finding probabilities using sample spaces
Possible Score: 57
Time Frame: 25–30 minutes

Lesson 3 Sample Spaces

Suppose you flip a penny and a dime. You can show all the possible outcomes in a table or in a tree diagram.

		dime	
		heads	tails
penny	heads	h, h	h, t
	tails	t, h	t, t

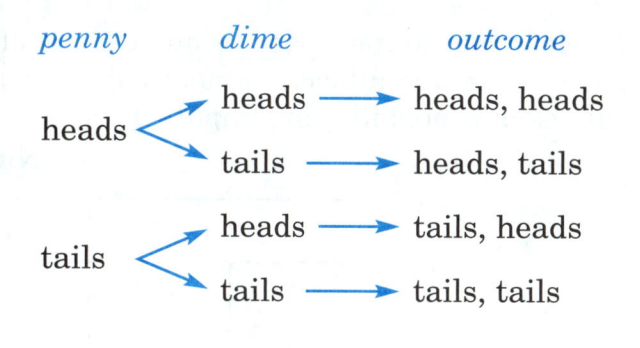

penny   dime   outcome

heads → heads → heads, heads
heads → tails → heads, tails
tails → heads → tails, heads
tails → tails → tails, tails

A list or a table of all the possible outcomes is called a **sample space**.

Use the sample spaces above to answer each question.

1. How many possible outcomes are there? __4__

2. What is the probability that both coins will land with heads up? __$\frac{1}{4}$__

3. What is the probability that one coin will land with heads up and the other will land with tails up? __$\frac{1}{2}$__

Suppose you toss a penny, a nickel, and a dime. Complete the sample space below to show the possible outcomes.

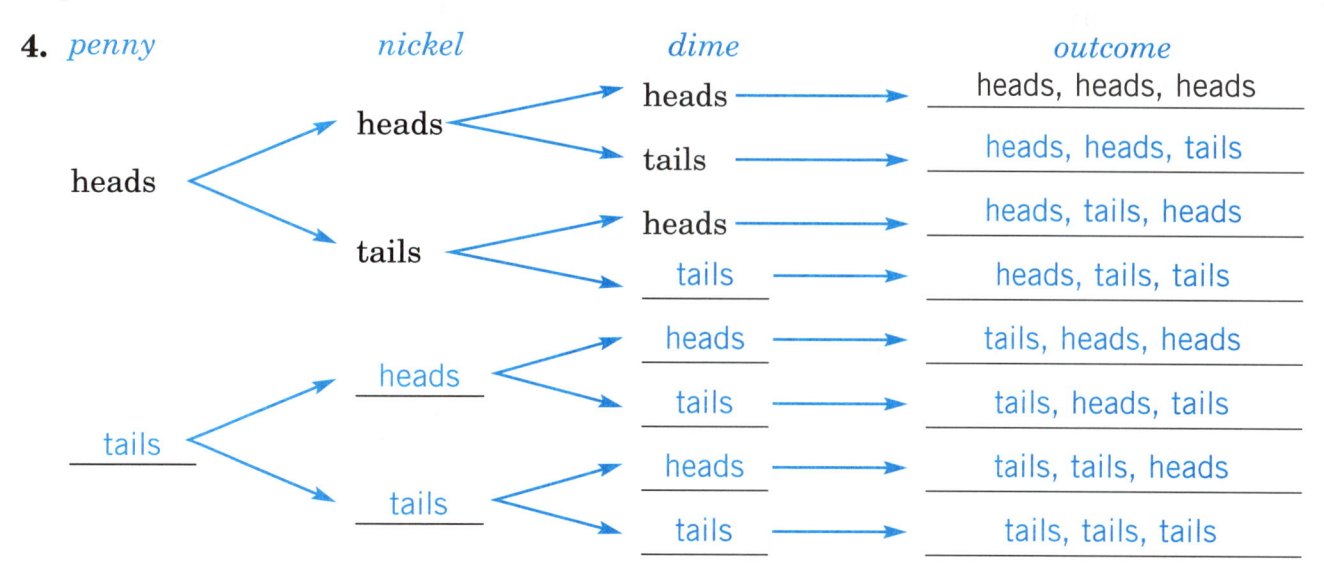

4. penny   nickel   dime   outcome

heads
– heads
  – heads → heads, heads, heads
  – tails → heads, heads, tails
– tails
  – heads → heads, tails, heads
  – tails → heads, tails, tails

tails
– heads
  – heads → tails, heads, heads
  – tails → tails, heads, tails
– tails
  – heads → tails, tails, heads
  – tails → tails, tails, tails

5. How many possible outcomes are there? __8__

6. What is the probability of all heads? __$\frac{1}{8}$__

7. What is the probability of two heads and one tail? __$\frac{3}{8}$__

8. What is the probability of at least one tail? __$\frac{7}{8}$__

Number Correct — LESSON FOLLOW-UP AND ERROR ANALYSIS

48–57: Have students write a sample space for flipping a coin and rolling a six-sided number cube.
37–47: On page 196, discuss that the chart numbers 1 to 6 are the numbers on each number cube.
Less than 37: Before they redo the pages, ask students to explain why it is important to count the total number of outcomes.

Lesson 3 Problem Solving

Complete the sample space below to show all the possible outcomes of rolling a blue number cube and a black number cube. Then use the sample space to solve each problem. Write each probability in simplest form.

Blue Number Cube

	1	2	3	4	5	6
1	1,1	2,1	3,1	4,1	5,1	6,1
2	1,2	2,2	3,2	4,2	5,2	6,2
3	1,3	2,3	3,3	4,3	5,3	6,3
4	1,4	2,4	3,4	4,4	5,4	6,4
5	1,5	2,5	3,5	4,5	5,5	6,5
6	1,6	2,6	3,6	4,6	5,6	6,6

Black Number Cube

1. What is the probability of rolling two 5s?

   The probability is $\frac{1}{36}$.

2. What is the probability of rolling the same number on both number cubes?

   The probability is $\frac{1}{6}$.

3. What is the probability of rolling 4,5 or 5,4?

   The probability is $\frac{1}{18}$.

4. What is the probability of rolling two number cubes that total 10?

   The probability is $\frac{1}{12}$.

5. What is the probability of rolling two number cubes that total 20?

   The probability is 0.

6. What is the probability of rolling two number cubes that total less than 13?

   The probability is 1.

7. What is the probability of rolling two number cubes that total 7?

   The probability is $\frac{1}{6}$.

8. What is the probability of rolling two different numbers?

   The probability is $\frac{5}{6}$.

CHAPTER 13
Probability

Lesson 3
Sample Spaces

Prerequisite Skills: finding probabilities

Lesson Focus: conducting probability experiments
Possible Score: 19
Time Frame: 25–30 minutes

Lesson 4   Probability Experiments

Try this experiment.

Flip a coin 10 times.
How many times did you get heads? _____

Flip a coin 10 more times.
Out of the 20 flips, did you
get heads exactly 10 times? _____

Use tally marks (/) to record the results.

heads	Answers will vary.
tails	See students' work.

Mathematical probability (what you have found in previous lessons) tells what is likely to happen. It does not tell what will actually happen. **Experimental probability** tells what happened during a particular experiment.

Try this experiment. Record your results. Use your results to answer each question.

1. Make paper cards like those shown at the right. Be sure the cards are all the same size. Draw one card without looking, record the result, and put the card back. Repeat the experiment 30 times.

Apple	Orange
Apple	Orange
Apple	Peach

Apple	Answers will vary.
Orange	See students' work.
Peach	

2. Based on your experiment, what is the probability of picking a card that says *Apple*?   _____

3. What is the mathematical probability of picking a card that says *Apple*? $\frac{3}{6}$ or $\frac{1}{2}$

4. Based on your experiment, what is the probability of picking a card that says *Orange*?   _____

5. What is the mathematical probability of picking a card that says *Orange*? $\frac{2}{6}$ or $\frac{1}{3}$

6. Based on your experiment, what is the probability of picking a card that says *Peach*?   _____

7. What is the mathematical probability of picking a card that says *Peach*? $\frac{1}{6}$

8. Repeat the experiment 30 more times. Did your experimental probability results come closer to the mathematical probability after more draws?   _____

9. Compare your results with other class members'. Did you all get exactly the same results?   _____

Number Correct

LESSON FOLLOW-UP AND ERROR ANALYSIS

16–19: Ask students to explain if the results of flipping a coin 50 times are any indication what the result of the next flip will be.
12–15: Ask students to compare mathematical probability and probability based on their own experiments.
Less than 12: Have students explain the importance of recording the results of their experiments accurately before they redo the pages.

Lesson 4   Problem Solving

Try each experiment. Record your results. Use your results to answer each question.

1. Make paper cards like those shown at the right. Be sure the cards are all the same size. Draw one card without looking, record the result, and put the card back. Repeat the experiment 100 times. (You could work with a friend and each person make 50 draws.)

Red	Blue
Red	Blue
Red	Blue
Red	Black
White	Black

Red	Answers will vary.
Blue	See students' work.
White	
Black	

2. Based on your experiment, what is the probability of picking a card that says *Red*? _____

3. What is the mathematical probability of picking a card that says *Red*? $\dfrac{2}{5}$

4. Based on your experiment, what is the probability of picking a card that says *Blue*? _____

5. What is the mathematical probability of picking a card that says *Blue*? $\dfrac{3}{10}$

6. Based on your experiment, what is the probability of picking a card that says *White*? _____

7. Based on your experiment, what is the probability of picking a card that says *Black*? _____

8. Flip a penny and a dime. Record the results with tally marks. Repeat the experiment 25 times.

Penny	Dime	
heads	heads	Answers will vary.
heads	tails	
tails	heads	
tails	tails	

9. Based on your experiment, what is the probability of both coins landing heads up? _____

10. Based on your experiment, what is the probability of at least one coin landing tails up? _____

CHAPTER 13
Probability

198

Prerequisite Skills: finding probabilities

Lesson Focus: writing probabilities as a percent
Possible Score: 20
Time Frame: 25–30 minutes

Lesson 5 Probability and Percent   PRE-ALGEBRA

You are to draw one marble without looking. The probability of drawing a blue marble is $\frac{5}{10}$ or $\frac{1}{2}$. You can write the probability as a percent in either of these two ways.

$$\frac{5}{10} = \frac{a}{100}$$

$$500 = 10a$$
$$50 = a$$

$$\begin{array}{r} 0.50 = 50\% \\ 10\overline{)5.00} \\ \underline{5\ 0} \\ 00 \\ \underline{00} \\ 0 \end{array}$$

$$\frac{5}{10} = \frac{50}{100} = 50\%$$

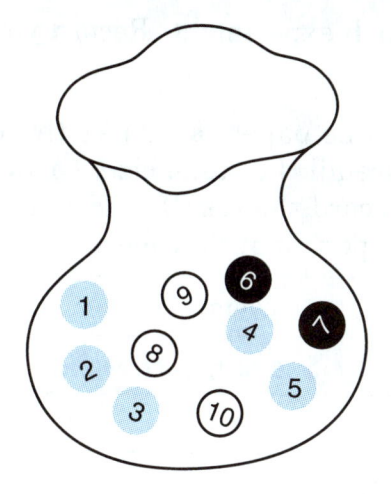

The probability of drawing a blue marble is 50%.

Solve each problem. Write each probability as a percent.

1. Using the bag of marbles at the top of the page, what is the probability of drawing a black marble?

    The probability is __20%__.

2. Using the bag of marbles at the top of the page, what is the probability of drawing a marble with a number on it?

    The probability is __100%__.

You are to choose one of the cards at the right without looking. Write each probability as a percent.

3. What is the probability of choosing *win*?

    The probability is __25%__.

4. What is the probability of **not** choosing *win*?

    The probability is __75%__.

5. What is the probability of choosing *lose*?

    The probability is __50%__.

6. What is the probability of choosing *draw again*?

    The probability is __25%__.

7. What is the probability of choosing *go home*?

    The probability is __0%__.

Number Correct	LESSON FOLLOW-UP AND ERROR ANALYSIS
17–20:	Have students write a probability problem using 75%.
13–16:	Ask students to explain how a probability can be expressed as a percent by changing the ratio to a fraction with a denominator of 100.
Less than 13:	Have students explain how to write the probability as a ratio and how to change it to a percent.

CHAPTER 13

199

Lesson 5 Problem Solving PRE-ALGEBRA

Try each experiment. Record your results. Write each probability as a percent.

1. Make paper cards like those shown. Be sure the cards are all the same size. Draw one card without looking, record the result, and put the card back. Repeat the experiment 20 times.

Red	Answers will vary.
Blue	See students' work.
White	
Black	

Red	Blue
White	Blue
White	Blue
Black	Black
Black	Black

2. Based on your experiment, what is the probability of picking a card that says *Red*? _____

3. What is the mathematical probability of picking a card that says *Red*?   __10%__

4. Based on your experiment, what is the probability of picking a card that says *Blue*? _____

5. What is the mathematical probability of picking a card that says *Blue*?   __30%__

6. Based on your experiment, what is the probability of picking a card that says *White*? _____

7. What is the mathematical probability of picking a card that says *White*?   __20%__

8. Based on your experiment, what is the probability of picking a card that says *Black*? _____

9. What is the mathematical probability of picking a card that says *Black*?   __40%__

10. What is the experimental probability of picking a card that does **not** say *Black*? _____

11. What is the mathematical probability of picking a card that does **not** say *Black*?   __60%__

12. What is the experimental probability of picking a card that says *Green*?   __0%__

13. What is the mathematical probability of picking a card that says *Green*?   __0%__

CHAPTER 13
Probability

Prerequisite Skills: finding probabilities

Lesson Focus: predicting outcomes using probability
Possible Score: 16
Time Frame: 25–30 minutes

Lesson 6   Predicting with Probability

You roll a number cube once. What is the possibility of getting a 5?

Suppose you roll a number cube 60 times. You can predict how many times you would expect to get a 5 as follows:

probability of getting a 5 ← → number of rolls

$\frac{1}{6} \times 60 = 10$ ← number of times you would expect to get a 5

A company finds that 2% of their calculators are defective. Predict how many calculators will be defective if they make 5000 calculators.

```
  5 0 0 0
× 0.0 2
---------
1 0 0.0 0
```

The company can expect 100 defective calculators.

Solve each problem. Write each probability as a fraction in simplest terms or as a percent.

1. You flip a coin once. What is the probability of getting heads?

   The probability is __$\frac{1}{2}$ or 50%__.

2. Suppose you flip a coin 200 times. Predict how many times you would expect to get tails.

   You should get tails about __100__ times.

3. A company knows that $\frac{1}{2}$% of the batteries they make are defective. If they produce 100 000 batteries, how many will be defective?

   __500__ batteries will be defective.

4. You spin the spinner at the right. What is the probability that the spinner will stop on *radio*?

   The probability is __$\frac{1}{4}$ or 25%__.

5. You spin the spinner at the right 20 times. Predict how many times the spinner will stop on *radio*.

   The spinner will stop on *radio* __5__ times.

6. You spin the spinner at the right. What is the probability that the spinner will stop on *pencil*?

   The probability is __$\frac{3}{4}$ or 75%__.

7. You spin the spinner at the right 60 times. Predict how many times the spinner will stop on *pencil*.

   The spinner will stop on *pencil* __45__ times.

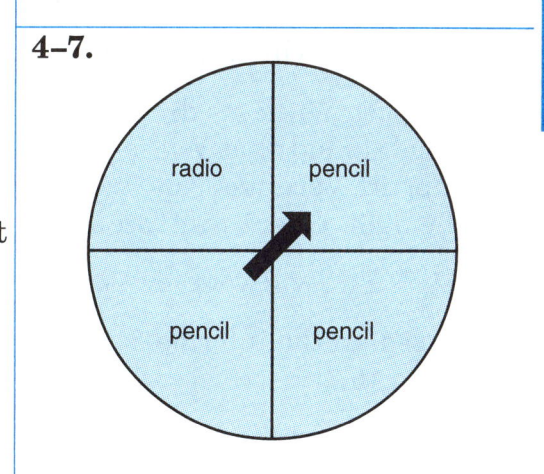

Number Correct

LESSON FOLLOW-UP AND ERROR ANALYSIS

13–16: Have students explain why a prediction based on probability may or may not be correct.

10–12: Check to see if students may have found the probability for one event, but forgot to multiply by the number of times the event occurs.

Less than 10: Before students redo page 202, they should realize that the percents are based on 100 people.

Lesson 6 Problem Solving

One hundred people were polled to see whom they preferred for union representative. The results are shown below. Use the results to solve each problem.

Candidate	Erickson	Nunez	Verdugo	Williams
Number of votes	34	27	25	14

1. What percent of those polled prefer Erickson?

   __34__ % prefer Erickson.

2. If 3000 people vote for union representative, predict how many will vote for Erickson.

   __1020__ will vote for Erickson.

3. What percent of those polled prefer Williams?

   __14__ % prefer Williams.

4. If 3000 people vote for union representative, predict how many will vote for Williams.

   __420__ will vote for Williams.

5. If 3000 people vote for union representative, predict how many will vote for Nunez.

   __810__ will vote for Nunez.

6. If 3000 people vote for union representative, predict how many will vote for Verdugo.

   __750__ will vote for Verdugo.

7. Suppose Williams drops out of the election. A poll found that Williams's supporters now support Verdugo. Now predict how many of the 3000 voters will vote for Verdugo.

   __1170__ will vote for Verdugo.

8. Based on the information in problem **7**, who will win?

   __Verdugo__ will win the election.

9. How many more votes will the winner get than Nunez will get?

   The winner will get __360__ more votes than Nunez.

CHAPTER 13 Probability

Lesson 6 Predicting with Probability

Prerequisite Skills: finding probabilities

Lesson Focus: conducting probability experiments
Possible Score: 14
Time Frame: 25–30 minutes

Lesson 7 More Probability Experiments

For some events it is hard to find a mathematical probability, so it is necessary to use experimental probability.

Make a cone from a piece of paper like this:

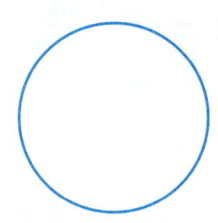

Trace around a circular object with a diameter between 3 and 6 inches.

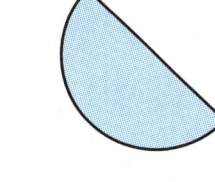

Cut out the circle. Fold it in half. Cut along the fold.

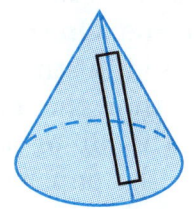

Tape the two edges of one semicircle together to make a cone.

Now toss the cone in the air so it lands on a hard surface. Record the results in the table at the right. Repeat the experiment 50 times.

Landing	Tallies (50 times)
(on base)	See students' work.
(on side)	

Use the experiment above. Write each probability as a percent.

1. Based on 50 tries, what is the probability that the cone will land on its base?   Answers will vary.

2. Based on 50 tries, what is the probability that the cone will land on its side? _____

3. Either do the experiment 50 more times or combine your results with those of another student. Record the totals for the 100 tosses in the table at the right.

(on base)	(on side)
(based on 100 tries)	

4. Based on 100 tries, what is the probability that the cone will land on its base? _____

5. Based on 100 tries, what is the probability that the cone will land on its side? _____

6. Are the results of **1** and **2** close to the results of **4** and **5**? _____

Number Correct
LESSON FOLLOW-UP AND ERROR ANALYSIS
12–14: Ask students if they can think of a way to influence the outcomes of the experiments in Problems 1–6.
9–11: Have students explain the importance of recording the results of their experiments accurately before they redo the pages.
Less than 9: Before students redo the pages, ask them to explain how to initially set up the experiments.

Lesson 7   Problem Solving

Do this experiment. Use your results to solve each problem. Write each probability as a percent.

1. Use a sheet of typing paper and a hair pin or needle. On the paper, draw parallel lines so that they are just slightly farther apart than the length of the pin or needle. Hold the pin or needle about 15 cm above the paper and let it drop. Record the results. If the pin or needle does not land on the paper, that turn does not count. Repeat the experiment 50 times.

Touched a line	Answers will vary.
Did not touch	See student's work.

2. After 50 tries, what is the probability that the pin or needle landed so it touched a line? _____

3. After 50 tries, what is the probability that the pin or needle landed so it did not touch a line? _____

4. Repeat the experiment 50 more times, or combine your results with those of another student.

Touched a line	
Did not touch	

5. After 100 tries, what is the probability that the pin or needle landed so it touched a line? _____

6. After 100 tries, what is the probability that the pin or needle landed so it did not touch a line? _____

Try the experiment again. This time draw the lines so they are only half as far apart. Record your results for 50 tries below.

Touched a line	
Did not touch	

7. After 50 tries, what is the probability that the pin or needle landed so it touched a line? _____

8. After 50 tries, what is the probability that the pin or needle landed so it did not touch a line? _____

Prerequisite Skills: finding probabilities

Lesson Focus: problem solving with probabilities
Possible Score: 9
Time Frame: 10–15 minutes

Lesson 8 Problem Solving

Solve each problem. Write each probability as a fraction in simplest form.

1. You choose one of the cards at the right without looking. What is the probability that the card will say *go*?

   The probability is __$\frac{1}{2}$__.

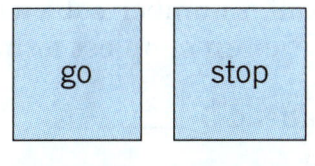

   | go | stop |

2. You choose one of the cards at the right without looking. What is the probability that the card will say *stop*?

   The probability is __$\frac{1}{3}$__.

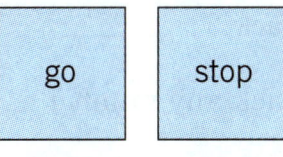

   | go | stop |

3. You choose one of the cards at the right without looking. What is the probability that the card will be *blue*?

   The probability is __1__.

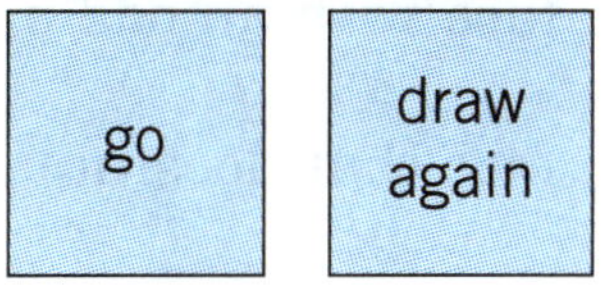

   | go | draw again |

Solve each problem. Write each probability as a percent.

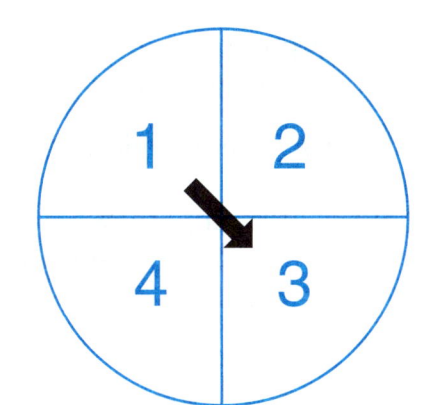

4. You spin the spinner at the right once. What is the probability that the spinner will stop on 4?

   The probability is __25%__.

5. You spin the spinner 80 times. Predict how many times the spinner will stop on 4.

   The spinner will stop on 4 __20__ times.

6. A company knows that 2% of their computer disks are defective. The company produced 400 000 computer disks. How many of those disks will be defective?

   __8000__ computer disks will be defective.

Complete this experiment. Flip a coin. Record the results. Repeat the experiment 19 more times. Write each probability as a percent.

7. Based on 20 flips, what is the probability that the coin landed *heads up*?

   The probability is __Answers will vary. They should be about 50%.__

8. Based on 20 flips, what is the probability that the coin landed *tails up*?

   The probability is _____.

9. What is the mathematical probability that a coin will land *heads up*?

   The probability is __50%__.

Number Correct	LESSON FOLLOW-UP AND ERROR ANALYSIS
8–9:	Have students set up, perform, and record their own probability experiment.
6–7:	Check to see if students found the probabilities correctly but forgot to change them to simplest form.
Less than 6:	Check to see if students found the probabilities correctly but forgot to change them to simplest form.

CHAPTER 13

205

For further evaluation, copy the Chapter Test on page 277.
For maintaining skills, use the Cumulative Review on pages 237–238.
For assessment of Chapters 1–13, use the Final Test on pages 209–212.

Possible Score: 17
Time Frame: 20–25 minutes

CHAPTER 13 PRACTICE TEST
Probability

You roll a number cube with faces marked as shown at the right. Write the probability as a fraction in simplest form that you will roll

1. ahead 5 ___$\frac{1}{6}$___

2. back 3 ___0___

3. back any number ___$\frac{1}{3}$___

4. a face with an odd number ___$\frac{1}{2}$___

5. a face that does **not** say *back* ___$\frac{2}{3}$___

| ahead 5 | back 4 | ahead 3 |
| lose a turn | back 1 | ahead 2 |

Solve each problem. Write each probability as a percent.

6. Suppose you flip a coin 150 times. Predict how many times you would expect to get tails.

    You should get tails about ___75___ times.

7. A company knows that 3% of their doodads are defective. If they produce 300 000 doodads, how many will be defective?

    ___9000___ doodads will be defective.

8. You spin the spinner at the right. What is the probability that the spinner will stop on *red*?

    The probability is ___40%___.

9. You spin the spinner at the right 20 times. Predict how many times the spinner will stop on *red*.

    The spinner will stop on *red* ___8___ times.

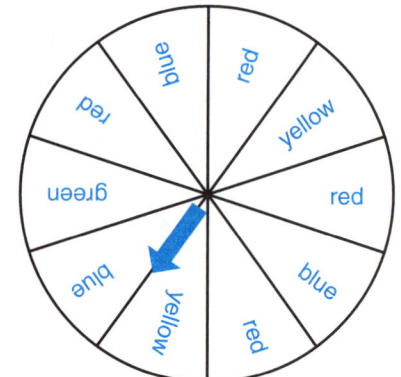

10. Complete the sample space for tossing a penny and a dime.

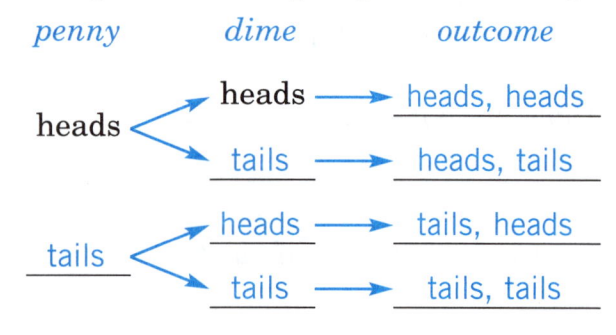

    penny   *dime*   *outcome*

    heads ⟨ heads ⟶ heads, heads
            tails ⟶ heads, tails

    ___tails___ ⟨ heads ⟶ tails, heads
            tails ⟶ tails, tails

Number Correct

LESSON FOLLOW-UP AND ERROR ANALYSIS

15–17: Have students set up, perform, and record their own probability experiment.
12–14: Check to see if students found the probabilities correctly but forgot to change them to simplest form.
Less than 12: Check to see if students found the probabilities correctly but forgot to change them to simplest form.

MID-TEST Chapters 1-6

Complete as indicated. Write each answer in simplest form.

	a	b	c	d	e
1.	83 920 +9 828 ――― 93 748	93 205 −28 591 ――― 64 614	49.235 +37.369 ――― 86.604	1.03 −0.9251 ――― 0.1049	145.5 643.56 9.25 +32.3 ――― 830.61
2.	5242 ×93 ――― 487 506	6.26 ×13 ――― 81.38	105 32)3360	1257.5 0.04)50.3	12.593 ×0.032 ――― 0.402976
3.	$\frac{5}{8}$ $\frac{7}{8}$ $+\frac{5}{8}$ ――― $2\frac{1}{8}$	$\frac{3}{4}$ $-\frac{1}{3}$ ――― $\frac{5}{12}$	$5\frac{1}{6}$ $-2\frac{7}{8}$ ――― $2\frac{7}{24}$	$5\frac{1}{4} \times 3\frac{1}{3} = 17\frac{1}{2}$	$1\frac{3}{5} \div 2\frac{2}{15} = \frac{3}{4}$

Solve each equation.

	a	b	c
4.	$4b = 24$ $b = 6$	$\frac{a}{8} = 13$ $a = 104$	$d + 29 = 120$ $d = 91$
5.	$h - 5 = 3 \times 7$ $h = 26$	$6j = 43 + 5$ $j = 8$	$\frac{s}{12} = 5 \times 6$ $s = 360$
6.	$7a + a = 80$ $a = 10$	$n + n - 1 = 11$ $n = 6$	$r + 3r + 23 = 51$ $r = 7$

PRISM MATHEMATICS
Purple Book

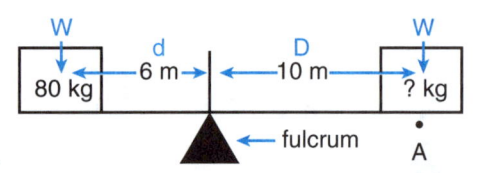

MID-TEST Chapters 1–6 (continued)

Solve each of the following.

	a	b	c
7.	$\frac{2}{5} = \frac{n}{15}$ n = 6	$\frac{9}{n} = \frac{1}{4}$ n = 36	$\frac{12}{18} = \frac{6}{n}$ n = 9

Complete each table. Write each fraction in simplest form.

a

fraction	decimal	percent
$\frac{1}{4}$	0.25	25%

b

fraction	decimal	percent
$\frac{1}{10}$	0.1	10%

Complete the following.

	a	b	c
9.	__6__ is 15% of 40	36 is 75% of __48__	43 is __50__ % of 86

Solve each problem.

10. A store sold 185 suits last month. Of those, 60% were women's suits. How many women's suits were sold?

    __111__ women's suits were sold.

11. Naomi borrowed $2000 for one year at 12% annual interest. How much interest did she pay?

    She paid $__240__ interest.

Write an equation for each problem. Solve each problem.

12. Anna made 12 more doodads than Aaron. Together they made 150 doodads. How many doodads did Anna make?

    Equation: __x + x + 12 = 150__

    Anna made __81__ doodads.

13. In the drawing at the right, how much mass would have to be applied at point A so the lever would be balanced?

    Equation: __80 × 6 = 10 × w__

    __48__ kg would have to be applied at point A.

PRISM MATHEMATICS
Purple Book

FINAL TEST Chapters 1-13

Complete as indicated. Write each answer in simplest form.

	a	b	c	d	e
1.	15 323 +5 628 ――― 20 951	8 6 8 5 −8 5 9 1 ――― 94	4 9.0 5 3 +5 8.3 6 ――― 107.413	5 1.5 −5.5 7 ――― 45.93	2.5 3 5 ×0.0 0 4 ――― 0.01014
2.	4 9 6 ×5 4 ――― 26 784	207 62)12 834	24 0.06)1.4 4	$1\frac{1}{2}$ $+3\frac{1}{2}$ ――― 5	$6\frac{1}{3}$ $-4\frac{1}{2}$ ――― $1\frac{5}{6}$
3.	$\frac{1}{5} \times 6\frac{2}{3} = 1\frac{1}{3}$	$1\frac{1}{2} \div 2\frac{1}{4} = \frac{2}{3}$	6 h 20 min +2 h 2 min ――― 8 h 22 min	3 h 10 min ×3 ――― 9 h 30 min	10 min 20 s −7 min 35 s ――― 2 min 45 s

Solve each of the following.

	a	b	c
4.	$6n = 24$ $n = 4$	$\frac{a}{3} = 43$ $a = 129$	$e + 9 = 12$ $e = 3$
5.	$h - 9 = 2 \times 5$ $h = 19$	$2j = 23 - 7$ $j = 8$	$\frac{s}{8} = 4 \times 3$ $s = 96$
6.	$2a + a = 30$ $a = 10$	$n + n - 5 = 15$ $n = 10$	$r + 7r + 14 = 30$ $r = 2$
7.	$\frac{2}{7} = \frac{n}{14}$ $n = 4$	$\frac{9}{n} = \frac{1}{7}$ $n = 63$	$\frac{12}{30} = \frac{6}{n}$ $n = 15$

PRISM MATHEMATICS
Purple Book

FINAL TEST   Chapters 1-13 (continued)

Complete each table. Write each fraction in simplest form.

8.

a

fraction	decimal	percent
$\frac{1}{2}$	0.5	50%

b

fraction	decimal	percent
$\frac{3}{4}$	0.75	75%

Complete the following.

	a	b	c
9.	__20__ is 25% of 80	105 is 75% of __140__	66 is __24__ % of 275
10.	6 m = __600__ cm	8 km = __8000__ m	100 L = __0.1__ kL
11.	100 mg = __0.1__ g	2.3 kg = __2300__ g	5600 mL = __5.6__ L
12.	60 cm = __60 000__ m	3 L = __3000__ mL	4 h 15 min = __255__ min
13.	2 t = __2000__ kg	2000 g = __2__ kg	3 kL = __3000__ L

Round as indicated.

		nearest thousand	nearest hundred	nearest ten
14.	32 546	33 000	32 500	32 550

Write an estimate for each exercise. Then find the answer.

15. 2 3 4 2   4 0 4 1 2   9 2 3
   +9 2 3   3000   − 9 3 4 2   30 000   × 3 1   27 000
   ─────   ───────   ─────
    3265   31 070   28 613

16. Water freezes at __0__ °C.

On the _____ before each name below, write the letter of the figure it describes.

17. __d__ line segment

18. __f__ circle

19. __e__ acute angle

20. __a__ right triangle

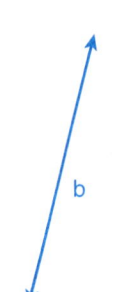

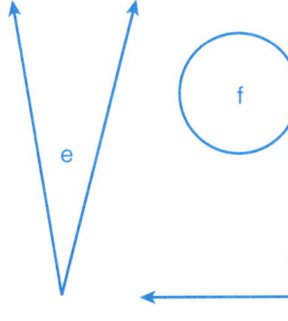

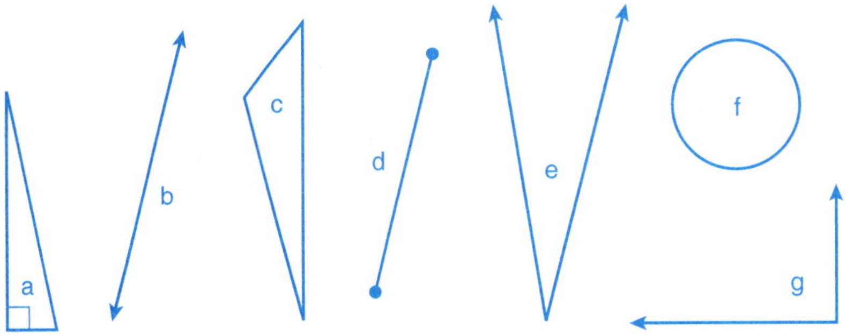

PRISM MATHEMATICS
Purple Book

FINAL TEST Chapters 1-13 (continued)

Use the similar triangles below to help you complete the following.

21. If $a = 16$, $b = 8$, and $d = 12$, then $e = \underline{\quad 6 \quad}$.

22. If $b = 3$, $c = 5$, and $e = 9$, then $f = \underline{\quad 15 \quad}$.

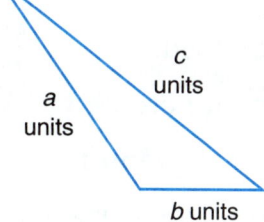

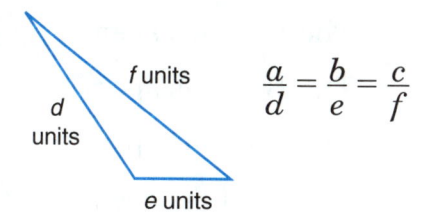

$$\frac{a}{d} = \frac{b}{e} = \frac{c}{f}$$

Find the area of each figure below. Use 3.14 for π.

23.

a. $\underline{\quad 18.75 \quad}$ cm²

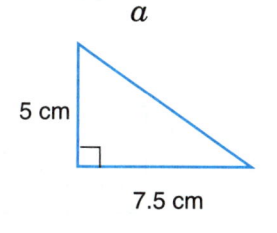

b. about $\underline{\quad 200.96 \quad}$ m²

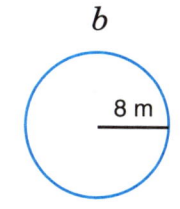

c. $\underline{\quad 68.87 \quad}$ m²

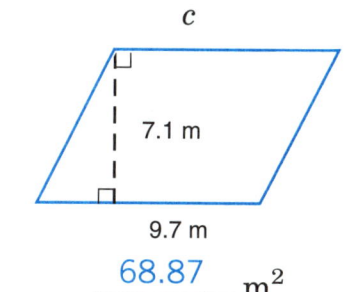

Find the surface area of each figure below. Use 3.14 for π.

24.

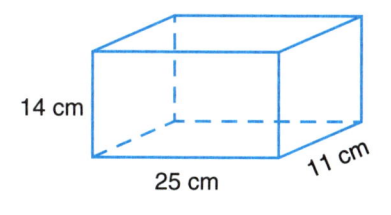

$\underline{\quad 1558 \quad}$ cm²

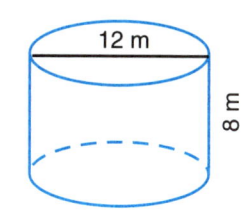

about $\underline{\quad 527.52 \quad}$ m²

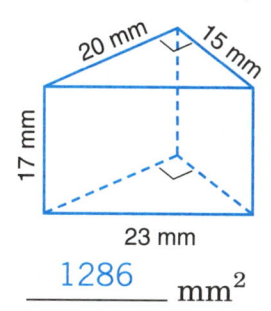

$\underline{\quad 1286 \quad}$ mm²

Find the volume of each figure below. Use 3.14 for π.

25.

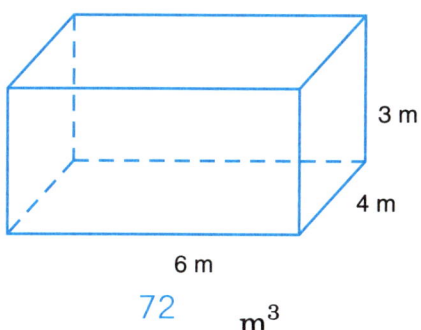

$\underline{\quad 72 \quad}$ m³

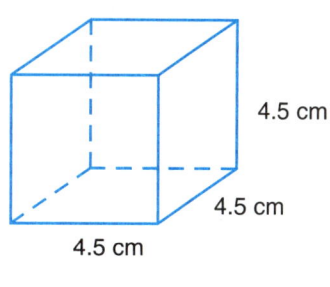

$\underline{\quad 91.125 \quad}$ cm³

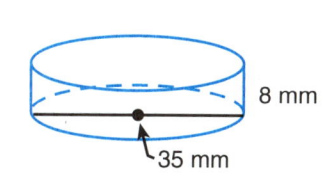

about $\underline{\quad 7700 \quad}$ mm³

FINAL TEST Chapters 1–13 (continued)

Use the circle graph to complete each sentence.

The Bentons' Vacation Expenses
- Hotel 36%
- Transportation 16%
- Food 29%
- Entertainment 12%
- Souvenirs 7%

26. The Bentons spent $2500 on vacation.

The Bentons spent $ __725__ on food.

The Bentons spent $ __900__ on a hotel.

The Bentons spent $ __175__ on souvenirs.

The Bentons spent $ __400__ on transportation.

The Bentons spent $ __300__ on entertainment.

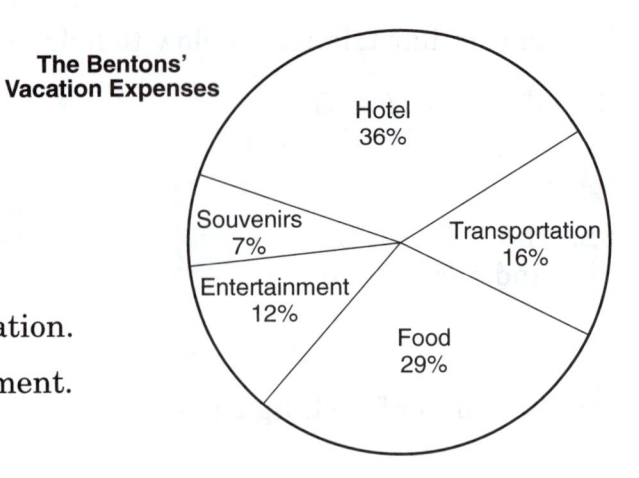

Use the cards at the right to solve each problem.

27. You draw one of the cards without looking. What is the probability of drawing a card that

has stripes? __$\frac{1}{3}$__

has dots? __$\frac{2}{3}$__

(Cards numbered 1–6: 1 dots, 2 dots, 3 stripes, 4 dots, 5 stripes, 6 dots)

28. You make 30 draws, replacing the card after each draw. Predict how many times you will draw each card that

is a multiple of 3. __10__

has dots and an odd number. __5__

Solve each problem.

29. A company made 40 000 batteries. It knows 2% are defective. How many defective batteries did the company make?

The company made __800__ defective batteries.

30. Jami borrowed $4000 for one year at 15% annual interest. How much interest did she pay?

Jami paid $ __600__ in interest.

31. Aliya is 2 years older than D'Marco. If you add their ages together, the sum is 40. How old is Aliya?

Equation: __x + x + 2 = 40__

Aliya is __21__ years old.

STOP

PRISM MATHEMATICS
Purple Book

212

FINAL TEST
Chapters 1–13

CHAPTER 1 CUMULATIVE REVIEW

Do each problem.
Find the correct answer.
Mark the space for the answer.

Part 1 Concepts

1. In which of these problems can the dividend be divided evenly by the divisor?
   - A 134 ÷ 8
   - ✓ B 135 ÷ 9
   - C 127 ÷ 7
   - D 136 ÷ 6

2. What number should replace the □ to make the following a true statement?

   10 462 = □ + 3219 + 4763
   - A 3490
   - B 7982
   - C 2472
   - ✓ D 2480

3. What difference is more than 4500 and less than 4700?
   - A 41 366 − 37 086
   - ✓ B 42 230 − 37 605
   - C 69 638 − 64 859
   - D 37 579 − 33 097

4. What digit is in the ten thousands place in the product 4579 × 746?
   - ✓ A 1
   - B 9
   - C 5
   - D 4

Part 2 Computation

5. 30 426
     891
   +6 055
   - ✓ A 37 372
   - B 38 662
   - C 38 364
   - D 39 462

6. 94 028
    −659
   - A 94 368
   - ✓ B 93 369
   - C 94 479
   - D 93 479

7. 7132
    ×101
   - A 737 766
   - B 720 433
   - C 719 423
   - ✓ D 720 332

8. 22)894
   - ✓ A 40 r14
   - B 41 r6
   - C 40 r21
   - D 39 r17

ANSWER ROW
1 Ⓐ ●Ⓒ Ⓓ   3 Ⓐ ●Ⓒ Ⓓ   5 ●Ⓑ Ⓒ Ⓓ   7 Ⓐ Ⓑ Ⓒ ●
2 Ⓐ Ⓑ Ⓒ ●   4 ●Ⓑ Ⓒ Ⓓ   6 Ⓐ ●Ⓒ Ⓓ   8 ●Ⓑ Ⓒ Ⓓ

PRISM MATHEMATICS
Purple Book

CHAPTER 1
CUMULATIVE REVIEW

CHAPTER 1 CUMULATIVE REVIEW (continued)

9. 16.937
    +4.584

    A 20.411
    B 22.612
    ✓ C 21.521
    D 23.481

10. 4.9018
    −2.9765

    A 2.0353
    ✓ B 1.9253
    C 1.8342
    D 2.1453

11. 52.62
    ×0.33

    A 18.0556
    B 17.4678
    C 16.8396
    ✓ D 17.3646

12. 2.4)8.832

    A 3.24
    ✓ B 3.68
    C 3.96
    D 4.12

Part 3 Applications

13. There are 56 943 people who live in the city where Candace works, and there are 39 017 people who live in the city where Candace was born. How many more people live in the city where Candace works than in the city where she was born?

    A 16 336
    B 16 535
    ✓ C 17 926
    D 18 016

14. If Daniel delivered 455 phone books each day for 4 days, how many phone books did he deliver altogether?

    ✓ A 1820   C 1455
    B 1660   D 1290

15. While testing tires at the R.J. Rubber Company, a new tire was rotated 1612 km for 26 hours. How fast was the tire rotating?

    A 55 km/h   C 60 km/h
    B 58 km/h   ✓ D 62 km/h

16. The thickness of Isaiah's piece of bread is 2.065 cm. The thickness of his stack of turkey is 1.158 cm. How thick will Isaiah's sandwich be when he combines the turkey with two pieces of bread?

    A 4.963 cm   C 4.399 cm
    ✓ B 5.288 cm   D 5.636 cm

17. Bianca and Lydia are counting the kilojoules (kJ) in what they eat. For dinner, they bought two roast beef subs with 2616 kJ each and two garden salads with 491 kJ each. How many calories were in the entire dinner?

    ✓ A 6214   C 6652
    B 7657   D 5439

18. Each jar of spaghetti sauce costs $2.55. How many jars can you buy with $20.40?

    ✓ A 8   C 10
    B 9   D 11

CHAPTER 2 CUMULATIVE REVIEW

Do each problem.
Find the correct answer.
Mark the space for the answer.

Part 1 Concepts

1. What is the least number that is evenly divisible by 16 and 24?
   - A 384
   - ✓ B 48
   - C 64
   - D 8

2. Which of these names the greatest number?
   - ✓ A two and four tenths
   - B nine tenths
   - C one and nine tenths
   - D two and nine hundredths

3. What number is in the numerator in the quotient $1\frac{6}{7} \div 2\frac{3}{8}$?
   - A 19
   - B 56
   - ✓ C 104
   - D 133

4. What is the reciprocal of $\frac{4}{9}$?
   - A $\frac{1}{9}$
   - ✓ B $\frac{9}{4}$
   - C $\frac{5}{9}$
   - D $\frac{1}{4}$

Part 2 Computation

5. 89 621
   48 306
   +5 774
   - A 142 810
   - B 152 741
   - C 140 355
   - ✓ D 143 701

6. 342
   ×56
   - A 18 810
   - ✓ B 19 152
   - C 19 096
   - D 23 892

7. 28.651
   4.009
   +17.560
   - ✓ A 50.220
   - B 48.210
   - C 51.020
   - D 49.110

8. 62.381
   −7.947
   - A 55.524
   - ✓ B 54.434
   - C 52.863
   - D 55.444

CHAPTER 2   CUMULATIVE REVIEW (continued)

9. 19.061
    ×24

    A 428.354
    B 487.554
    ✓C 457.464
    D 381.224

10. 1.6)39.536

    A 25.34
    B 21.62
    C 28.95
    ✓D 24.71

11. $1\frac{4}{5}$
    $-\frac{2}{3}$

    ✓A $1\frac{2}{15}$
    B $1\frac{4}{15}$
    C $1\frac{1}{5}$
    D $\frac{11}{15}$

12. $2\frac{1}{6} \times 3\frac{2}{5}$

    A $6\frac{1}{15}$
    B $6\frac{1}{10}$
    ✓C $7\frac{11}{30}$
    D $7\frac{7}{30}$

Part 3 Applications

13. Desiree's car averages 6.9 km per litre of gasoline in the city. How many miles would she be able to travel on 33.1 L of gasoline?

    ✓A 228.39   C 220.39
    B 228.28   D 220.29

14. At the stadium, there are 84 243 seats. At Tuesday night's game, 59 567 seats were filled. How many seats were empty?

    A 22 636   C 26 215
    ✓B 24 676   D 27 897

15. If you were to travel from Ottawa, to Berlin, Germany, you would travel 7050 km. From Berlin to Hong Kong, it is 10 100 km. How many kilometres would you travel from Ottawa to Hong Kong, by way of Berlin?

    A 16 690   ✓C 17 150
    B 16 080   D 14 650

16. On Saturday, Jasmine spent $1\frac{2}{3}$ h driving her son to and from a soccer game, $\frac{3}{5}$ h driving her dog to the veterinarian, and $1\frac{2}{15}$ h driving her daughter to and from a softball game. How long did Jasmine drive on Saturday?

    A 3 h   C $2\frac{11}{15}$ h
    B $4\frac{1}{5}$ h   ✓D $3\frac{2}{5}$ h

17. The boys' record for the high jump at Piston High School is 1.95 m. The girls' record is 1.61 m. How much more distance do the girls need to tie the boys' record?

    A 0.26 m   C 0.53 m
    ✓B 0.34 m   D 0.56 m

18. Hal's football practice runs $2\frac{1}{4}$ h for 4 days a week. How many hours does Hal practise each week?

    A $8\frac{1}{4}$ ✓C 9
    B $8\frac{1}{2}$ D $9\frac{3}{4}$

CHAPTER 3 CUMULATIVE REVIEW

Do each problem.
Find the correct answer.
Mark the space for the answer.

Part 1 Concepts

1. Which of the following products has a 7 in the thousands place?
   - A 417 × 335
   - B 908 × 47
   - ✓ C 387 × 149
   - D 216 × 816

2. Which of these is less than $\frac{4}{9}$?
   - ✓ A $\frac{1}{3}$
   - B $\frac{1}{2}$
   - C $\frac{3}{4}$
   - D $\frac{5}{8}$

3. Which of the following expressions represents the phrase "18 more than a number"?
   - A $18 \times n$
   - B $18 - n$
   - C $n \div 18$
   - ✓ D $n + 18$

4. Which equation has a solution of 23?
   - ✓ A $40 = x + 17$
   - B $154 = 7x$
   - C $x - 9 = 16$
   - D $x + 36 = 55$

Part 2 Computation

5. 68 542
   −59 763
   - ✓ A 8779
   - B 19 889
   - C 10 078
   - D 9869

6. $314.98
   +78.28
   - A $381.37
   - B $376.86
   - ✓ C $393.26
   - D $382.16

7. 146.3
   ×17.8
   - A 2319.64
   - ✓ B 2604.14
   - C 2163.76
   - D 2111.24

8. $4\frac{1}{6}$
   $3\frac{3}{8}$
   $+5\frac{1}{2}$
   - A $13\frac{5}{6}$
   - B $12\frac{1}{8}$
   - C $14\frac{5}{24}$
   - ✓ D $13\frac{1}{24}$

PRISM MATHEMATICS
Purple Book

CHAPTER 3
CUMULATIVE REVIEW

217

CHAPTER 3 CUMULATIVE REVIEW (continued)

9. $2\frac{3}{4} \times 4\frac{1}{4}$

   A $8\frac{3}{16}$
  ✓B $11\frac{11}{16}$
  C $9\frac{1}{2}$
  D $10\frac{5}{16}$

10. $4\frac{1}{5} \div 3\frac{2}{3}$

  A $1\frac{3}{10}$
  B $2\frac{1}{15}$
  ✓C $1\frac{8}{55}$
  D $1\frac{2}{15}$

11. $\frac{z}{6} = 7$

  ✓A $z = 42$
  B $z = 49$
  C $z = 54$
  D $z = 48$

12. $250 - 70 = x + 100$

  A $x = 70$
  ✓B $x = 80$
  C $x = 180$
  D $x = 95$

Part 3 Applications

13. When Lonnie won the lottery, he decided to divide his $50 000 evenly among eight charities. How much money will each charity receive?

  A $8175
  B $7325
  C $5950
  ✓D $6250

14. In the previous problem, if Lonnie had decided to buy a new car for $27 487.95, how much money would he have left from his winnings?

  A $33 623.15
  ✓B $22 512.05
  C $26 534.05
  D $20 423.05

15. The garden centre parking lot holds 325 cars. On Saturday evening, the lot was $\frac{3}{5}$ full. How many cars were in the parking lot?

  ✓A 195
  B 210
  C 225
  D 240

16. Celeste lost 7 kg in $3\frac{1}{2}$ months. How many kilograms did she lose per month?

  A 9 kg
  ✓B 2 kg
  C 10 kg
  D 5 kg

17. Josh and Logan are filling plastic eggs with candy pieces to give out at school. If Josh fills 110 of the 225 eggs, how many eggs will Logan need to fill?

  A 95
  B 100
  C 105
  ✓D 115

18. Twelve rooms were reserved yesterday at the Townsend Hotel. If that is $\frac{1}{6}$ of all the rooms, how many rooms are there at the hotel?

  A 65
  B 68
  ✓C 72
  D 76

CHAPTER 4 CUMULATIVE REVIEW

Do each problem.
Find the correct answer.
Mark the space for the answer.

Part 1 Concepts

1. What number is 10 000 more than 23 971 056?

    A 23 961 056
    B 23 972 056
    ✓ C 23 981 056
    D 24 081 056

2. What digit is in the hundredths place in the quotient $0.034\overline{)0.21012}$?

    A 1
    B 6
    C 3
    ✓ D 8

3. What is the value of d in the following equation?

    $7d + 2d + 13 = 85$

    ✓ A 8
    B 9
    C 10.3
    D 72

4. If $a = 7$, then $3a - 6 = $ ____?

    A 4
    B 21
    ✓ C 15
    D 18

Part 2 Computation

5. $27\overline{)32\,481}$

    ✓ A 1203
    B 1217 r18
    C 1226
    D 1196 r6

6. 10.748
   -0.967

    A 10.881
    B 9.671
    ✓ C 9.781
    D 11.715

7. $8\frac{3}{10} - 6\frac{2}{5}$

    A $2\frac{1}{5}$
    B $2\frac{1}{10}$
    ✓ C $1\frac{9}{10}$
    D $1\frac{2}{5}$

8. $4\frac{1}{9} \times 2\frac{1}{2}$

    A $8\frac{1}{9}$
    ✓ B $10\frac{5}{18}$
    C $8\frac{1}{18}$
    D $9\frac{7}{18}$

CHAPTER 4   CUMULATIVE REVIEW (continued)

9. $15m = 255$
   - A $m = 14$
   - B $m = 15$
   - C $m = 16$
   - ✓ D $m = 17$

10. $\dfrac{x}{3} = 65 + 25$
    - A $x = 90$
    - ✓ B $x = 270$
    - C $x = 30$
    - D $x = 180$

11. $9e - 3e =$
    - ✓ A $6e$
    - B $27e$
    - C 6
    - D $11e$

12. $2s + s + 16 = 49$
    - A $s = 6$
    - B $s = 9$
    - ✓ C $s = 11$
    - D $s = 14$

Part 3 Applications

13. If each ticket in a bundle of 1800 tickets costs 65¢, what is the cost of the bundle?
    - A $1075
    - B $1250
    - ✓ C $1170
    - D $1105

14. Ricky spent $2\frac{1}{3}$ h doing homework on Monday, $1\frac{3}{4}$ h on Tuesday, and $2\frac{1}{6}$ h on Wednesday. How long did he spend doing homework in those three days?
    - ✓ A $6\frac{1}{4}$ h
    - B $5\frac{5}{12}$ h
    - C $5\frac{1}{2}$ h
    - D $6\frac{2}{3}$ h

15. An electrician has just finished $\frac{6}{7}$ of a 77-h job. How many hours has he finished?
    - A 55
    - B 62
    - C 59
    - ✓ D 66

16. Katrina has earned 52 stars in school this year for good behaviour. She has earned four times as many as Brooklyn. How many stars has Brooklyn earned?
    - A 11
    - ✓ B 13
    - C 208
    - D 18

17. William is twice as old as his son, Elijah. Their combined age is 78. How old is Elijah?
    - A 20
    - B 23
    - ✓ C 26
    - D 30

18. In the previous problem, how old is William?
    - A 60
    - ✓ B 52
    - C 46
    - D 40

CHAPTER 5 CUMULATIVE REVIEW

Do each problem.
Find the correct answer.
Mark the space for the answer.

Part 1 Concepts

1. Which of the following fractions is not equivalent to $\frac{3}{7}$?

   A $\frac{9}{21}$

   ✓ B $\frac{12}{27}$

   C $\frac{6}{14}$

   D $\frac{15}{35}$

2. $1.05 =$

   A $\frac{10}{15}$

   B $\frac{15}{10}$

   C $\frac{100}{105}$

   ✓ D $\frac{105}{100}$

3. Which of the following phrases represents the expression $n \div 3$?

   A three divided by a number
   B three less than a number
   ✓ C a number divided by three
   D a number increased by three

4. Which of the following decimals is greater than $3\frac{2}{7}$?

   ✓ A 3.35
   B 3.25
   C 2.7
   D 3.099

Part 2 Computation

5. 29 601
       357
    +8 273

   A 39 241
   B 41 321
   C 36 142
   ✓ D 38 231

6. 1342.05
    ×34.6

   A 47 851.341
   B 49 342.821
   ✓ C 46 434.930
   D 44 562.620

7. $5\frac{1}{3} \div 2\frac{1}{6}$

   A $2\frac{1}{6}$
   ✓ B $2\frac{6}{13}$
   C $2\frac{5}{6}$
   D $3\frac{1}{11}$

8. $8 \times 5 = \frac{r}{4}$

   ✓ A $r = 160$
   B $r = 175$
   C $r = 40$
   D $r = 205$

CHAPTER 5 CUMULATIVE REVIEW (continued)

9. $4x + x + 22 = 62$
   - A $x = 20$
   - B $x = 5$
   - C $x = 200$
   - ✓ D $x = 8$

10. $t + (2t + 7) = 16$
    - A $t = 4$
    - B $t = 6$
    - ✓ C $t = 3$
    - D $t = 5$

11. $\frac{2}{5} = \frac{v}{125}$
    - ✓ A $v = 50$
    - B $v = 75$
    - C $v = 25$
    - D $v = 100$

12. 35 is 20% of ___.
    - A 190
    - ✓ B 175
    - C 168
    - D 160

Part 3 Applications

13. Joaquin needed four new tires for his car. The tires cost $86.97 each. How much did Joaquin pay for his tires?
    - ✓ A $347.88
    - B $386.75
    - C $456.98
    - D $412.08

14. Angela's trip was 605 km and took 11 h. She drove the same distance each hour. What was her average travelling speed?
    - ✓ A 55 km/h
    - B 50 km/h
    - C 61 km/h
    - D 64 km/h

15. Last Year, Renee could run $3\frac{2}{3}$ laps around the track in 10 min. This year she can run $4\frac{7}{9}$ laps. How many more laps can she run this year than last year?
    - A $1\frac{1}{3}$
    - ✓ B $1\frac{1}{9}$
    - C $1\frac{1}{2}$
    - D $1\frac{8}{9}$

16. Derek has noticed that the temperature has risen 6 degrees since he woke up this morning. The thermometer now reads 27°C. What was the temperature when Derek woke up?
    - A 33°C
    - B 29°C
    - C 24°C
    - ✓ D 21°C

17. The goalie for the red hockey team had 12 saves out of 15 shots last week. At that rate, how many saves will she have if there are 25 shots?
    - A 18
    - B 19
    - ✓ C 20
    - D 21

18. The grade 4s at Current Elementary School raised $3250 for new playground equipment. They needed $195 of that total to pay for expenses. What percent of the total was used for expenses?
    - A 4%
    - ✓ B 6%
    - C 7%
    - D 9%

CHAPTER 6 CUMULATIVE REVIEW

Do each problem.
Find the correct answer.
Mark the space for the answer.

Part 1 Concepts

1. What number should replace the □ to make the following a true statement?

   54 709 = □ + 34 855 + 1648

   A 18 200
   B 36 503
   C 33 207
   ✓ D 18 206

2. What number is 500 000 less than 362 745 309?

   ✓ A 362 245 309
   B 362 695 309
   C 357 745 309
   D 362 145 309

3. How would you write 37% as a fraction?

   A $\frac{3}{7}$
   ✓ B $\frac{37}{100}$
   C $\frac{.37}{100}$
   D $\frac{37}{100}$

4. Which formula can you use to find the interest on a loan?

   A $i = \frac{p \times r}{t}$
   B $i = p + r + t$
   ✓ C $i = p \times r \times t$
   D $i = \frac{r \times t}{p}$

Part 2 Computation

5. 52 013
    −9 874

    A 61 887
    B 53 249
    ✓ C 42 139
    D 47 658

6. $3\frac{1}{9}$
    $4\frac{2}{3}$
    $+5\frac{7}{18}$

    A $13\frac{5}{6}$
    ✓ B $13\frac{1}{6}$
    C $12\frac{11}{18}$
    D $14\frac{2}{9}$

7. $33 = \frac{f}{9}$

   ✓ A $f = 297$
   B $f = 3.667$
   C $f = 231$
   D $f = 363$

8. $2x + 4x + 6 = 42$

   A $x = 4$
   ✓ B $x = 6$
   C $x = 8$
   D $x = 10$

ANSWER ROW
1 Ⓐ Ⓑ Ⓒ ● 　3 Ⓐ ● Ⓒ Ⓓ 　5 Ⓐ Ⓑ ● Ⓓ 　7 ● Ⓑ Ⓒ Ⓓ
2 ● Ⓑ Ⓒ Ⓓ 　4 Ⓐ Ⓑ ● Ⓓ 　6 Ⓐ ● Ⓒ Ⓓ 　8 Ⓐ ● Ⓒ Ⓓ

PRISM MATHEMATICS
Purple Book

CHAPTER 6
CUMULATIVE REVIEW

CHAPTER 6 CUMULATIVE REVIEW (continued)

9. ___ is 41% of 600.
   - A 258
   - B 14.63
   - C 683
   - ✓ D 246

10. $\frac{1}{8}$ is ___% of $1\frac{1}{4}$.
    - A 5
    - B 20
    - ✓ C 10
    - D 18

11. What is the rate on a loan if interest = $72, principal = $200, and time = 2 years?
    - A 26%
    - B 31%
    - C 22%
    - ✓ D 18%

12. What is the time on a loan if interest = $3780, principal = $21 000, and rate = 9%?
    - A 1 year
    - ✓ B 2 years
    - C $1\frac{1}{2}$ years
    - D 3 years

Part 3 Applications

13. Aliana found a wall oven she would like to buy for $365.95. The clerk told her it had a small dent in the back of it, so they would take $18.30 off of the price. How much did Aliana pay for the oven?
    - ✓ A $347.65
    - B $317.55
    - C $321.75
    - D $357.65

14. A recipe to make bread requires $1\frac{1}{4}$ packets of yeast. Sandra is making $2\frac{1}{2}$ times the amount of bread that the recipe calls for. How much yeast will she need?
    - A $2\frac{3}{4}$ packets
    - C $2\frac{7}{8}$ packets
    - B $3\frac{1}{2}$ packets
    - ✓ D $3\frac{1}{8}$ packets

15. Last month, Geraldine delivered three times as many newspapers as Ralph. They delivered a total of 2088 newspapers. How many newspapers did Ralph deliver?
    - A 436
    - ✓ C 522
    - B 481
    - D 579

16. When Elsie and Trish were 5 years old, Elsie's mass was 5 kg more than Trish. Their combined mass was 49 kg. What was Elsie's mass?
    - A 21 kg
    - C 24 kg
    - B 22 kg
    - ✓ D 27 kg

17. Keenan saved and deposited $10\frac{1}{2}$% of $1200 he earned mowing lawns. How much did he deposit?
    - ✓ A $126
    - C $198
    - B $160
    - D $210

18. At an annual interest rate of $6\frac{1}{4}$%, Luigi earned $56 in 2 years. How much did he deposit to do this?
    - A $422
    - C $506
    - ✓ B $448
    - D $524

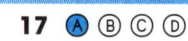

CHAPTER 7   CUMULATIVE REVIEW

Do each problem.
Find the correct answer.
Mark the space for the answer.

Part 1 Concepts

1. Which of these is another name for six and fifteen thousandths?

   A 6.15
   B 6.000 15
   C 6015
   ✓ D 6.015

2. What is the value of n in the following proportion?

   $\frac{27}{n} = \frac{108}{136}$

   ✓ A 34
   B 21
   C 31
   D 544

3. $17b + 5b - 9b =$

   A $22b$
   B $8b$
   ✓ C $13b$
   D $31b$

4. What operation do you perform to convert grams to milligrams?

   A multiply by 100
   B divide by 100
   ✓ C multiply by 1000
   D divide by 1000

Part 2 Computation

5. $3.6 \overline{)432.036}$

   A 104.31
   B 152.10
   ✓ C 120.01
   D 134.11

6. $2\frac{1}{5} \times 3\frac{1}{3}$

   A $6\frac{1}{15}$
   ✓ B $7\frac{1}{3}$
   C $6\frac{1}{5}$
   D $7\frac{2}{15}$

7. $82 - 16 = 11c$

   A $c = 10$
   B $c = 8$
   C $c = 5$
   ✓ D $c = 6$

8. $x + 3x + 7 = 79$

   ✓ A $x = 18$
   B $x = 21$
   C $x = 19$
   D $x = 15$

CHAPTER 7 CUMULATIVE REVIEW (continued)

9. $5\frac{3}{4}\% =$
   - A 5.75
   - B 0.575
   - ✓ C 0.0575
   - D 0.00575

10. What is the interest on a loan if principal = $700, rate = 6%, and time = 2 years?
    - A $42
    - B $79
    - C $82
    - ✓ D $84

11. In problem 10, if interest was compounded annually, what is the total amount at the end of two years?
    - A $784.00
    - ✓ B $786.52
    - C $742.00
    - D $744.52

12. 0.639 kL =
    - ✓ A 639 L
    - B 6.39 L
    - C 63.9 L
    - D 0.0639 L

Part 3 Applications

13. Keesha borrowed $1200 for one year. Interest on the first $500 of the loan was 12%. It was 9% interest on the rest. How much interest did she pay?
    - A $188
    - B $154
    - ✓ C $123
    - D $96

14. In a certain village, $\frac{3}{5}$ of the men are married. If 85 men live in the village, how many are married?
    - A 39
    - B 46
    - ✓ C 51
    - D 60

15. The Davidson triplets and the Johnson twins were absent on Thursday. They make up $\frac{1}{5}$ of the entire class. How many students are in the class?
    - A 20
    - ✓ B 25
    - C 30
    - D 35

16. At Rosemont Junior High School, 1 out of every 6 children rides their bikes to and from school. There are 438 children who attend Rosemont. How many ride their bikes?
    - ✓ A 73
    - B 58
    - C 102
    - D 82

17. The president of a large corporation spent $1071.00 on baseball tickets for his employees. If each ticket costs $12.75, how many employees does he have?
    - A 75
    - ✓ B 84
    - C 91
    - D 106

18. Yasmine's father said she could not go swimming until the temperature reached 25°C. Yasmine read the thermometer and it said 23°C. How many degrees does the temperature have to rise for her to be allowed to go swimming?
    - A 5°C
    - B 4°C
    - C 3°C
    - ✓ D 2°C

CHAPTER 8 CUMULATIVE REVIEW

Do each problem.
Find the correct answer.
Mark the space for the answer.

Part 1 Concepts

1. What difference is more than 11 000 and less than 11 500?

    A 47 541 − 35 891
    B 33 564 − 22 619
    ✓C 75 930 − 64 580
    D 59 797 − 48 267

2. If $y = 11$, then $4y - 22 = $ ___?

    A 44
    ✓B 22
    C 20
    D 37

3. How many complete minutes are in 675 s?

    A 13
    B 12
    ✓C 11
    D 22

4. Which of the following is a measurement of mass?

    A grams
    B kilolitres
    ✓C Celsius
    D decimetres

Part 2 Computation

5. 3650.2×16

    A 54 753.2
    ✓B 58 403.2
    C 58 563.2
    D 58 404.8

6. $3\frac{5}{6} - 2\frac{1}{3}$

    A $\frac{5}{6}$
    B $1\frac{2}{3}$
    C $1\frac{1}{6}$
    ✓D $1\frac{1}{2}$

7. $\frac{x}{110} = 18 - 13$

    A $x = 440$
    B $x = 330$
    ✓C $x = 550$
    D $x = 660$

8. $f + 5f = 126$

    A $f = 18$
    ✓B $f = 21$
    C $f = 23$
    D $f = 27$

CHAPTER 8   CUMULATIVE REVIEW (continued)

9. 124 is 25% of ____.
   ✓ A 496
   B 330
   C 422
   D 478

10. What is the principal on a loan if interest = $312, rate = 13%, and time = 3 years?
    A $1200
    B $1400
    C $600
    ✓ D $800

11. 14.5 g =
    ✓ A 14 500 mg
    B 1450 mg
    C 145 mg
    D 1.45 mg

12. 2 h 20 min
       ×4
    ─────────
    A 8 h 40 min
    B 8 h 20 min
    ✓ C 9 h 20 min
    D 9 h 40 min

Part 3 Applications

13. Keith ran $16\frac{3}{4}$ laps around the track. Harold ran $14\frac{5}{6}$ laps. How many more laps did Keith run than Harold?
    A $1\frac{1}{6}$
    C $2\frac{1}{4}$
    B $1\frac{7}{12}$
    ✓ D $1\frac{11}{12}$

14. After their dart game, Jason had 48 points which was $\frac{1}{4}$ the number of points that Derek had. How many points did Derek have?
    A 164
    ✓ B 192
    C 12
    D 238

15. Marie and her friends are watching the city pool being filled. The pool holds 162 000 L. If the pool is 75% full, how many gallons are in it?
    A 113 400 L
    ✓ C 121 500 L
    B 129 600 L
    D 137 700 L

16. Ricardo deposited $400 in a two-year account at 8%. If this account were compounded annually, how much interest would be in the account after two years?
    ✓ A $66.56
    C $32.00
    B $64.00
    D $34.56

17. A recipe for chocolate fudge calls for 60 g of cocoa powder. How many milligrams would that be?
    A 20 000 mg
    C 40 000 mg
    ✓ B 60 000 mg
    D 30 000 mg

18. It takes 20 h 30 min to drive from Halifax to Toronto. It takes 10 h 45 min to drive from Halifax to Quebec City. How long does it take to drive from Quebec City to Toronto?
    A 10 h 45 min
    C 30 h 15 min
    ✓ B 31 h 15 min
    D 9 h 45 min

STOP

CHAPTER 9 CUMULATIVE REVIEW

Do each problem.
Find the correct answer.
Mark the space for the answer.

Part 1 Concepts

1. What is the least number that is evenly divisible by 12 and 32?

   A 64
   ✓B 96
   C 128
   D 192

2. What digit is in the thousandths place in the product 1.0678×3.14?

   ✓A 2
   B 7
   C 5
   D 8

3. What operation do you perform to convert litres to kilolitres?

   A multiply by 1000
   ✓B divide by 1000
   C multiply by 100
   D divide by 10

4. What type of triangle has angle measures 98°, 46°, and 36°?

   A acute
   B right
   C isosceles
   ✓D obtuse

Part 2 Computation

5. $\begin{array}{r} 740.12 \\ 3.48 \\ +96.75 \\ \hline \end{array}$

   A 728.35
   B 739.25
   C 839.45
   ✓D 840.35

6. $8\frac{1}{3} \div 2\frac{1}{4}$

   A $4\frac{1}{3}$
   ✓B $3\frac{19}{27}$
   C $4\frac{3}{4}$
   D $3\frac{1}{12}$

7. $a + a + 17 = 55$

   ✓A $a = 19$
   B $a = 22$
   C $a = 18$
   D $a = 16$

8. $\frac{m}{7} = \frac{72}{84}$

   A $m = 8.16$
   B $m = 4$
   C $m = 864$
   ✓D $m = 6$

ANSWER ROW   1 Ⓐ ●　Ⓒ Ⓓ   3 Ⓐ ● Ⓒ Ⓓ   5 Ⓐ Ⓑ Ⓒ ●   7 ● Ⓑ Ⓒ Ⓓ
             2 ● Ⓑ Ⓒ Ⓓ   4 Ⓐ Ⓑ Ⓒ ●   6 Ⓐ ● Ⓒ Ⓓ   8 Ⓐ Ⓑ Ⓒ ●

PRISM MATHEMATICS
Purple Book

CHAPTER 9   CUMULATIVE REVIEW (continued)

9. What is the time of a loan if interest = $162, principal = $900, and rate = 8%?

   A 3 years
   B $2\frac{3}{4}$ years
   ✓ C $2\frac{1}{4}$ years
   D $1\frac{3}{4}$ years

14. During the $3\frac{1}{4}$ h soccer camp, the players worked on dribbling, passing, and shooting activities. How much time was spent on each activity, if the time was divided evenly?

   ✓ A $1\frac{1}{12}$ h   C $1\frac{1}{2}$ h
   B $\frac{11}{12}$ h   D $1\frac{1}{4}$ h

10. 13 dm =

   A 0.13 m
   ✓ B 1.3 m
   C 130 m
   D 0.013 m

15. The 14th hole at the Winterwax Golf Course is 190 m long. How long would it be on a model with a scale of 2 cm to 76 m?

   A 4 cm   C 6 cm
   ✓ B 5 cm   D 7 cm

11. 18 kg =

   A 10 800 g
   B 1800 g
   ✓ C 18 000 g
   D 1.8 g

16. Mr. Newman has $1300 in a savings account that pays 6% interest compounded annually. How much will Mr. Newman have in his account after two years?

   A $1378.00   C $1535.21
   B $1498.09   ✓ D $1460.68

12. A(n) _____ triangle never has sides the same length.

   A isosceles
   B equilateral
   ✓ C scalene
   D straight

17. In Warehouse #5, Chandler counted 52 boxes with a mass of 113 kg each. If his boss needs an estimate of the mass for all the boxes, what would Chandler tell him?

   ✓ A 5500 kg   C 550 kg
   B 5650 kg   D 5720 kg

Part 3 Applications

13. Mrs. Sarma made three deposits for the home and school association last month. The first deposit was $325.65, next was $519.43, and the last deposit was $272.18. How much money did Mrs. Sarma deposit all together?

   A $1003.36   ✓ C $1117.26
   B $1096.14   D $1204.15

18. If you put a right angle and an angle that measures 36° together with one common ray, what would be the angle measure they form?

   A 81°   ✓ C 126°
   B 156°   D 111°

CHAPTER 10 CUMULATIVE REVIEW

Do each problem.
Find the correct answer.
Mark the space for the answer.

Part 1 Concepts

1. In which of these problems can the dividend be evenly divided by the divisor?
   - A $3481 \div 19$
   - B $2149 \div 23$
   - C $3318 \div 32$
   - ✓ D $2523 \div 29$

2. What number does x represent to make the equation true?

   $\frac{7}{9} - \frac{x}{9} = \frac{2}{9}$

   - ✓ A 5
   - B 1
   - C 8
   - D 9

3. The prefix 'hecto' means
   - A 0.1
   - B 10
   - ✓ C 100
   - D 1000

4. The value of $\sqrt{27}$ is between _____.
   - A 6 and 7
   - B 20 and 30
   - C 2 and 14
   - ✓ D 5 and 6

Part 2 Computation

5. $\begin{array}{r} \$2978.10 \\ -\ 1989.63 \\ \hline \end{array}$
   - A $1099.57
   - B $1989.66
   - ✓ C $988.47
   - D $899.47

6. $9\frac{4}{5} + 8\frac{5}{6}$
   - ✓ A $18\frac{19}{30}$
   - B $17\frac{2}{15}$
   - C $17\frac{3}{10}$
   - D $18\frac{1}{6}$

7. $d = 310.5 \times 8$
   - A $d = 2412$
   - ✓ B $d = 2484$
   - C $d = 2564$
   - D $d = 2173.5$

8. ____ is 145% of 120.
   - A 177
   - B 170
   - C 168
   - ✓ D 174

CHAPTER 10   CUMULATIVE REVIEW (continued)

9. $112 - 17 = r + 45$

    A $r = 38$
    B $r = 42$
    C $r = 47$
    ✓ D $r = 50$

10. What is the interest on a loan if principal = $400, rate = $9\frac{1}{2}\%$, and time = $3\frac{1}{2}$ years?

    ✓ A $133
    B $126
    C $114
    D $119

11. 3 h 45 min
    +4 h 45 min
    ──────────

    A 7 h 45 min
    ✓ B 8 h 30 min
    C 8 h 15 min
    D 8 h 45 min

12. $21^2 =$

    A 529
    B 42
    C 484
    ✓ D 441

Part 3 Applications

13. Last week, Branham ran $42\frac{1}{2}$ times around the block. The block is 2 km around. How many kilometres did Branham run?

    A 40 km   C 21 km
    ✓ B 85 km   D 2 km

14. Leah answered 34 out of 40 questions correct on her driver's test. What percent of the test did she get right?

    A 81%   C 92%
    B 88%   ✓ D 85%

15. What kind of rate will Mr. Nardiello receive if he'll pay $714 in interest after $3\frac{1}{2}$ years if he borrows $3400?

    A 4%   ✓ C 6%
    B 5%   D 7%

16. For the carnival, eight tables were placed end to end. Each table is 2 m long. What is the total length of the tables?

    A 4 m   C 10 m
    ✓ B 16 m   D 6 m

17. What is the square root of 9?

    ✓ A 3   C 87
    B 6   D 12

18. The fire department ladder leans against a building. The bottom of the ladder rests 5 m from the building, and the top of his ladder rests 12 m up on the building wall. How long is the ladder?

    A 14 m   C 16 m
    B 15 m   ✓ D 13 m

STOP

CHAPTER 11 CUMULATIVE REVIEW

Do each problem.
Find the correct answer.
Mark the space for the answer.

Part 1 Concepts

1. Which of the following expressions represents the phrase "34 decreased by twice a number"?
   - A $34 + 2 + n$
   - ✓ B $34 - 2n$
   - C $2n - 34$
   - D $34 \div 2n$

2. Which of these is greater than $5\frac{3}{8}$?
   - ✓ A $5\frac{6}{11}$
   - B $5\frac{1}{4}$
   - C $5\frac{2}{7}$
   - D $5\frac{7}{20}$

3. Two angles that have the same size are called _____ angles.
   - A isosceles
   - B acute
   - ✓ C congruent
   - D equilateral

4. What is the formula for the area of a triangle?
   - A $A = bh$
   - ✓ B $A = \frac{1}{2}bh$
   - C $A = \frac{1}{2} + b + h$
   - D $A = \pi r^2$

Part 2 Computation

5. $47 \overline{)11\,938}$
   - ✓ A 254
   - B 303 r18
   - C 286
   - D 269 r39

6. $\frac{15}{16} - \frac{3}{4}$
   - A $\frac{1}{4}$
   - B $\frac{7}{16}$
   - ✓ C $\frac{3}{16}$
   - D $\frac{1}{2}$

7. $9 \times 6 = 18p$
   - A $p = 5$
   - B $p = 2$
   - C $p = 4$
   - ✓ D $p = 3$

8. $18\frac{1}{4}\% =$
   - A 0.018 25
   - B 1.825
   - C 18.25
   - ✓ D 0.1825

PRISM MATHEMATICS
Purple Book

CHAPTER 11
CUMULATIVE REVIEW

233

CHAPTER 11   CUMULATIVE REVIEW (continued)

9. 7 min 40 s
   ×3

   A 22 min 20 s
   ✓B 23 min
   C 22 min
   D 21 min 40 s

10. $\sqrt{4356} =$

    ✓A 66
    B 62
    C 55
    D 68

11. What is the perimeter of a square if s = 11 cm?

    A 22 cm
    B 60.5 cm
    ✓C 44 cm
    D 33 cm

12. What is the approximate surface area of a cylinder if height = 3 cm and radius = 2 cm? Use 3.14 for π.

    A 12.56 cm²
    ✓B 62.80 cm²
    C 51.40 cm²
    D 31.40 cm²

Part 3 Applications

13. When Mrs. Jackson was sick, her daughter, Laura, drove back and forth to see her five times in one month. If the distance was 116 km from Laura's house to her mother's, how many kilometres did Laura drive to see her mother in that one month?

    A 955 km
    B 580 km
    ✓C 1160 km
    D 660 km

14. Diedri bought 4.5 kg of mixed nuts. The store clerk said she was guaranteed to find 25% of that mixture to be cashews. At least how many kilogram of cashews did Diedri get?

    ✓A 1.125 kg
    B 1.25 kg
    C 2.125 kg
    D 0.9 kg

15. For how many years did Alfonzo have a loan of $900 at 5% interest if the interest is $45?

    A 4
    B 3
    C 2
    ✓D 1

16. Mr. Omar showed a movie on physics to his science students. The first day, they watched 47 min. The second day, they watched the remaining 73 min. If Mr. Omar showed the movie to four classes, how many total hours did the movie play?

    A 6
    ✓B 8
    C 4
    D 10

17. Myra and her friends left the dock in their sailboat and sailed 4 km west and then 3 km south. How far was the sailboat from the dock then?

    ✓A 5 km
    B 6 km
    C 7 km
    D 8 km

18. Tony is responsible for mowing the vacant lot across the street. The lot is rectangular and measures 155 m by 36 m. How many square metre does Tony mow?

    A 382 m²
    B 2451 m²
    ✓C 5580 m²
    D 7320 m²

CHAPTER 12 CUMULATIVE REVIEW

Do each problem.
Find the correct answer.
Mark the space for the answer.

Part 1 Concepts

1. What number is in the numerator when you write the product of $8\frac{3}{11} \times 6\frac{1}{5}$ as a mixed numeral?
   - A 55
   - B 2821
   - C 51
   - ✓ D 16

2. What type of triangle has angle measures 67°, 58°, and 55°?
   - ✓ A acute
   - B right
   - C isosceles
   - D obtuse

3. Perpendicular lines form _____ angles.
   - A 120°
   - ✓ B 90°
   - C 45°
   - D 60°

4. Which of the following sector measurements represent a whole circle?
   - A 170°, 120°, 60°, 30°
   - B 150°, 90°, 45°, 25°, 20°
   - C 168°, 90°, 72°, 36°
   - ✓ D 130°, 90°, 80°, 60°

Part 2 Computation

5. $3251.06
   −973.48
   - A $2387.58
   - B $3388.68
   - C $2267.77
   - ✓ D $2277.58

6. $8\frac{9}{11} =$
   - ✓ A $\frac{97}{11}$
   - B $\frac{109}{11}$
   - C $\frac{95}{11}$
   - D $\frac{171}{11}$

7. $\frac{2}{x} = \frac{170}{425}$
   - A $x = 7$
   - B $x = 0.8$
   - ✓ C $x = 5$
   - D $x = 2.5$

8. 146.8 m =
   - ✓ A 14 680 cm
   - B 1468 cm
   - C 14.68 cm
   - D 1.468 cm

GO

ANSWER ROW
1 Ⓐ Ⓑ Ⓒ ●   3 Ⓐ ● Ⓒ Ⓓ   5 Ⓐ Ⓑ Ⓒ ●   7 Ⓐ Ⓑ ● Ⓓ
2 ● Ⓑ Ⓒ Ⓓ   4 Ⓐ Ⓑ Ⓒ ●   6 ● Ⓑ Ⓒ Ⓓ   8 ● Ⓑ Ⓒ Ⓓ

PRISM MATHEMATICS
Purple Book

CHAPTER 12
CUMULATIVE REVIEW

235

CHAPTER 12 CUMULATIVE REVIEW (continued)

9. $4\frac{3}{4} \div 2\frac{1}{2}$
   - A $2\frac{1}{4}$
   - ✓ B $1\frac{9}{10}$
   - C $2\frac{3}{8}$
   - D $1\frac{3}{16}$

10. $29^2 =$
    - A 729
    - B 58
    - C 116
    - ✓ D 841

11. What is the approximate area of a circle if the radius = 30 cm? Use 3.14 for π.
    - A 188.4 cm²
    - B 706.5 cm²
    - ✓ C 2826 cm²
    - D 3017.5 cm²

12. 55% of a circular region =
    - A 162°
    - ✓ B 198°
    - C 216°
    - D 234°

Part 3 Applications

13. In three days, Georgina got 18 h of sleep. At this rate, how many h of sleep would she get in 7 days?
    - A 31 h
    - B 36 h
    - C 39 h
    - ✓ D 42 h

14. Mrs. Fay has 22 more papers to mark than Mr. Spencer. Together, they have 216 papers to mark. How many papers does Mrs. Fay have to mark?
    - A 101
    - ✓ B 119
    - C 127
    - D 97

Games Won by the Two Teams (Baseball, Soccer by Month: March, April, May, June)

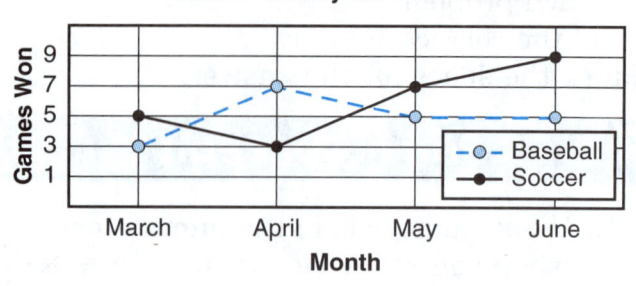

15. In which month did the soccer team win seven games?
    - ✓ A May
    - B June
    - C March
    - D April

16. There are 24 pain reliever tablets in a bottle. Each tablet contains 200 mg of an active ingredient called ibuprofen. How many grams of ibuprofen are there in the bottle?
    - A 320 g
    - B 4800 g
    - ✓ C 4.8 g
    - D 48 g

17. The Nicholsons have an above-ground circular swimming pool in their backyard. The diameter of the pool is 7 m. What is the approximate circumference of the pool? Use 3.14 for π.
    - A about 10.99 m
    - ✓ B about 21.98 m
    - C about 14 m
    - D about 43.96 m

18. Monique is making a circle graph to represent how she spends her day. She spends 35% of a day sleeping. What should be the measure of the sector that represents the sleep?
    - ✓ A 126°
    - B 103°
    - C 97.2°
    - D 75°

CHAPTER 13   CUMULATIVE REVIEW

Do each problem.
Find the correct answer.
Mark the space for the answer.

Part 1 Concepts

1. How much would the value of 825 910 be decreased by if you replaced the 8 with a 7?
   - A 10 000
   - ✓ B 100 000
   - C 1000
   - D 1

2. Which of these is not equal to the others?
   - A 65%
   - B $\frac{65}{100}$
   - ✓ C 0.065
   - D 65 ÷ 100

3. What is the formula for the volume of a cylinder?
   - ✓ A $V = Bh$
   - B $V = 2\pi rh + 2\pi r^2$
   - C $V = \frac{1}{3}Bh$
   - D $V = 2\pi r^2 h$

4. How many possible outcomes are there if you flip a coin and roll a number cube?
   - A 6
   - B 8
   - ✓ C 12
   - D 24

Part 2 Computation

5. $\frac{5}{6} \div 1\frac{2}{3}$
   - A $\frac{1}{3}$
   - B $\frac{1}{6}$
   - C $\frac{2}{9}$
   - ✓ D $\frac{1}{2}$

6. $81 + 19 = \frac{300}{x}$
   - A $x = 2$
   - B $x = 10$
   - ✓ C $x = 3$
   - D $x = 30$

7. $3\frac{3}{4}$ is ___% of 5.
   - A 50
   - B 62
   - C 71
   - ✓ D 75

8. 1.650 km =
   - ✓ A 1650 m
   - B 165 m
   - C 16.5 m
   - D 0.165 m

ANSWER ROW
1 Ⓐ ● Ⓒ Ⓓ   3 ● Ⓑ Ⓒ Ⓓ   5 Ⓐ Ⓑ Ⓒ ●   7 Ⓐ Ⓑ Ⓒ ●
2 Ⓐ Ⓑ ● Ⓓ   4 Ⓐ Ⓑ ● Ⓓ   6 Ⓐ Ⓑ ● Ⓓ   8 ● Ⓑ Ⓒ Ⓓ

PRISM MATHEMATICS
Purple Book

CHAPTER 13   CUMULATIVE REVIEW (continued)

9. 6 h 2 min
   +4 h 2 min

   A 10 h 2 min
   ✓ B 11 h 1 min
   C 11 h
   D 10 h 1 min

10. $11^2 =$

   A 22
   B 110
   ✓ C 121
   D 132

11. What is the approximate area of a circle if the radius = 10 cm? Use 3.14 for π.

   A 31.4 cm²
   ✓ B 314 cm²
   C 78.5 cm²
   D 62.8 cm²

12. If you roll a number cube 30 times, how many times would you expect to roll a 3?

   A 10
   B 6
   C 4
   ✓ D 5

Part 3 Applications

13. Food prices are expected to rise 4% in the next year. If this happens, how much will a cart of groceries that cost $175 today cost in one year?

   A $198
   B $189
   ✓ C $182
   D $180

14. Samantha drew a triangle with different length measurements on all three sides. What type of triangle did Samantha draw?

   A isosceles
   B equilateral
   ✓ C scalene
   D pyramid

15. When the Rodriquez family flew to Aruba, the flight there took 3 h 48 min and the flight home took 3 hours 55 min. How long did the Rodriquez family spend flying on their vacation?

   ✓ A 7 h 43 min   C 6 h 33 min
   B 7 h 15 min   D 6 h 58 min

16. Marcus spent $1\frac{1}{2}$ h on his math homework, $1\frac{3}{4}$ h on his science project, and $1\frac{2}{3}$ h on his essay. How did Marcus spend on schoolwork in all?

   A $3\frac{3}{4}$ h   C $4\frac{1}{4}$ h
   ✓ B $4\frac{11}{12}$ h   D $4\frac{1}{2}$ h

17. In a circle graph, Roger's farm is divided into percentages of different farm animals. Roger has 15% cows, 25% chickens, 35% horses, 15% pigs, and 10% goats. If Roger has 160 farm animals in all, how many chickens does he have?

   A 24   C 56
   ✓ B 40   D 16

18. At Carmen's Ice Cream Shoppe, the owner said that 12% of the total dips of ice cream sold are rocky road. If 225 dips were sold tomorrow, how many would you predict to be rocky road?

   A 17   C 23
   B 20   ✓ D 27

ALGEBRA READINESS
Variables, Expressions, Equations

In algebra,

- a **variable** is a symbol, usually a letter of the alphabet, that stands for an unknown number. x
- an **algebraic expression** is a combination of variables, numbers, and at least one operation. $x + 6$
- an **equation** is a sentence that contains an equal sign. $x + 6 = 13$

Write *expression* or *equation* for each of the following.

	a	b	c
1.	$n + 13 = 20$ __equation__	$6ab$ __expression__	$25 + x$ __expression__
2.	$9 \times n = 63$ __equation__	$8 + x - y$ __expression__	$34 + 79 = 113$ __equation__

Translate each phrase into an algebraic expression.

	a	b
3.	ten more than x __$x + 10$__	7 decreased by n __$7 - n$__
4.	twelve less than a __$a - 12$__	the product of 15 and 34 __15×34__
5.	the sum of five and six __$5 + 6$__	a number divided by 13 __$n \div 13$__

Translate each sentence into an equation.

6. Eleven times a number is 132. __$11 \times n = 132$__

7. Twenty minus fourteen equals six. __$20 - 14 = 6$__

8. Ten less than a number equals forty. __$n - 10 = 40$__

Write the following in words. **Answers may vary. Typical answers given.**

9. $n + 5$ five more than a number

10. $6 - a$ six decreased by a number

11. $35 \times 25 = 875$ Thirty-five times twenty-five equals eight hundred seventy-five.

PRISM MATHEMATICS
Purple Book

ALGEBRA READINESS
Variables, Expressions, Equations

ALGEBRA READINESS
Properties of Numbers

Commutative Properties of Addition and Multiplication
The order in which numbers are added does not change the sum. $\quad a + b = b + a$
The order in which numbers are multiplied does not change the product. $\quad x \times y = y \times x$

Associative Properties of Addition and Multiplication
The grouping of addends does not change the sum. $\quad (a + b) + c = a + (b + c)$
The grouping of factors does not change the product. $\quad (x \times y) \times z = x \times (y \times z)$

Identity Properties of Addition and Multiplication
The sum of an addend and zero is that addend. $\quad a + 0 = a$
The product of a factor and one is that factor. $\quad a \times 1 = a$

Properties of Zero
The product of a factor and zero is zero. $\quad a \times 0 = 0$
The quotient of zero and any non-zero number is zero. $\quad 0 \div a = 0$

Name the property shown by each statement.

	a	b
1.	$x \times 1 = x$ __identity__	$(12 \times a) \times b = 12 \times (a \times b)$ __associative__
2.	$54m + n = n + 54m$ __commutative__	$0 \div 3xy = 0$ __property of zero__
3.	$(7a + b) + 5 = (b + 7a) + 5$ __commutative__	$\frac{15x}{y} \times 0 = 0$ __property of zero__
4.	$(w + x) + 0 = (w + x)$ __identity__	$\frac{1}{3}c + \frac{2}{5}d = \frac{2}{5}d + \frac{1}{3}c$ __commutative__

Rewrite each expression using the property indicated.

	a	b
5.	property of zero: $8a \times 0 =$ __0__	commutative: $7d + 13e =$ __$13e + 7d$__
6.	identity: $1 \times (3x + 11) =$ __$(3x + 11)$__	associative: $(x \times 2y) \times z =$ __$x \times (2y \times z)$__
7.	identity: $\frac{3}{5}w + 0 =$ __$\frac{3}{5}w$__	commutative: $m \times 2n =$ __$2n \times m$__
8.	associative: $a + (4b + c) =$ __$(a + 4b) + c$__	property of zero: $0 \div (8xy) =$ __0__

ALGEBRA READINESS
The Distributive Property

Distributive Property
If one factor in a product is a sum, multiplying each addend by the other factor before adding does not change the product.

$a \times (b + c) = (a \times b) + (a \times c)$ For example, $4 \times (15 + 9) = (4 \times 15) + (4 \times 9)$
$\phantom{a \times (b + c) = (a \times b) + (a \times c) \quad \text{For example,}} 4 \times 24 = 60 + 36$
$\phantom{a \times (b + c) = (a \times b) + (a \times c) \quad \text{For example,} 4 \times (15 +} 96 = 96$

Rewrite each expression using the distributive property.

	a		b
1.	$x \times (y + 15) =$ __(x × y) + (x × 15)__	$(35 \times 4x) + (35 \times 6y) =$	__35 × (4x + 6y)__
2.	$(d \times 7) + (d \times 2e) =$ __d × (7 + 2e)__	$z \times (23 + 5y) =$	__(z × 23) + (z × 5y)__
3.	$j \times (3k + m) =$ __(j × 3k) + (j × m)__	$(46 \times b) + (46 \times c) =$	__46 × (b + c)__
4.	$(17 \times s) + (17 \times t) =$ __17 × (s + t)__	$(42 \times x) + (42 \times y) =$	__42 × (x + y)__
5.	$132 \times (a + d) =$ __(132 × a) + (132 × d)__	$(x + y) \times z =$	__(x × z) + (y × z)__

Replace each w with 11, x with 0, y with 7, and z with 20.
Then evaluate each expression.

	a		b
6.	$z \times (w + x) =$ __220__	$(x \times w) + (x \times z) =$	__0__
7.	$y \times (w + x) =$ __77__	$(y \times w) + (y \times z) =$	__217__
8.	$w \times (x + y) =$ __77__	$(w \times z) + (w \times y) =$	__297__
9.	$(w \times z) + (w \times x) =$ __220__	$x \times (z + y) =$	__0__
10.	$(z \times y) + (z \times z) =$ __540__	$(z \times x) + (z \times w) =$	__220__
11.	$w \times (z + x + y) =$ __297__	$(z \times w) + (z \times x) + (z \times y) =$	__360__

PRISM MATHEMATICS
Purple Book

ALGEBRA READINESS
The Distributive Property

241

ALGEBRA READINESS
Evaluating Expressions

Algebraic expressions can be evaluated using the rules called **Order of Operations.**

1. Do all operations within parentheses.
2. Do all multiplications and divisions from left to right.
3. Do all additions and subtractions from left to right.

$(3 + 6) \times 3 = 27$
$5 \times 4 + 2 = 22$
$12 - 3 + 5 = 14$

Name the operation that should be done first. Then find the value.

a

1. $16 - (4 \times 2)$ __multiply__; __8__
2. $9 \times 6 - 3$ __multiply__; __51__
3. $(3 + 4) \times (6 - 3)$ __subtract__; __21__

b

$8 + 6 \div 3$ __divide__; __10__
$4 + 6 \times 7 - 1$ __multiply__; __45__
$8 \div 2 + (3 - 1)$ __add or subtract__; __6__

Evaluate each expression if $a = 8$, $b = 4$, and $c = 2$.

a

4. $b \times c - a$ __0__
5. $4 + b - c$ __6__
6. $8 \times (b + c)$ __48__
7. $(a + a) \div c$ __8__
8. $9 - (a \div b)$ __7__
9. $b \div c + a - b$ __6__
10. $c \times (a + b)$ __24__

b

$a \div b + c$ __4__
$3 \times a \div 4$ __6__
$a + a \div c$ __12__
$(a + b) \div c$ __6__
$(b + c) \times a$ __48__
$(b + c) \times (a + b)$ __72__
$(c \times a) + (c \times b)$ __24__

Write true or false.

a

11. $8 + 24 \div 4 - 2 = 12$ __true__
12. $24 - 10 - 3 \times 4 = 2$ __true__

b

$18 \div 3 + (5 - 2) = 3$ __false__
$42 \div 7 \times 6 = 1$ __false__

ALGEBRA READINESS
Solving Equations Using Addition and Subtraction

Subtraction Property of Equality
If you subtract the same number from each side of an equation, the two sides remain equal.

$$x + 8 = 14$$

To undo the addition of 8, subtract 8.

$$x + 8 - 8 = 14 - 8$$
$$x + 0 = 6$$
$$x = 6$$

Addition Property of Equality
If you add the same number to each side of an equation, the two sides remain equal.

$$n - 6 = 7$$

To undo the subtraction of 6, add 6.

$$n - 6 + 6 = 7 + 6$$
$$n - 0 = 13$$
$$n = 13$$

Write the operation that would undo the operation in the equation.

	a		b
1.	$x - 16 = 20$ __addition__		$24 + n = 38$ __subtraction__
2.	$14 = n - 32$ __addition__		$a + 50 = 84$ __subtraction__

Solve each equation.

	a			b	
3.	$n - 7 = 12$	__19__	$x + 17 = 25$	__8__
4.	$a - 11 = 6$	__17__	$32 + b = 40$	__8__
5.	$x + 9 = 18$	__9__	$n - 45 = 90$	__135__
6.	$16 + a = 54$	__38__	$12 + x = 24$	__12__
7.	$b - 15 = 0$	__15__	$83 + n = 83$	__0__
8.	$16 + b = 32$	__16__	$52 = a - 5$	__57__
9.	$35 = n + 15$	__20__	$x + 18 = 19$	__1__

Write and solve an equation for each situation.

10. A total of 97 students tried out for the debate team. If 45 of the students were girls, how many were boys? __$45 + n = 97$; 52__

11. Three members left the debate team during the year. If 12 members remained, how many were on the team originally? __$n - 3 = 12$; 15__

ALGEBRA READINESS
Solving Equations Using Multiplication and Division

Division Property of Equality
If you divide each side of an equation by the same nonzero number, the two sides remain equal.

$$3 \times n = 15$$

To undo multiplication by 3, divide by 3.

$$\frac{3 \times n}{3} = \frac{15}{3}$$
$$n = 5$$

Multiplication Property of Equality
If you multiply each side of an equation by the same number, the two sides remain equal.

$$\frac{a}{3} = 9$$

To undo division by 3, multiply by 3.

$$\frac{a}{3} \times 3 = 9 \times 3$$
$$a = 27$$

Write the operation that would undo the operation in the equation.

	a			b	
1.	$6 \times a = 24$	division	$\frac{x}{4} = 16$	multiplication
2.	$4 = \frac{n}{3}$	multiplication	$42 = 7 \times a$	division
3.	$x \times 8 = 56$	division	$\frac{a}{8} = 16$	multiplication

Solve each equation.

	a			b	
4.	$\frac{x}{3} = 4$	12	$6 \times a = 54$	9
5.	$x \times 12 = 144$	12	$\frac{n}{6} = 16$	96
6.	$\frac{x}{8} = 24$	192	$9 \times n = 81$	9
7.	$54 = x \times 6$	9	$8 = \frac{n}{7}$	56
8.	$72 = 9 \times a$	8	$n \times 16 = 160$	10
9.	$356 \times n = 356$	1	$34 \times a = 544$	16
10.	$\frac{n}{15} = 38$	570	$x \times 53 = 3445$	65

ALGEBRA READINESS
Solving Two-Step Equations

A **two-step equation** is solved by undoing each operation in the equation.

$$4n + 5 = 17$$

To undo the addition of 5, subtract 5.

$$4n + 5 - 5 = 17 - 5$$
$$4n = 12$$

To undo the multiplication of 4, divide by 4.

$$\frac{4n}{4} = \frac{12}{4}$$
$$n = 3$$

$$\frac{n}{4} - 1 = 2$$

To undo the subtraction of 1, add 1.

$$\frac{n}{4} - 1 + 1 = 2 + 1$$
$$\frac{n}{4} = 3$$

To undo the division by 4, multiply by 4.

$$\frac{n}{4} \times 4 = 3 \times 4$$
$$n = 12$$

Solve each equation.

	a	b	c
1.	$2x + 5 = 11$ __3__	$3a - 5 = 7$ __4__	$6n + 8 = 50$ __7__
2.	$2b - 9 = 7$ __8__	$5x + 15 = 35$ __4__	$\frac{a}{5} - 3 = 0$ __15__
3.	$\frac{n}{6} + 12 = 15$ __18__	$7 + 3x = 28$ __7__	$2n - 4 = 6$ __5__
4.	$\frac{a}{12} - 10 = 2$ __144__	$\frac{n}{10} - 9 = 1$ __100__	$6n - 12 = 18$ __5__
5.	$\frac{n}{6} - 12 = 0$ __72__	$\frac{a}{7} - 3 = 1$ __28__	$4 + 10x = 74$ __7__
6.	$8a - 50 = 6$ __7__	$\frac{a}{3} - 6 = 6$ __36__	$12 = 9x - 15$ __3__
7.	$\frac{n}{9} - 9 = 0$ __81__	$\frac{a}{12} - 15 = 3$ __216__	$18a - 6 = 30$ __2__

Write the equation. Then solve. Typical equations given.

8. Seven more than two times a number is 23. _____$7 + 2n = 23$; 8_____

9. Three times a number, increased by 4, equals 31. _____$3n + 4 = 31$; 9_____

10. Eight less than five times a number is 27. _____$5n - 8 = 27$; 7_____

11. Twice a number, decreased by 16, is 54. _____$2n - 16 = 54$; 35_____

PRISM MATHEMATICS
Purple Book

ALGEBRA READINESS
Solving Two-Step Equations

245

ALGEBRA READINESS
Solving Equations

Some equations contain multiple steps.

$2 + 6 + 4x = 80$

Combine $2 + 6 = 8$.

$8 + 4x = 80$
$8 - 8 + 4x = 80 - 8$
$4x = 72$
$\dfrac{4x}{4} = \dfrac{72}{4}$
$x = 18$

$\dfrac{a}{4+6} - 3 = 11$

Simplify the denominator.

$\dfrac{a}{10} - 3 = 11$
$\dfrac{a}{10} - 3 + 3 = 11 + 3$
$\dfrac{a}{10} = 14$
$10 \times \dfrac{a}{10} = 14 \times 10$
$a = 140$

Solve each equation.

                         a   b

1. $\dfrac{n}{15-8} + 31 = 45$ $n = \underline{\ 98\ }$ $7 + 18 + 3x = 34$ $x = \underline{\ 3\ }$

2. $\dfrac{x}{11-3} + 7 = 16$ $x = \underline{\ 72\ }$ $5d + 15 + 5 = 45$ $d = \underline{\ 5\ }$

3. $6a - 37 = 3 + 2$ $a = \underline{\ 7\ }$ $8 + 4b + 21 = 33$ $b = \underline{\ 1\ }$

4. $7 + \dfrac{u}{24-18} = 12$ $u = \underline{\ 30\ }$ $33 - 15 + 3z = 57$ $z = \underline{\ 13\ }$

5. $8c + 108 - 95 = 45$ $c = \underline{\ 4\ }$ $\dfrac{h}{34-17} - 27 = 3$ $h = \underline{\ 510\ }$

6. $\dfrac{w}{8-5} - 21 = 14$ $w = \underline{\ 105\ }$ $11y + 53 - 30 = 78$ $y = \underline{\ 5\ }$

7. $27 + 23 + 10d = 60$ $d = \underline{\ 1\ }$ $49 - 44 + 13x = 96$ $x = \underline{\ 7\ }$

8. $123 + \dfrac{r}{7+9} = 131$ $r = \underline{\ 128\ }$ $85 - 67 = 9 + \dfrac{w}{14}$ $w = \underline{\ 126\ }$

9. $\dfrac{m}{36-19} - 11 = 6$ $m = \underline{\ 289\ }$ $24 - 11 = \dfrac{n}{3} + 6$ $n = \underline{\ 21\ }$

10. $15 + 37 - 8 + 9b = 98$ $b = \underline{\ 6\ }$ $39 + \dfrac{z}{26+8-11} = 58$ $z = \underline{\ 437\ }$

PRISM MATHEMATICS
Purple Book

ALGEBRA READINESS
Solving Inequalities

An **inequality** is a mathematical sentence that contains an inequality symbol ($>, <, \geq, \leq$).
$>$ means *is greater than*. $<$ means *is less than*.
$\geq$ means *is greater than or equal to*. $\leq$ means *is less than or equal to*.
An inequality is solved the same way an equation is solved.

$x - 3 > 10$ $\qquad\qquad\qquad\qquad\qquad$ $a + 5 \leq 8$
Add 3 to both sides of the inequality. $\qquad$ *Subtract 5 from both sides of the inequality.*
$x - 3 + 3 > 10 + 3$ $\qquad\qquad\qquad\qquad$ $a + 5 - 5 \leq 8 - 5$
$x > 13$ $\qquad\qquad\qquad\qquad\qquad\qquad$ $a \leq 3$

An inequality can have more than one solution.

x is any number greater than 13. $\qquad\qquad$ a is any number less than or equal to 3.

Write *true* or *false*.

	a		b		c
1.	$7 > 2$ __true__		$5 < 3$ __false__		$4 \geq 2$ __true__
2.	$6 \leq 5$ __false__		$0 > 2$ __false__		$9 \leq 9$ __true__

Use the given value to tell if each inequality is *true* or *false*.

	a		b
3.	$n + 2 \geq 7$ if $n = 6$ __true__		$14 \geq x + 6$ if $x = 4$ __true__
4.	$3a \geq 7$ if $a = 0$ __false__		$2 < 2x - 5$ if $x = 3$ __false__

Give a value for the variable in each inequality.

	a		b
5.	$n + 4 > 5$ __a number greater than 1__		$x - 3 < 7$ __a number less than 10__
6.	$a + 8 < 11$ __a number less than 3__		$n - 5 > 3$ __a number greater than 8__
7.	$x + 6 > 8$ __a number greater than 2__		$a - 6 < 9$ __a number less than 15__
8.	$n \leq 5$ __a number 5 or less__		$x \geq 12$ __a number 12 or more__
9.	$a \geq 3$ __a number 3 or greater__		$n \leq 10$ __a number 10 or less__
10.	$a + 1 > 9$ __a number 9 or more__		$n - 1 \leq 7$ __a number 8 or less__

ALGEBRA READINESS
Inequalities on a Number Line

You can graph the solution of an inequality on a number line.

The following graphs compare x and 5.

An open dot means that 5 is not a solution.

$x > 5$

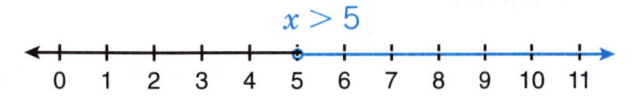

$x < 5$

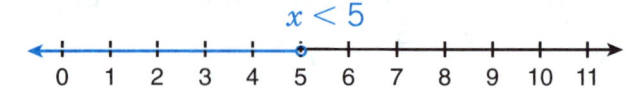

A closed dot means that 5 is a solution.

$x \geq 5$

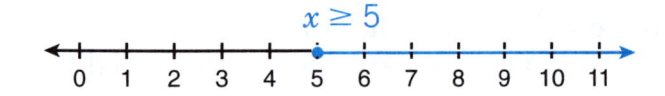

$x \leq 5$

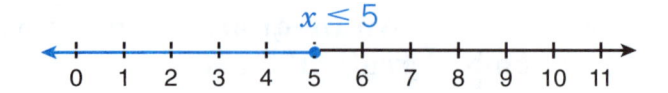

Graph each inequality on a number line.

                   a b

1. $d > 3$ $y \leq 8$

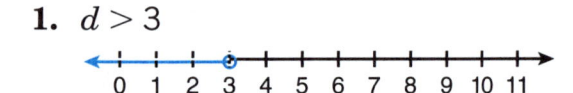

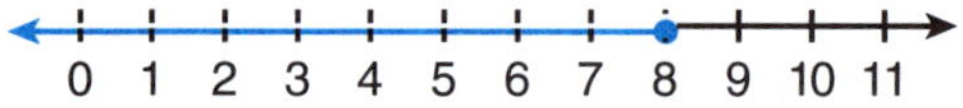

2. $x > 11$ 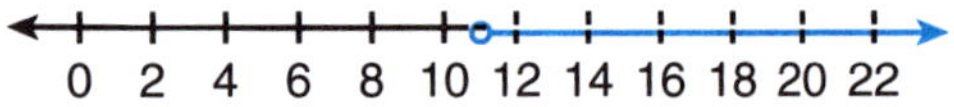 $n < 4$

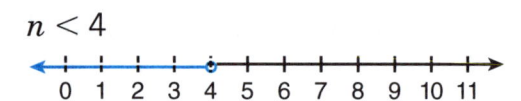

3. $h \geq 0$ 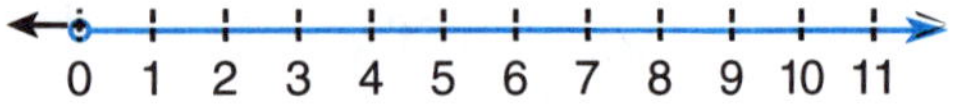 $p > 15$

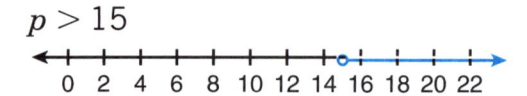

Solve each inequality. Graph the solution on a number line.

                   a b

4. $a - 3 \geq 5$ $a \geq 8$ $g + 11 < 20$ $g < 9$

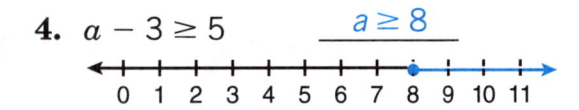

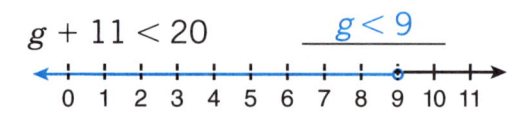

5. $3 + u \leq 6$ $u \leq 3$ 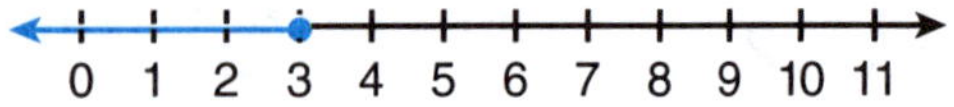 $7 + x < 15$ $x < 8$

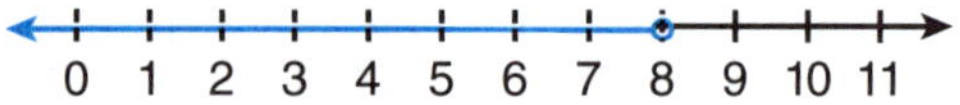

6. $j + 7 > 7$ $j > 0$ 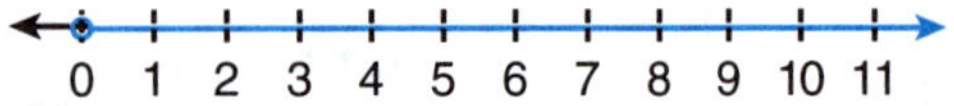 $p + 18 - 9 \geq 20$ $p \geq 11$

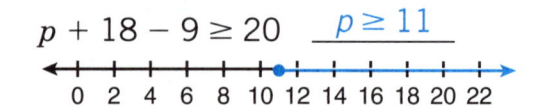

ALGEBRA READINESS
Integers

Negative and positive whole numbers are called **integers**.

Integers are often shown on a number line with zero as a starting point.

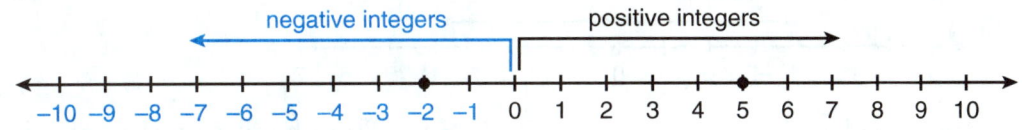

The greater of two integers is always the one farther to the right on a number line.

Say: −2 is less than 5. Say: 5 is greater than −2.
Write: −2 < 5 Write: 5 > −2

Use integers to name each point on a number line.

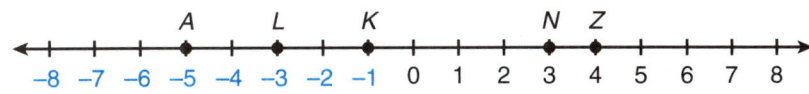

1. N __3__   L __−3__   Z __4__   K __−1__   A __−5__

Graph each point on the number line below.

2. B, −7   F, 1   M, 4   P, −4   S, 5

Write < or > in each ☐.

	a	b	c	d
3.	−1 > −3	4 > 2	0 < 5	0 > −1
4.	−4 < −2	−8 < 0	4 > −4	−1 > −7
5.	−6 < 1	2 > −6	−5 < 0	−7 > −8

List each set of integers in order from least to greatest.

 a b

6. 4, 0, −2, −1   __−2, −1, 0, 4__   −6, −1, 1, −5   __−6, −5, −1, 1__

7. 1, 0, −1, −7, −3   __−7, −3, −1, 0, 1__   −2, 2, 0, −3, 3   __−3, −2, 0, 2, 3__

PRISM MATHEMATICS
Purple Book

ALGEBRA READINESS
Integers

ALGEBRA READINESS
Absolute Value

The **absolute value** of a number is the distance that number is from zero on the number line. The absolute value of a number is always positive.

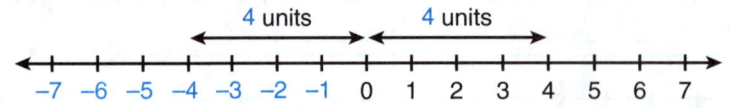

Say: The absolute value of −4 is 4.
Write: $|-4| = 4$

Say: The absolute value of 4 is 4.
Write: $|4| = 4$

Write the absolute value of each number.

	a	b	c
1.	$|-7| = $ __7__	$|14| = $ __14__	$|0| = $ __0__
2.	$|25| = $ __25__	$|-16| = $ __16__	$|-33| = $ __33__
3.	$|-78| = $ __78__	$|118| = $ __118__	$|-250| = $ __250__

Write < or > in each ☐.

	a	b	c
4.	$\|-6\|$ > $\|4\|$	$\|5\|$ > $\|-4\|$	$\|9\|$ < $\|-13\|$
5.	$\|0\|$ < $\|-5\|$	$\|-6\|$ > $\|-3\|$	$\|11\|$ < $\|15\|$
6.	$\|-25\|$ > $\|-23\|$	$\|-10\|$ > $\|0\|$	$\|-7\|$ < $\|-9\|$
7.	$\|35\|$ < $\|47\|$	$\|55\|$ > $\|-45\|$	$\|-34\|$ < $\|37\|$
8.	$\|-84\|$ > $\|-81\|$	$\|103\|$ > $\|-98\|$	$\|-138\|$ < $\|-157\|$

List in order from least to greatest.

	a	b
9.	−5, 7, $\|-9\|$, 0 ___−5, 0, 7, $\|-9\|$___	$\|-3\|$, −8, 5, $\|-7\|$ ___−8, $\|-3\|$, 5, $\|-7\|$___
10.	0, $\|5\|$, −7, $\|-6\|$ ___−7, 0, $\|5\|$, $\|-6\|$___	−11, 10, $\|-9\|$, 11 ___−11, $\|-9\|$, 10, 11___

ALGEBRA READINESS
Adding and Subtracting Integers

The sum of two positive integers is a **positive** integer. $3 + 2 = 5$

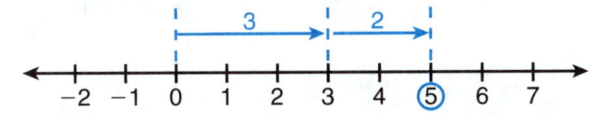

The sum of two negative integers is a **negative** integer. $-4 + (-2) = -6$

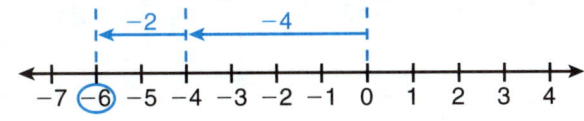

To add integers with different signs, **subtract** their absolute values. Give the result the same sign as the integer with the greatest absolute value. $6 + (-2) = 4$

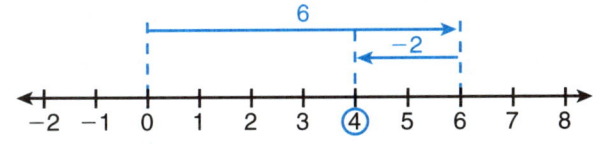

To subtract an integer, **add** its opposite.
The subtraction problem $-8 - 3 = -11$ can be rewritten as the addition problem $-8 + (-3) = -11$. -3 is the opposite of 3.

Add.

	a	b	c	d
1.	$7 + (-3) = \underline{4}$	$5 + 3 = \underline{8}$	$-9 + 4 = \underline{-5}$	$-6 + (-2) = \underline{-8}$
2.	$-12 + 9 = \underline{-3}$	$-4 + (-6) = \underline{-10}$	$3 + 18 = \underline{21}$	$3 + (-9) = \underline{-6}$
3.	$-1 + (-6) = \underline{-7}$	$12 + 14 = \underline{26}$	$8 + (-6) = \underline{2}$	$-4 + 8 = \underline{4}$
4.	$-12 + 0 = \underline{-12}$	$-14 + (-2) = \underline{-16}$	$0 + (-1) = \underline{-1}$	$14 + (-14) = \underline{0}$
5.	$68 + (-42) = \underline{26}$	$-97 + 38 = \underline{-59}$	$-16 + (-16) = \underline{-32}$	$48 + 52 = \underline{100}$

Subtract.

	a	b	c	d
6.	$8 - (-4) = \underline{12}$	$10 - 6 = \underline{4}$	$-8 - 5 = \underline{-13}$	$9 - (-6) = \underline{15}$
7.	$21 - 15 = \underline{6}$	$18 - (-9) = \underline{27}$	$10 - (-5) = \underline{15}$	$-6 - (-5) = \underline{-1}$
8.	$-4 - 9 = \underline{-13}$	$-8 - 6 = \underline{-14}$	$-12 - (-7) = \underline{-5}$	$5 - 11 = \underline{-6}$
9.	$16 - 31 = \underline{-15}$	$-8 - 12 = \underline{-20}$	$-4 - 0 = \underline{-4}$	$-5 - 2 = \underline{-7}$
10.	$2 - (-15) = \underline{17}$	$-8 - (-18) = \underline{10}$	$9 - (-17) = \underline{26}$	$0 - 8 = \underline{-8}$

PRISM MATHEMATICS
Purple Book

ALGEBRA READINESS
Multiplying and Dividing Integers

The product of two integers with **like** signs is **positive**.
The product of two integers with **unlike** signs is **negative**.

$3 \times 5 = 15$ $-3 \times (-5) = 15$
$-6 \times 3 = -18$ $6 \times (-3) = -18$

The quotient of two integers with **like** signs is **positive**.
The quotient of two integers with **unlike** signs is **negative**.

$8 \div 4 = 2$ $-8 \div (-4) = 2$
$6 \div (-3) = -2$ $-6 \div 3 = -2$

State whether each answer is positive or negative.

	a	b	c
1.	$18 \times (-7) =$ negative	$6 \times (-48) =$ negative	$-12 \times (-15) =$ positive
2.	$-18 \div (-9) =$ positive	$54 \div (-6) =$ negative	$-56 \div 7 =$ negative

Multiply or divide.

	a	b	c
3.	$8 \times (-9) =$ −72	$-9 \times (-6) =$ 54	$-12 \times 8 =$ −96
4.	$-56 \div (-7) =$ 8	$-54 \div 9 =$ −6	$96 \div (-8) =$ −12
5.	$11 \times (-8) =$ −88	$72 \div 9 =$ 8	$10 \times (-10) =$ −100
6.	$63 \div (-9) =$ −7	$-35 \div 5 =$ −7	$126 \times (-1) =$ −126
7.	$7 \times (-7) =$ −49	$235 \div (-1) =$ −235	$-634 \times 0 =$ 0
8.	$-64 \div (-8) =$ 8	$0 \div (-147) =$ 0	$-12 \times (-12) =$ 144

Write *true* or *false*. If false, state the reason.

9. The product of two positive integers is never negative. _____true_____

10. The product of two negative integers is always negative. _false, it is always positive_

11. The quotient of two negative integers is always positive. _____true_____

PRISM MATHEMATICS
Purple Book

ALGEBRA READINESS
Powers and Exponents

Numbers can be expressed in different ways. $10\,000 = 10 \times 10 \times 10 \times 10$
A shorter way to express 10 000 is by using **exponents**. $10\,000 = 10^4$
An exponent tells how many times a number, called the **base**, is used as a factor.

$$\text{base} \rightarrow 10^{6} = 10 \times 10 \times 10 \times 10 \times 10 \times 10 = 1\,000\,000$$
$$2^4 = 2 \times 2 \times 2 \times 2 = 16$$

(exponent labels the 6 in 10^6)

Numbers that are expressed using exponents are called **powers**.

Write each power as the product of the same factor.

	a		b
1.	8^4 $8 \times 8 \times 8 \times 8$	9^3	$9 \times 9 \times 9$
2.	23^2 23×23	1^6	$1 \times 1 \times 1 \times 1 \times 1 \times 1$
3.	10^5 $10 \times 10 \times 10 \times 10 \times 10$	2^7	$2 \times 2 \times 2 \times 2 \times 2 \times 2 \times 2$
4.	6^3 $6 \times 6 \times 6$	49^4	$49 \times 49 \times 49 \times 49$

Use exponents to express the following.

	a		b	
5.	$3 \times 3 \times 3 \times 3$	3^4	$9 \times 9 \times 9$	9^3
6.	15×15	15^2	$2 \times 2 \times 2 \times 2 \times 2 \times 2$	2^6
7.	$1 \times 1 \times 1 \times 1$	1^4	$4 \times 4 \times 4 \times 4 \times 4$	4^5
8.	$12 \times 12 \times 12$	12^3	$10 \times 10 \times 10$	10^3

Evaluate each expression.

	a		b	
9.	x^3 if $x = 5$	125	n^2 if $n = 9$	81
10.	a^5 if $a = 2$	32	b^7 if $b = 10$	10 000 000
11.	x^2 if $x = 15$	225	n^4 if $n = 3$	81

PRISM MATHEMATICS
Purple Book

ALGEBRA READINESS
Negative Exponents

Numbers between 0 and 1 can be expressed using **negative exponents.**

Any nonzero number raised to a negative power is the same as 1 divided by that number raised to the absolute value of the power.

$$x^{-a} = \frac{1}{x^a}$$

$$10^{-3} = \frac{1}{10^3} = \frac{1}{10 \times 10 \times 10} = \frac{1}{1000} \qquad 3^{-5} = \frac{1}{3^5} = \frac{1}{3 \times 3 \times 3 \times 3 \times 3} = \frac{1}{243}$$

Rewrite each expression using a positive exponent. Then write it in expanded form.

                                                  a   *b*

1. $10^{-2} = \dfrac{1}{10^2} = \dfrac{1}{10 \times 10}$ $8^{-4} = \dfrac{1}{8^4} = \dfrac{1}{8 \times 8 \times 8 \times 8}$

2. $6^{-3} = \dfrac{1}{6^3} = \dfrac{1}{6 \times 6 \times 6}$ $11^{-4} = \dfrac{1}{11^4} = \dfrac{1}{11 \times 11 \times 11 \times 11}$

3. $5^{-5} = \dfrac{1}{5^5} = \dfrac{1}{5 \times 5 \times 5 \times 5 \times 5}$ $18^{-3} = \dfrac{1}{18^3} = \dfrac{1}{18 \times 18 \times 18}$

4. $2^{-7} = \dfrac{1}{2^7} = \dfrac{1}{2 \times 2 \times 2 \times 2 \times 2 \times 2 \times 2}$ $12^{-5} = \dfrac{1}{12^5} = \dfrac{1}{12 \times 12 \times 12 \times 12 \times 12}$

Use negative exponents to rewrite the following.

                                                  a   *b*

5. $\dfrac{1}{5 \times 5 \times 5}$ 5^{-3} $\dfrac{1}{3 \times 3 \times 3 \times 3}$ 3^{-4}

6. $\dfrac{1}{14 \times 14 \times 14 \times 14}$ 14^{-4} $\dfrac{1}{8 \times 8 \times 8}$ 8^{-3}

7. $\dfrac{1}{10 \times 10 \times 10 \times 10 \times 10}$ 10^{-5} $\dfrac{1}{2 \times 2 \times 2 \times 2 \times 2 \times 2 \times 2 \times 2}$ 2^{-8}

8. $\dfrac{1}{24 \times 24 \times 24 \times 24}$ 24^{-4} $\dfrac{1}{15 \times 15 \times 15 \times 15 \times 15 \times 15}$ 15^{-6}

Evaluate each expression.

                                       a   *b*

9. a^{-2} if $a = 3$ $\dfrac{1}{9}$ x^{-4} if $x = 2$ $\dfrac{1}{16}$

10. b^{-3} if $b = 5$ $\dfrac{1}{125}$ m^{-4} if $m = 4$ $\dfrac{1}{256}$

PRISM MATHEMATICS
Purple Book

ALGEBRA READINESS
Negative Exponents

ALGEBRA READINESS
Multiplying and Dividing Powers

To **multiply** powers *that have the same base*, **add** the exponents.

$a^m \times a^n = a^{m+n}$
$10^3 \times 10^2 = 10^{3+2} = 10^5$

To **divide** powers *that have the same base*, **subtract** the exponents.

$a^m \div a^n = a^{m-n}$
$10^3 \div 10^2 = 10^{3-2} = 10^1$

Find each product.

	a		b		c	
1.	$5^3 \times 5^6$	5^9	$3^2 \times 3^4$	3^6	$n^6 \times n^2$	n^8
2.	$9^3 \times 9^1$	9^4	$x \times x$	x^2	$10^4 \times 10^4$	10^8
3.	$12^3 \times 12^4$	12^7	$a \times a^5$	a^6	$15^5 \times 15^3$	15^8

Verify each product by replacing the powers with their values.

	a		b	
4.	$3^3 \times 3^2 = 3^5$	$27 \times 9 = 243$	$2^2 \times 2^3 = 2^5$	$4 \times 8 = 32$
5.	$3 \times 3^4 = 3^5$	$3 \times 81 = 243$	$5 \times 5 = 5^2$	$5 \times 5 = 25$
6.	$2^4 \times 2^2 = 2^6$	$16 \times 4 = 64$	$3^2 \times 3^2 = 3^4$	$9 \times 9 = 81$

Find each quotient.

	a		b		c	
7.	$7^7 \div 7^2$	7^5	$a^4 \div a^2$	a^2	$8^3 \div 8^1$	8^2
8.	$9^5 \div 9^2$	9^3	$6^{12} \div 6^6$	6^6	$5^8 \div 5^3$	5^5
9.	$4^4 \div 4$	4^3	$7^6 \div 7^5$	7	$15^4 \div 15^3$	15

Verify each quotient by replacing the powers with their values.

	a		b	
10.	$3^4 \div 3^2 = 3^2$	$81 \div 9 = 9$	$2^5 \div 2^3 = 2^2$	$32 \div 8 = 4$
11.	$4^3 \div 4 = 4^2$	$64 \div 4 = 16$	$5^2 \div 5 = 5$	$25 \div 5 = 5$
12.	$3^3 \div 3 = 3^2$	$27 \div 3 = 9$	$10^5 \div 10^2 = 10^3$	$100\,000 \div 100 = 1000$

PRISM MATHEMATICS
Purple Book

ALGEBRA READINESS
Multiplying and Dividing Powers

ALGEBRA READINESS
Scientific Notation

A number written in **scientific notation** is shown as the product of a factor between 1 and 10 and a power of 10.

30 000 — Move the decimal point 4 places to the left. Multiply by 10^4.
3×10^4
$5\,780\,000 = 5.78 \times 10^6$

0.0003 — Move the decimal point 4 places to the right. Multiply by 10^{-4}.
3×10^{-4}
$0.006\,23 = 6.23 \times 10^{-3}$

Express each of the following in scientific notation.

	a		b		c	
1.	6300	6.3×10^3	7000	7×10^3	540	5.4×10^2
2.	0.5	5×10^{-1}	0.006	6×10^{-3}	0.0007	7×10^{-4}
3.	690	6.9×10^2	0.20	2×10^{-1}	50 000	5×10^4
4.	0.0017	1.7×10^{-3}	0.064	6.4×10^{-2}	8 000 000	8×10^6
5.	0.609	6.09×10^{-1}	0.003	3×10^{-3}	0.0852	8.52×10^{-2}

Express each scientific notation as indicated.

	a		b		c	
6.	7.5×10^2	750	3×10^3	3000	9×10^4	90 000
7.	5×10^{-3}	0.005	8×10^{-1}	0.8	4×10^{-2}	0.04
8.	6.5×10^2	650	9.04×10^3	9040	7×10^{-1}	0.7
9.	6.47×10^2	647	1.2×10^3	1200	5.8×10^{-2}	0.058
10.	2×10^{-2}	0.02	0.2×10^3	200	8.1×10^3	8100

ALGEBRA READINESS
Ordered Pairs

The location of any point on a grid can be indicated by an **ordered pair** of numbers. Point A on the grid at the right is indicated by the ordered pair (2, 4) because it is located at 2 on the horizontal scale x, and at 4 on the vertical scale y. The number on the horizontal scale x is always named first in an ordered pair. (0, 0) is called the **origin.**

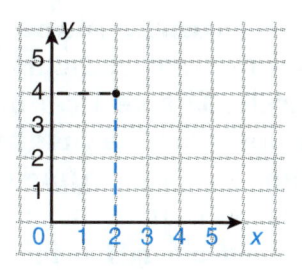

Grid 1

Use Grid 1 to name the point for each ordered pair.

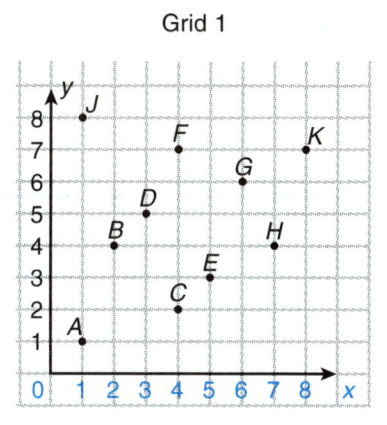

	a		b	
1.	(4, 2)	C	(7, 4)	H
2.	(8, 7)	K	(2, 4)	B
3.	(6, 6)	G	(3, 5)	D
4.	(5, 3)	E	(1, 1)	A
5.	(4, 7)	F	(1, 8)	J

Use Grid 2 to find the ordered pair for each labelled point.

Grid 2

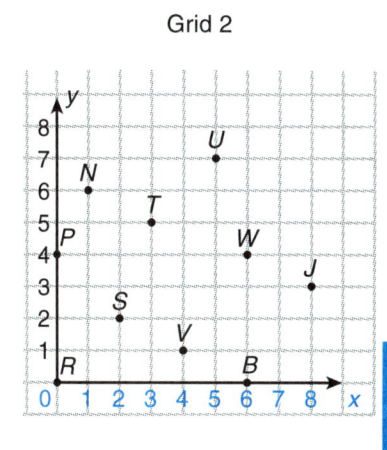

	a		b	
6.	J	(8, 3)	N	(1, 6)
7.	S	(2, 2)	W	(6, 4)
8.	R	(0, 0)	B	(6, 0)
9.	T	(3, 5)	V	(4, 1)
10.	U	(5, 7)	P	(0, 4)

Locate four points on the grid and name each ordered pair.

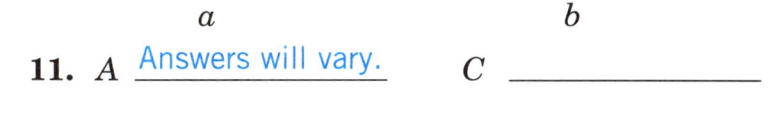

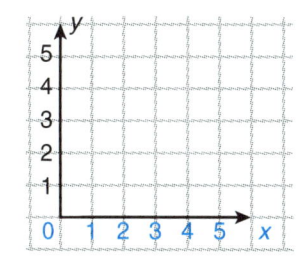

	a		b	
11.	A	Answers will vary.	C	
12.	Z		R	

PRISM MATHEMATICS
Purple Book

ALGEBRA READINESS
Ordered Pairs

257

ALGEBRA READINESS
Graphing in Four Quadrants

A **coordinate plane** is formed by two number lines that are perpendicular. The horizontal line is the ***x*-axis.** The vertical line is the ***y*-axis.** The axes intersect at the **origin.** The axes divide the coordinate plane into four **quadrants.** The first number in an ordered pair is the ***x*-coordinate.** The second number is the ***y*-coordinate.**

To plot the point $(6, -3)$ on a coordinate plane, start at 0 and move 6 units right then 3 units down.

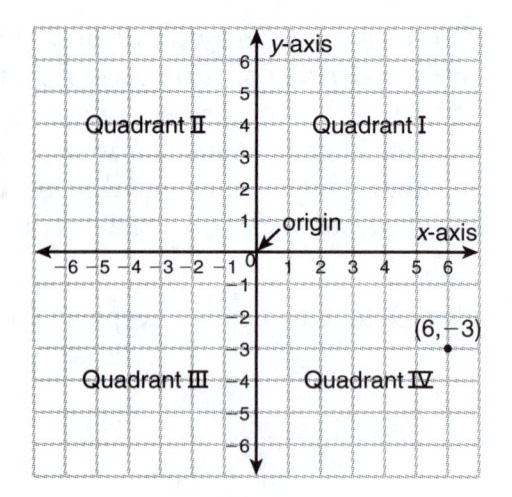

Use Grid 1 to name the point for each ordered pair.

	a		b	
1.	$(1, -6)$	B	$(-3, -7)$	G
2.	$(-5, 0)$	D	$(4, 2)$	F
3.	$(-7, -3)$	E	$(5, 0)$	A
4.	$(-6, 1)$	H	$(-4, 3)$	C

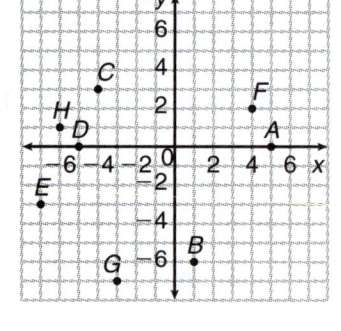

Plot each ordered pair on Grid 2.

	a	b
5.	$T(6, -5)$	$R(-4, 0)$
6.	$U(-3, 1)$	$P(-7, 2)$
7.	$S(0, -7)$	$W(1, -2)$
8.	$Q(0, 0)$	$V(2, 7)$

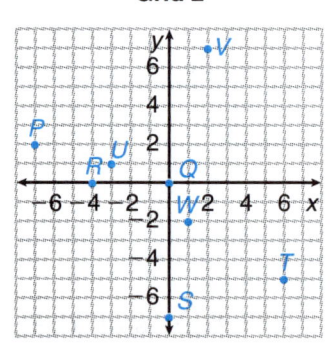

State the quadrant in which each ordered pair would be located.

	a		b		c	
9.	$(-9, 4)$	Quadrant II	$(5, -1)$	Quadrant IV	$(-3, -3)$	Quadrant III
10.	$(18, -33)$	Quadrant IV	$(12, 20)$	Quadrant I	$(-34, 42)$	Quadrant II

PRISM MATHEMATICS
Purple Book

ALGEBRA READINESS
Making Function Tables

A **function** is a rule that states for each value of one variable that there is exactly one related value for the other variable.

For example, $y = 3x - 6$ is a function.

A **function table** organizes values of a function.

Each x-value and its corresponding y-value can be thought of as ordered pairs.

x	y
1	-3
2	0
3	3
4	6

$y = 3x - 6$
let $x = 1, 2, 3, 4$
$y = 3(1) - 6$
$y = -3$

Make a function table for each function and the given values of x.

              a   b   c

1. $y = 8 + 2x$ $y = \frac{3x}{2}$ $y = 12 - 8x$

let $x = -4, -2, 0, 2, 4$ let $x = -2, -1, 0, 1, 2$ let $x = 0, 1, 2, 3, 4$

x	y
-4	0
-2	4
0	8
2	12
4	16

x	y
-2	-3
-1	-1.5
0	0
1	1.5
2	3

x	y
0	12
1	4
2	-4
3	-12
4	-20

2. $y = 5x - 15$ $y = \frac{x}{4} - 5$ $y = \frac{x}{3} + 4$

let $x = -5, 0, 1, 3, 8$ let $x = -8, -4, 0, 4, 8$ let $x = -9, -3, 0, 6, 12$

x	y
-5	-40
0	-15
1	-10
3	0
8	25

x	y
-8	-7
-4	-6
0	-5
4	-4
8	-3

x	y
-9	1
-3	3
0	4
6	6
12	8

Write the function that is represented by each function table.

3. $y = x - 7$ $y = -3x$ $y = 2x + 1$

x	y
-2	-9
-1	-8
0	-7
1	-6
2	-5

x	y
0	0
2	-6
4	-12
6	-18
8	-24

x	y
-1	-1
0	1
1	3
2	5
3	7

ALGEBRA READINESS
Graphing Linear Functions

A **linear function** is one that can be represented on a coordinate plane as a straight line.

To graph a linear function, create a function table with at least two ordered pairs. Then plot these ordered pairs on a coordinate plane and draw a line through the points.

Graph the linear function $y = 6 - 2x$.

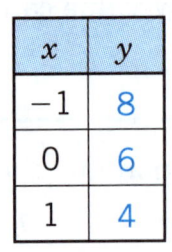

x	y
−1	8
0	6
1	4

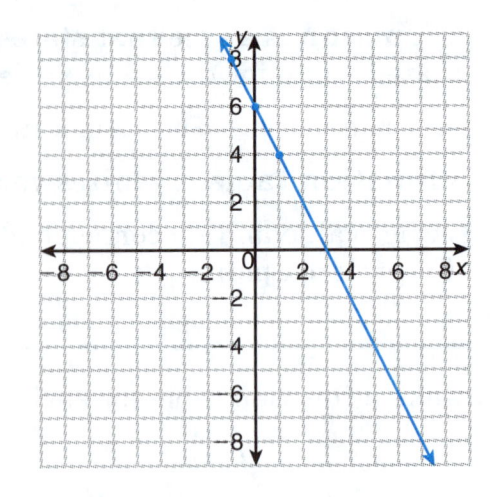

Graph each linear function.

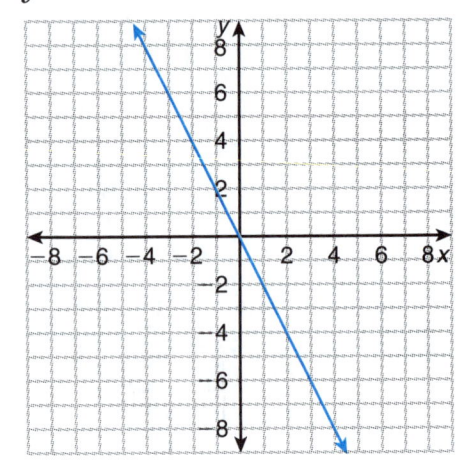

        a   *b*

1. $y = -2x$ $y = 7 - \dfrac{x}{2}$

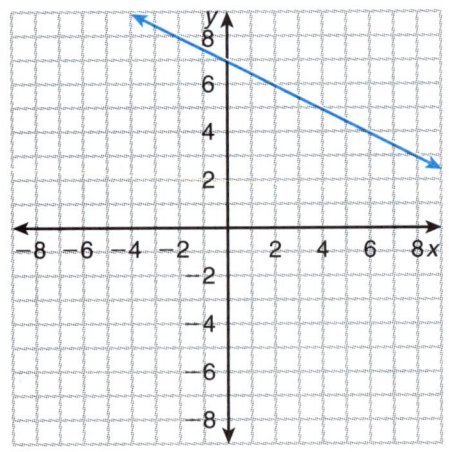

2. $y = 5x - 4$ 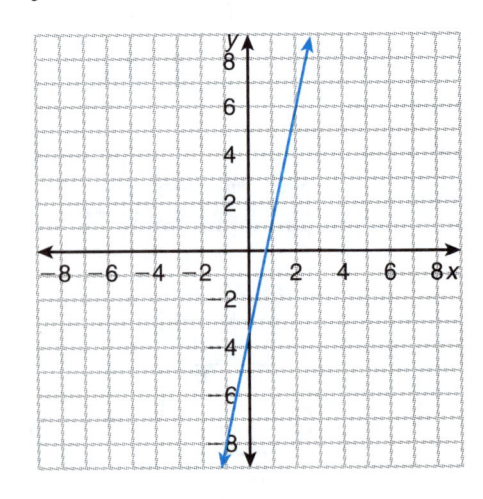 $y = \dfrac{3}{4}x + 5$

PRISM MATHEMATICS
Purple Book

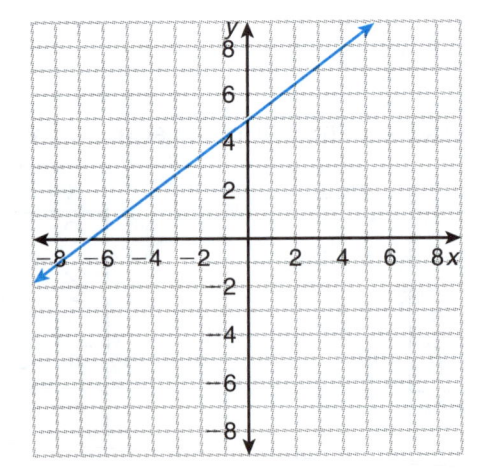

ALGEBRA READINESS
Slope

The slope of a line is the ratio of the change in y to the corresponding change in x.

$$\text{slope} = \frac{\text{change in } y}{\text{change in } x}$$

In Quadrant I, the change in y is 2 and the corresponding change in x is 3. Therefore, the slope of the line is $\frac{2}{3}$.

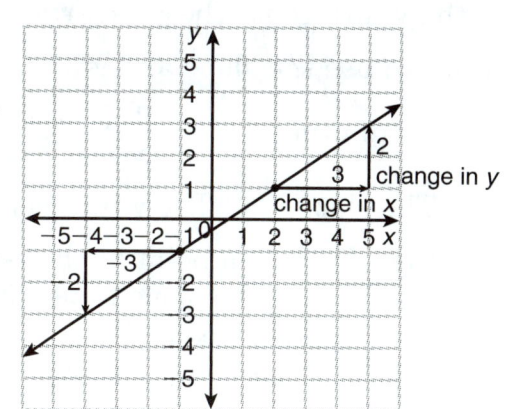

The slope of the line is the same in Quadrant III.

$$\frac{\text{change in } y}{\text{change in } x} = \frac{-2}{-3} \text{ or } \frac{2}{3}$$

To find the slope of a line when given two ordered pairs on that line: find the ratio of the difference in the y-coordinates and the difference in the x-coordinates.

Find the slope of the line passing through (6, −4) and (3, 2).

$$\text{Slope} = \frac{-4 - 2}{6 - 3} = \frac{-6}{3} = \frac{-2}{1}$$

Find the slope of each graphed line.

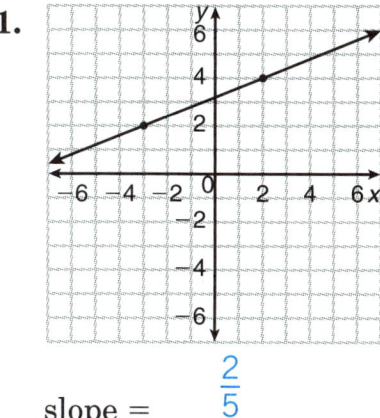

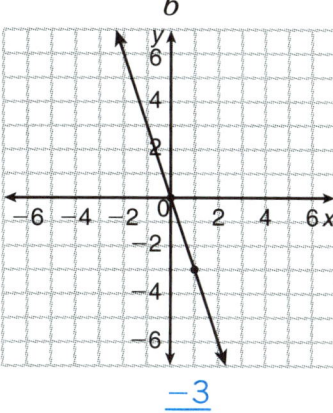

	a	b	c
1.	slope = $\frac{2}{5}$	slope = $\frac{-3}{1}$	slope = $\frac{1}{2}$

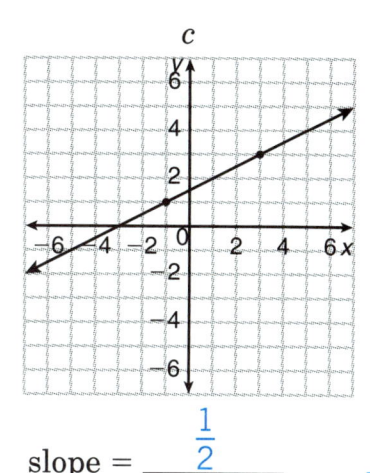

Find the slope of the line passing through each pair of points.

	a	b	c
2.	(5, −7), (3, 2)	(3, −1), (−3, −4)	(−4, −2), (8, −6)
	slope = $-\frac{9}{2}$	slope = $\frac{1}{2}$	slope = $-\frac{1}{3}$
3.	(6, −5), (7, −3)	(−4, 1), (0, 0)	(1, −3), (4, −1)
	slope = $\frac{2}{1}$	slope = $-\frac{1}{4}$	slope = $\frac{2}{3}$

PRISM MATHEMATICS
Purple Book

ALGEBRA READINESS
Slope

261

ALGEBRA READINESS
Slope-Intercept Form

The **slope-intercept form** of a linear equation is $y = mx + b$, where m is the slope and b is the y-intercept. The **y-intercept** of a line is the point where the line crosses the y-axis.

You can use the slope and y-intercept to graph a line.

Graph the line $y = \frac{2}{3}x + 2$.

The slope is $\frac{2}{3}$. The y-intercept is 2.

Step 1: Place a point at the y-intercept, 2.
Step 2: Use the slope to plot another point. The slope is $\frac{2}{3}$.
Step 3: Draw a line through the two points.

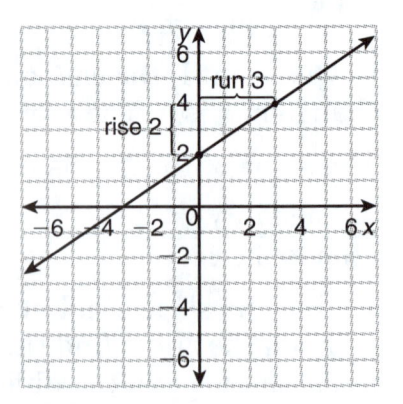

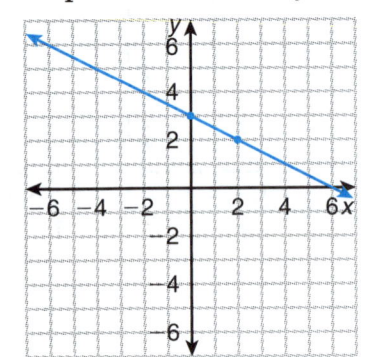

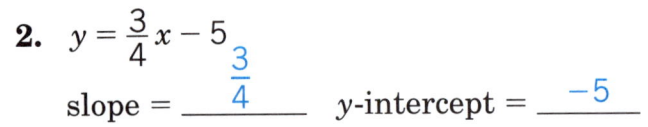

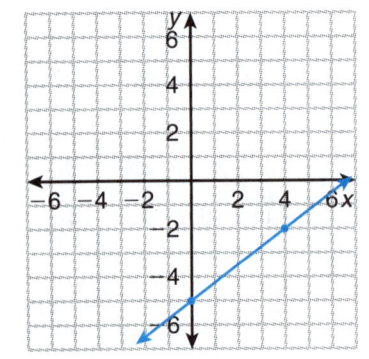

Name the slope and y-intercept of each line. Then graph the line.

                     a   b

1. $y = -\frac{1}{2}x + 3$　　　　　　　　$y = 3x - 2$

   slope = $\frac{-1}{2}$　　y-intercept = 3　　　　slope = $\frac{3}{1}$　　y-intercept = -2

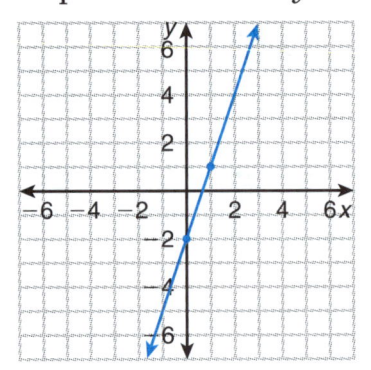

2. $y = \frac{3}{4}x - 5$　　　　　　　　$y = -4x + 1$

   slope = $\frac{3}{4}$　　y-intercept = -5　　　　slope = $\frac{-4}{1}$　　y-intercept = 1

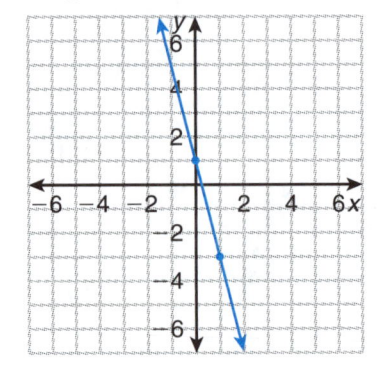

Prism Math Assessment

INFORMAL ASSESSMENT

✓ Daily Lessons

Performance on the daily lessons is informal assessment to chart progress and understanding. The daily lesson scores can be averaged over the course of a grading period to provide one part of a student's evaluation.

✓ Lesson Follow-up

Suggestions for addressing student performance on the daily lesson can be found at the end of each lesson in the Teacher's Edition.

✓ Problem-Solving

The problem-solving exercises that follow the daily lessons provide specific insight into whether students can apply the skill they have just practiced. The ability to apply the skill in another context is a clear sign of understanding.

FORMAL ASSESSMENT

✓ Readiness Check

The Readiness Checks at the beginning of **Levels Red–Purple** offer insight into student skill level and placement.

✓ Chapter Pretests

The Chapter Pretests at the beginning of each chapter provides insight so you can plan your instruction according to the needs of the student.

✓ Chapter Tests

The Chapter Tests in blackline master form in the back of each Teacher's Edition provide for a formal assessment of each chapter's content.

- Using the **Chapter Tests**

  Administer the Chapter Test after students complete the Cumulative Review at the end of each chapter. Record the scores as part of the math evaluation.

ASSESSMENT

Prism Math Assessment

SELF ASSESSMENT

✓ **Chapter Practice Tests**

The Chapter Practice Tests give students an opportunity to evaluate their knowledge of the chapter content before taking the Chapter Test. Students can record the scores on the Chapter Practice Test chart as part of their math evaluation.

✓ **Self Assessment**

The students can keep track of their progress by recording their scores from the Chapter Practice Tests on the chart located on the inside back cover.

✓ **Cumulative Review**

The Cumulative Review for each chapter in the Student Edition offers insight into student ability to maintain skills taught from the beginning of the book.

CHAPTER 1 TEST

NAME _____

Add or subtract.

	a	b	c	d
1.	6159 241 +398	64.31 22.85 +9.74	45 628 13 105 +24 359	47.611 3.298 +39.1
2.	85.32 −63.78	14 360 −12 819	9.401 −3.695	50 101 −18 957
3.	157 823 21 425 +21 428	56.73 4.18 +3.14	7.2 −3.4	6752 −974

Multiply or divide.

4.	751 ×38	6819 ×315	72.3 ×18.2	2.703 ×0.014
5.	639 ×582	72.631 ×0.155	23)18 321	84)563
6.	57)43 605	0.7)4.48	4.3)10.105	3.25)0.585

PRISM MATHEMATICS
Purple Book

CHAPTER 1 TEST

CHAPTER 2 TEST

NAME _____

Write each fraction in simplest form.

	a	b	c	d
1.	$\dfrac{5}{10} =$	$\dfrac{30}{36} =$	$\dfrac{18}{27} =$	$\dfrac{75}{100} =$

Rename.

2. $\dfrac{7}{8} = \dfrac{70}{}$ $\quad 4 = \dfrac{}{6}$ $\quad \dfrac{7}{9} = \dfrac{63}{}$ $\quad 4\dfrac{3}{5} = \dfrac{}{5}$

Write each sum, difference, product, or quotient in simplest form.

	a	b	c	d
3.	$\dfrac{2}{5}$ $+\dfrac{4}{5}$	$4\dfrac{1}{2}$ $+6\dfrac{7}{12}$	$2\dfrac{3}{4}$ $+3\dfrac{2}{3}$	$7\dfrac{7}{15}$ $5\dfrac{1}{6}$ $+1\dfrac{2}{3}$
4.	$\dfrac{12}{13}$	$3\dfrac{1}{3}$ $-\dfrac{3}{4}$	$9\dfrac{5}{7}$ $-2\dfrac{9}{10}$	$3\dfrac{1}{3}$ $-2\dfrac{2}{3}$
5.	$\dfrac{7}{8} \times \dfrac{2}{7} =$	$\dfrac{3}{10} \times \dfrac{5}{9} =$	$2\dfrac{5}{8} \times \dfrac{4}{7} =$	$4\dfrac{2}{5} \times 9\dfrac{1}{11} =$
6.	$\dfrac{6}{7} \div \dfrac{3}{4} =$	$7 \div \dfrac{1}{3} =$	$3\dfrac{3}{4} \div 1\dfrac{1}{2} =$	$2\dfrac{4}{5} \div 1\dfrac{1}{3} =$

PRISM MATHEMATICS
Purple Book

CHAPTER 3 TEST

NAME _____

Solve each equation.

	a	b	c
1.	$5t = 45$	$39 - 25 = 2x$	$72 = 8p$
2.	$16 = 6 + c$	$a + 4 = 29$	$13 + 19 = z + 5$
3.	$\dfrac{y}{5} = 7$	$12 = \dfrac{x}{12}$	$\dfrac{d}{3} = 15 + 9$
4.	$43 = b - 12$	$m - 8 = 14$	$e - 5 = 17 + 13$
5.	$33 + 21 = s + 14$	$68 + 9 = \dfrac{t}{11}$	$42 + 11 = x - 15$
6.	$12 = \dfrac{f}{4}$	$13e = 169$	$10n = 20 \times 5$

Write an equation for the problem. Solve.

7. There are 12 empty seats on the ride. This is $\frac{1}{3}$ of the total number of seats. How many seats are there?

Equation: _____   There are _____ seats.

PRISM MATHEMATICS
Purple Book

CHAPTER 4 TEST

NAME _____

Solve each problem.

1. Maggie collected four times as many fossils as Henry. They collected a total of 240 fossils. How many fossils did Henry collect?

   Equation: _____

   Henry collected _____ fossils.

2. Brittany scored 7 more points than Sharla. Together they scored 53 points. How many points did Brittany score?

   Equation: _____

   Brittany scored _____ points.

3. Maya has seven less than three times as many magnets than Cindy. If they have a total of 33, how many magnets does each girl have?

   Equation: _____

   Maya has _____ magnets.

   Cindy has _____ magnets.

4. At 31 km per hour, how far can Julia bike in 4 h?

   Julia can bike _____ km in 4 h.

5. Paul has twice as many CDs as Jeriel and Alexis combined. Paul has 28 CDs and Alexis has 5 CDs. How many CDs does Jeriel have?

   Jeriel has _____ CDs.

PRISM MATHEMATICS
Purple Book

CHAPTER 4 TEST

CHAPTER 5 TEST

Express each of the following as a ratio in two ways as shown.

	a	b

1. 22 baskets in 2 games _____ _____

2. 7 pencils to 10 students _____ _____

3. 3 leaves to 1 clover _____ _____

4. 8 boys to 12 girls _____ _____

Solve the following.

      a   b

5. $\dfrac{3}{6} = \dfrac{n}{12}$ $\dfrac{5}{n} = \dfrac{15}{24}$

6. $\dfrac{7}{9} = \dfrac{n}{72}$ $\dfrac{2}{8} = \dfrac{14}{n}$

7. $\dfrac{8}{n} = \dfrac{20}{45}$ $\dfrac{n}{15} = \dfrac{9}{27}$

Complete the following.

      a   b

8. _____ is 25% of 32.   17.4 is 30% of _____.

9. 380 is 95% of _____.   _____ is 8.5% of 200.

10. $1\dfrac{3}{4}$ is _____% of $2\dfrac{1}{2}$.   8.5 is _____% of 68.

11. 600 is _____% of 300.   _____ is 40% of 40.

PRISM MATHEMATICS
Purple Book

CHAPTER 5 TEST

CHAPTER 6 TEST

Complete the following for simple interest.

	principal	rate	time	interest
1.	220	20%	4 years	
2.	$500	6.5%	5 years	
3.	$750		$\frac{1}{2}$ year	$67.50
4.	$920	25%		$977.50
5.		$8\frac{1}{2}$%	3 years	$1083.75
6.	$1500		2 years	$390

Interest is to be compounded in each account below. Find the total amount that will be in each account after the period of time indicated.

	principal	rate	time	compounded	total amount
7.	$200	5%	2 years	annually	
8.	$500	8%	3 years	annually	
9.	$700	4%	2 years	semiannually	
10.	$300	6%	$\frac{1}{3}$ year	monthly	

PRISM MATHEMATICS
Purple Book

CHAPTER 7 TEST

Measure each line segment to the nearest unit as indicated.

1. _____ cm
2. _____ mm

Complete the following.

	a		b
3.	35 cm = _____ mm		520 cm = _____ m
4.	13 m = _____ cm		300 mm = _____ m
5.	8 km = _____ m		6.2 m = _____ mm
6.	4 m = _____ km		7.1 mm = _____ cm
7.	14 L = _____ mL		3.5 kL = _____ L
8.	0.011 L = _____ mL		12.9 mL = _____ L
9.	1800 L = _____ kL		670 L = _____ kL
10.	200 g = _____ kg		92 kg = _____ g
11.	5.4 kg = _____ g		3000 mg = _____ g
12.	80 mg = _____ g		0.4 g = _____ mg

13. Water freezes at _____ ° Celsius.
14. Water boils at _____ ° Celsius.

Solve each problem.

15. Jenny's pot of chili had a mass of 3.2 kg. Meghan's pot of chili had a mass of 2800 g. Whose pot of chili had a greater mass?

    _____ pot of chili had a mass of _____ kg more.

16. Jaxon has to walk 3.5 km to get home. He has already walked 1900 m. How many more kilometres does he need to walk to get home? How many metres is that?

    Jaxon needs to walk _____ more kilometres.

    That is _____ m.

CHAPTER 8 TEST

NAME _____

Complete the following.

	a	b
1.	4 m = _____ cm	2 m = _____ mm
2.	150 min = _____ h	5 kg = _____ g
3.	7 kg = _____ g	3 h 14 min = _____ min
4.	144 h = _____ days	3 km = _____ m
5.	4 L = _____ mL	8 L = _____ mL

Add, subtract, or multiply.

	a	b	c
6.	4 m −2 cm	5 L +3 mL	3 min 21 s ×2
7.	7 g −3 mg	4 kg ×5	8 h 45 min +6 h 37 min

Round as indicated.

	a nearest ten	b nearest hundred	c nearest thousand
8. 6851	_____	_____	_____
9. 74 883	_____	_____	_____

Write an estimate for each exercise. Then find the answer.

10. 6254   8159   485
    +3867   −6901   ×64

PRISM MATHEMATICS
Purple Book

272

CHAPTER 8 TEST

CHAPTER 9 TEST

Use the figures below to answer the questions that follow.

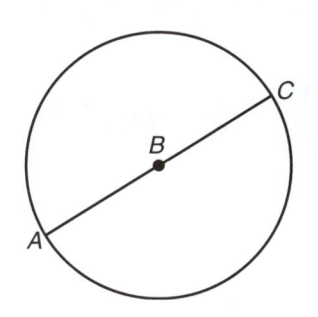

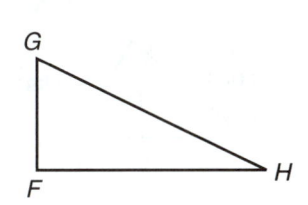

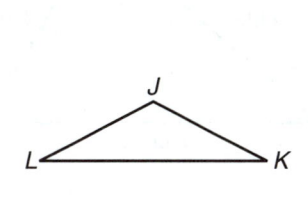

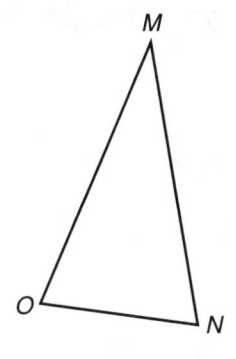

1. Name the diameter of circle B. _____   Name a radius of circle B. _____

2. Which figures are isosceles triangles?   Which figure is an acute triangle?
   _____   _____

3. Which figure is a right triangle?   Which figure is a scalene triangle?
   _____   _____

Use a protractor to measure each angle. Then describe each angle by writing *acute*, *obtuse*, or *right*.

           a   *b*   *c*

4. _____°, _____   _____°, _____   _____°, _____

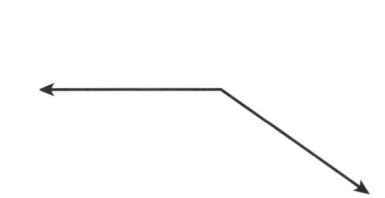

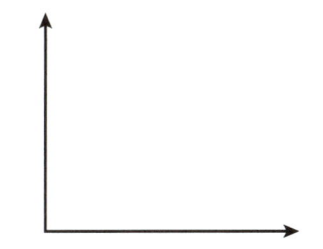

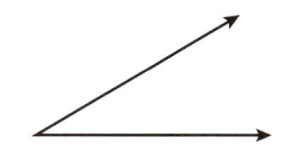

Tell whether each pair of lines is *parallel* or *perpendicular*.

5. _____   _____   _____

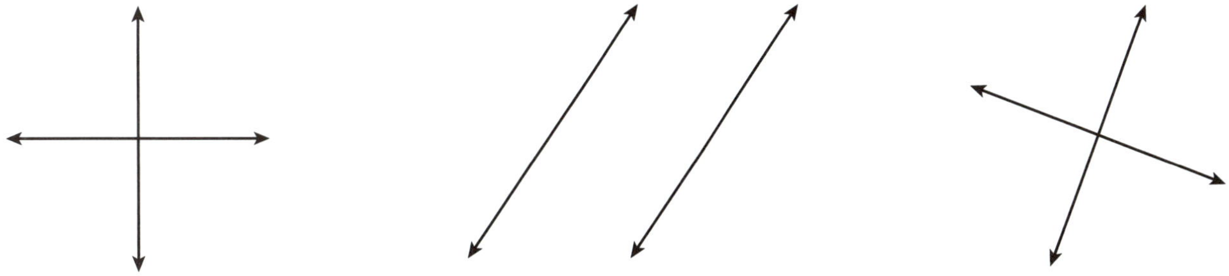

CHAPTER 10 TEST

Use the triangles below to help you complete the following.

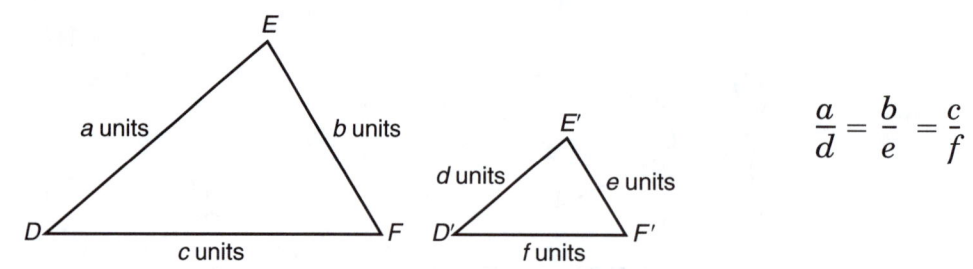

$$\frac{a}{d} = \frac{b}{e} = \frac{c}{f}$$

1. If $a = 5$, $d = 10$, $c = 7$, then $f =$ _____.

2. If $b = 45$, $e = 15$, $a = 21$, then $d =$ _____.

3. If $c = 2$, $f = 3$, $d = 9$, then $a =$ _____.

4. If $d = 15$, $a = 12$, $f = 10$, then $c =$ _____.

Use the triangle below and the table on page **149** to help you complete the following.

5. If $a = 3$ and $b = 4$, then $c =$ _____.

6. If $a = 9$ and $c = 15$, then $b =$ _____.

7. If $b = 10$ and $c = 15$, then $a \approx$ _____.

8. If $a = 5$ and $c = 9$, then $b \approx$ _____.

$$c^2 = a^2 + b^2 \text{ or } c^2 = b^2 + a^2$$

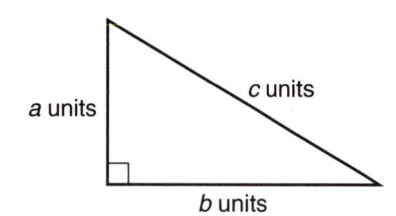

Solve each of the following.

9. A building and a person cast shadows as shown below. What is the height of the building?

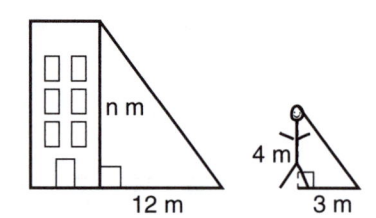

   The height of the building is _____ m.

10. A rectangular mirror is 2 m by 3 m. What is the distance between opposite corners of the mirror?

    The distance is about _____ m.

PRISM MATHEMATICS
Purple Book

CHAPTER 11 TEST

Find the perimeter and area of each figure.

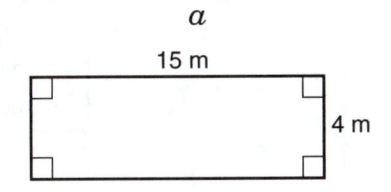

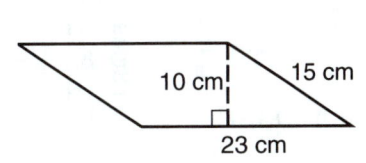

　　　　　　　　a　　　　　　　　　　　　　　b　　　　　　　　　　　　　　c

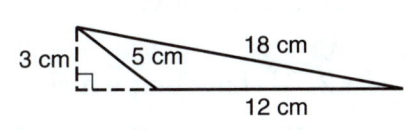

1.

perimeter: _____ m　　　　　_____ cm　　　　　_____ cm
area: 　　_____ m²　　　　　_____ cm²　　　　_____ cm²

Complete the table below. Use $3\frac{1}{7}$ for π. Find the approximate circumference and area.

	diameter	radius	approximate circumference	approximate area
2.	12 cm	cm	cm	cm²
3.	m	3 m	m	m²

Find the surface area of each figure. Use 3.14 for π.

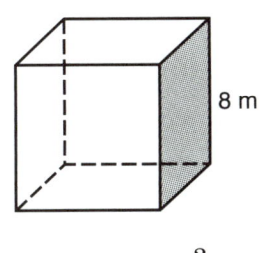

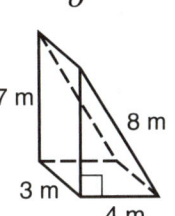

　　　　　　　　a　　　　　　　　　　　　　　b　　　　　　　　　　　　　　c

4.

_____ m²　　　　　_____ m²　　　　about _____ cm²

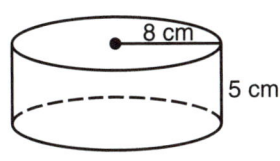

Find the volume of each figure. Use 3.14 for π.

5.

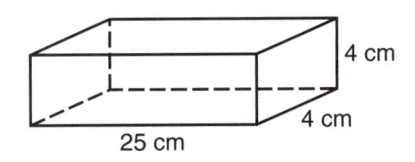

_____ cm³　　　　　_____ cm³　　　　about _____ m³

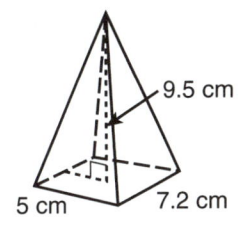

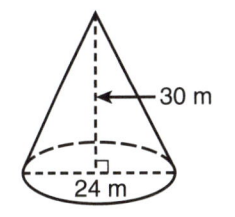

PRISM MATHEMATICS
Purple Book

CHAPTER 11 TEST

275

CHAPTER 12 TEST

NAME _____

Use the line graph to answer each question.

1. In 2000, how many months had an increase in sales? _____

2. In 2001, between which two months was the largest decline in sales? _____

3. In what month were the sales in 2000 equal to the sales in 2001? _____

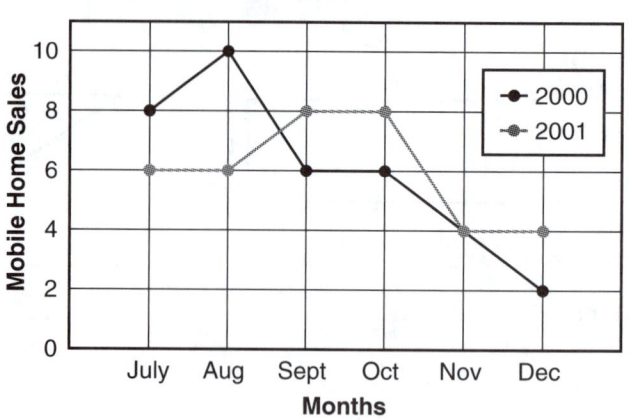

Mobile Home Sales

Use the bar graph to determine how this graph is misleading.

4. By looking at the bars, how many times greater do you expect the population of Ada to be than the population of Cass?
   _____

5. What is incorrect that brings you to reach the conclusion from question 4?
   _____

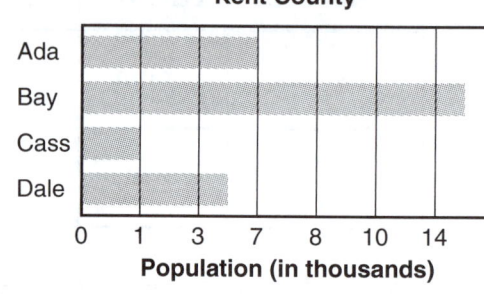

Population of Towns in Kent County

6. Use the chart to determine the number of degrees for each meal time.

   breakfast = _____

   mid-morning snack = _____

   lunch = _____

   dinner = _____

   late-night snack = _____

Daily Calorie Breakdown	
	Percent
breakfast	30%
mid-morning snack	10%
lunch	20%
dinner	35%
late night snack	5%

Daily Calorie Breakdown

7. Create a circle graph using the information from question 6.

PRISM MATHEMATICS
Purple Book

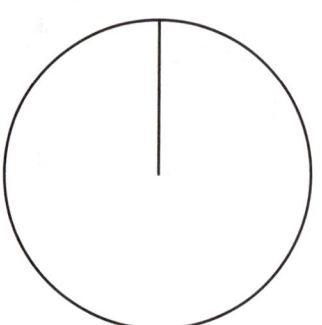

CHAPTER 13 TEST

NAME _____

Pretend that, without looking, you select a shape from the bag. Write the probability of each selection as a fraction in simplest form.

1. a blue shape _____

2. a circle _____

3. a hexagon _____

4. a white or blue shape _____

5. a white circle _____

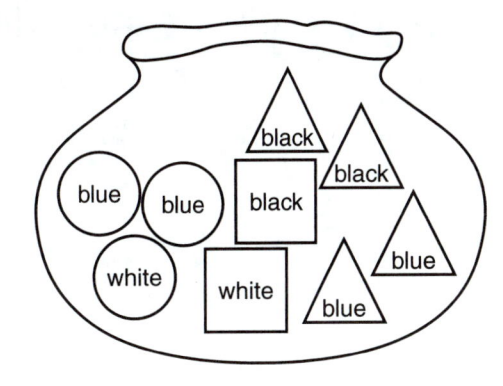

Solve each problem. Write each probability as a percent.

6. Suppose you flip a coin 300 times. Predict how many times you would expect to get tails.

   You should get tails about _____ times.

7. A company knows that 4% of their T-shirts are defective. The company produced 500 000 T-shirts. How many will be defective?

   _____ T-shirts will be defective.

8. According to the wheel, what is the probability that the spinner will stop on an even number?

   The probability is _____.

9. You spin the spinner 80 times. Predict how many times the spinner will stop on a multiple of 3.

   The spinner will stop on a multiple of 3 _____ times.

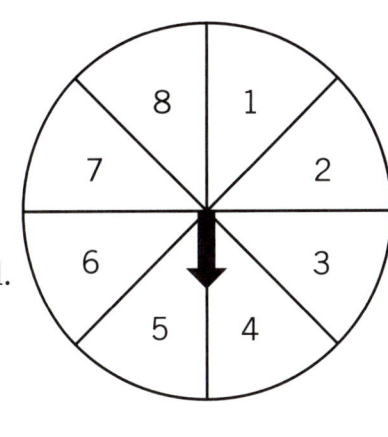

10. Complete the sample space for tossing a dime and a nickel.

dime	nickel	outcome
heads	tails	

PRISM MATHEMATICS
Purple Book

CHAPTER 13 TEST

277

CHAPTER 1 TEST

Possible Score: 20
Time Frame: 15–20 minutes

Add or subtract.

	a	b	c	d
1.	6159 241 +398 **6798**	64.31 22.85 +9.74 **96.9**	45 628 13 105 +24 359 **83 092**	47.611 3.298 +39.1 **90.009**
2.	85.32 −63.78 **21.54**	14 360 −12 819 **1541**	9.401 −3.695 **5.706**	50 101 −18 957 **31 144**
3.	157 823 21 425 +21 428 **200 676**	56.73 4.18 +3.14 **64.05**	7.2 −3.4 **3.8**	6752 −974 **5778**

Multiply or divide.

	a	b	c	d
4.	751 ×38 **25 538**	6819 ×315 **2 147 985**	72.3 ×18.2 **1 315.86**	2.703 ×0.014 **0.037 842**
5.	639 ×582 **371 898**	72.631 ×0.155 **11.257 805**	23)18 321 **796 r13**	84)563 **6 r59**
6.	57)43 605 **765**	0.7)4.48 **6.4**	4.3)10.105 **2.35**	3.25)0.585 **0.18**

PRISM MATHEMATICS
Purple Book

CHAPTER 1 TEST
265

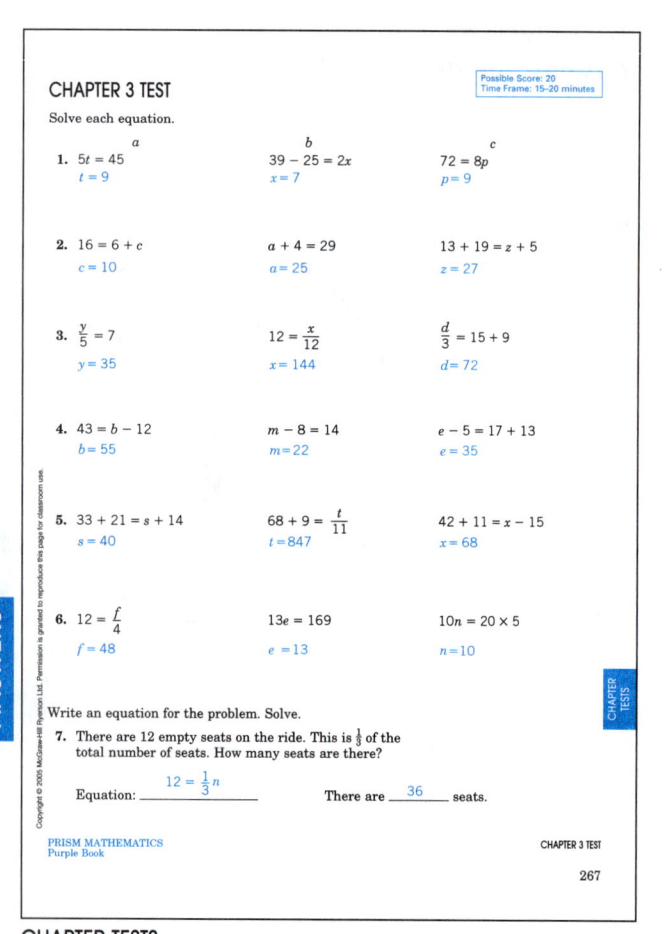

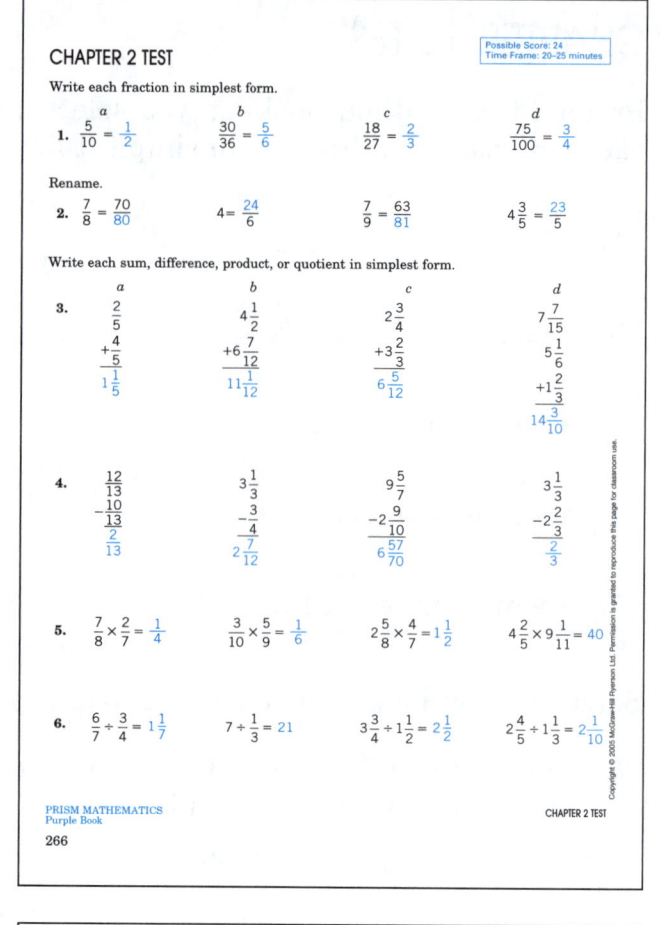

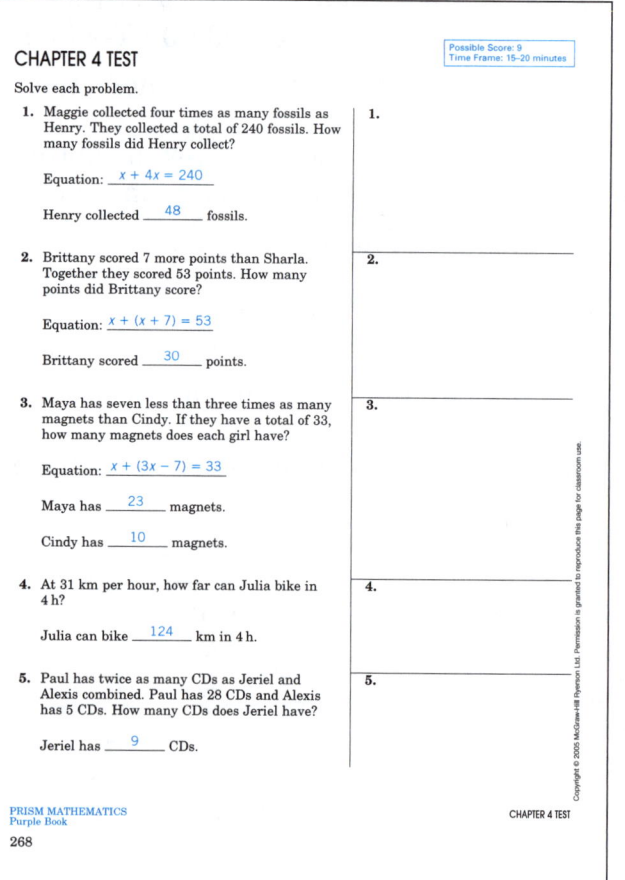

CHAPTER 2 TEST

Possible Score: 24
Time Frame: 20–25 minutes

Write each fraction in simplest form.

	a	b	c	d
1.	$\frac{5}{10} = \frac{1}{2}$	$\frac{30}{36} = \frac{5}{6}$	$\frac{18}{27} = \frac{2}{3}$	$\frac{75}{100} = \frac{3}{4}$

Rename.

| 2. | $\frac{7}{8} = \frac{70}{80}$ | $4 = \frac{24}{6}$ | $\frac{7}{9} = \frac{63}{81}$ | $4\frac{3}{5} = \frac{23}{5}$ |

Write each sum, difference, product, or quotient in simplest form.

	a	b	c	d
3.	$\frac{2}{5} + \frac{4}{5} = 1\frac{1}{5}$	$4\frac{1}{2} + 6\frac{7}{12} = 11\frac{1}{12}$	$2\frac{3}{4} + 3\frac{2}{3} = 6\frac{5}{12}$	$7\frac{7}{15} + 5\frac{1}{6} + 1\frac{2}{3} = 14\frac{3}{10}$
4.	$\frac{12}{13} - \frac{10}{13} = \frac{2}{13}$	$3\frac{1}{3} - \frac{3}{4} = 2\frac{7}{12}$	$9\frac{5}{7} - 2\frac{9}{10} = 6\frac{57}{70}$	$3\frac{1}{3} - 2\frac{2}{3} = \frac{2}{3}$
5.	$\frac{7}{8} \times \frac{2}{7} = \frac{1}{4}$	$\frac{3}{10} \times \frac{5}{9} = \frac{1}{6}$	$2\frac{5}{8} \times \frac{4}{7} = 1\frac{1}{2}$	$4\frac{2}{5} \times 9\frac{1}{11} = 40$
6.	$\frac{6}{7} \div \frac{3}{4} = 1\frac{1}{7}$	$7 \div \frac{1}{3} = 21$	$3\frac{3}{4} \div 1\frac{1}{2} = 2\frac{1}{2}$	$2\frac{4}{5} \div 1\frac{1}{3} = 2\frac{1}{10}$

PRISM MATHEMATICS
Purple Book

CHAPTER 2 TEST
266

CHAPTER 3 TEST

Possible Score: 20
Time Frame: 15–20 minutes

Solve each equation.

	a	b	c
1.	$5t = 45$ $t = 9$	$39 − 25 = 2x$ $x = 7$	$72 = 8p$ $p = 9$
2.	$16 = 6 + c$ $c = 10$	$a + 4 = 29$ $a = 25$	$13 + 19 = z + 5$ $z = 27$
3.	$\frac{y}{5} = 7$ $y = 35$	$12 = \frac{x}{12}$ $x = 144$	$\frac{d}{3} = 15 + 9$ $d = 72$
4.	$43 = b − 12$ $b = 55$	$m − 8 = 14$ $m = 22$	$e − 5 = 17 + 13$ $e = 35$
5.	$33 + 21 = s + 14$ $s = 40$	$68 + 9 = \frac{t}{11}$ $t = 847$	$42 + 11 = x − 15$ $x = 68$
6.	$12 = \frac{f}{4}$ $f = 48$	$13e = 169$ $e = 13$	$10n = 20 \times 5$ $n = 10$

Write an equation for the problem. Solve.

7. There are 12 empty seats on the ride. This is $\frac{1}{3}$ of the total number of seats. How many seats are there?

Equation: $12 = \frac{1}{3}n$ There are __36__ seats.

PRISM MATHEMATICS
Purple Book

CHAPTER 3 TEST
267

CHAPTER 4 TEST

Possible Score: 9
Time Frame: 15–20 minutes

Solve each problem.

1. Maggie collected four times as many fossils as Henry. They collected a total of 240 fossils. How many fossils did Henry collect?

   Equation: $x + 4x = 240$

   Henry collected __48__ fossils.

2. Brittany scored 7 more points than Sharla. Together they scored 53 points. How many points did Brittany score?

   Equation: $x + (x + 7) = 53$

   Brittany scored __30__ points.

3. Maya has seven less than three times as many magnets than Cindy. If they have a total of 33, how many magnets does each girl have?

   Equation: $x + (3x − 7) = 33$

   Maya has __23__ magnets.

   Cindy has __10__ magnets.

4. At 31 km per hour, how far can Julia bike in 4 h?

   Julia can bike __124__ km in 4 h.

5. Paul has twice as many CDs as Jeriel and Alexis combined. Paul has 28 CDs and Alexis has 5 CDs. How many CDs does Jeriel have?

   Jeriel has __9__ CDs.

PRISM MATHEMATICS
Purple Book

CHAPTER 4 TEST
268

CHAPTER TESTS
Answers

278

CHAPTER 5 TEST

Possible Score: 20
Time Frame: 20-25 minutes

Express each of the following as a ratio in two ways as shown.

	a	b
1. 22 baskets in 2 games	22 to 2	$\frac{22}{2}$
2. 7 pencils to 10 students	7 to 10	$\frac{7}{10}$
3. 3 leaves to 1 clover	3 to 1	$\frac{3}{1}$
4. 8 boys to 12 girls	8 to 12	$\frac{8}{12}$

Solve the following.

a
5. $\frac{3}{6} = \frac{n}{12}$ $n = 6$
6. $\frac{7}{9} = \frac{n}{72}$ $n = 56$
7. $\frac{8}{n} = \frac{20}{45}$ $n = 18$

b
$\frac{5}{n} = \frac{15}{24}$ $n = 8$
$\frac{2}{8} = \frac{14}{n}$ $n = 56$
$\frac{n}{15} = \frac{9}{27}$ $n = 5$

Complete the following.

a
8. __8__ is 25% of 32.
9. 380 is 95% of __400__.
10. $1\frac{3}{4}$ is __70__% of $2\frac{1}{2}$.
11. 600 is __200__% of 300.

b
17.4 is 30% of __58__.
__17__ is 8.5% of 200.
8.5 is __12.5__% of 68.
__16__ is 40% of 40.

269

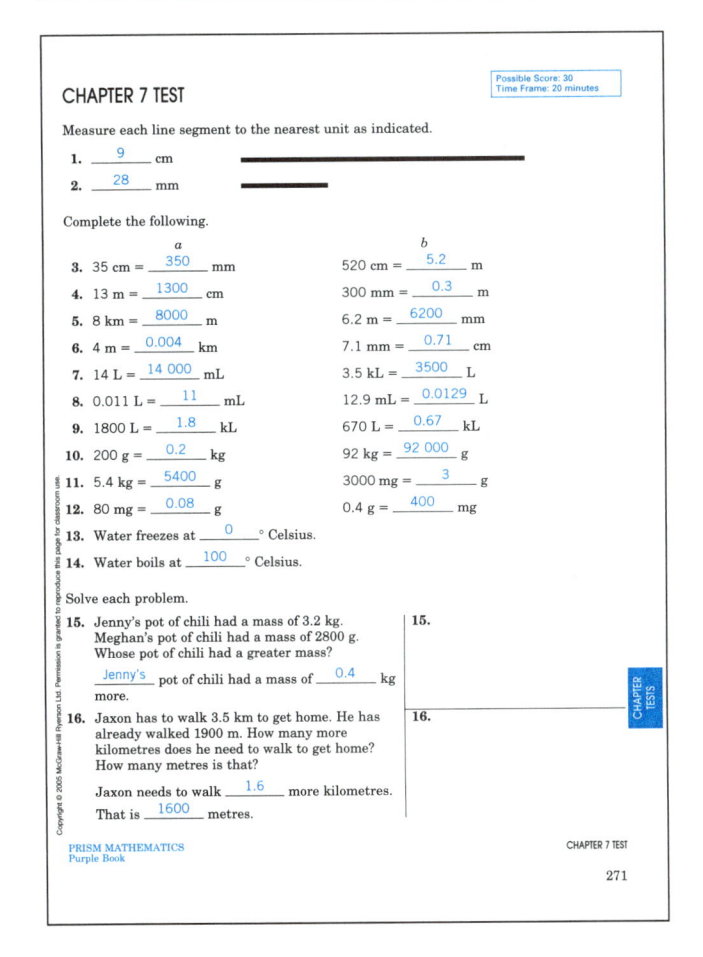

CHAPTER 6 TEST

Possible Score: 20
Time Frame: 15-20 minutes

Complete the following for simple interest.

	principal	rate	time	interest
1.	220	20%	4 years	$176
2.	$500	6.5%	5 years	$162.50
3.	$750	18%	$\frac{1}{2}$ year	$67.50
4.	$920	25%	$4\frac{1}{4}$ years	$977.50
5.	$4250	$8\frac{1}{2}$%	3 years	$1083.75
6.	$1500	13%	2 years	$390

Interest is to be compounded in each account below. Find the total amount that will be in each account after the period of time indicated.

	principal	rate	time	compounded	total amount
7.	$200	5%	2 years	annually	$220.50
8.	$500	8%	3 years	annually	$629.86
9.	$700	4%	2 years	semiannually	$818.90
10.	$300	6%	$\frac{1}{3}$ year	monthly	$378.74

270

CHAPTER 7 TEST

Possible Score: 30
Time Frame: 20 minutes

Measure each line segment to the nearest unit as indicated.

1. __9__ cm
2. __28__ mm

Complete the following.

a
3. 35 cm = __350__ mm
4. 13 m = __1300__ cm
5. 8 km = __8000__ m
6. 4 m = __0.004__ km
7. 14 L = __14 000__ mL
8. 0.011 L = __11__ mL
9. 1800 L = __1.8__ kL
10. 200 g = __0.2__ kg
11. 5.4 kg = __5400__ g
12. 80 mg = __0.08__ g
13. Water freezes at __0__ ° Celsius.
14. Water boils at __100__ ° Celsius.

b
520 cm = __5.2__ m
300 mm = __0.3__ m
6.2 m = __6200__ mm
7.1 mm = __0.71__ cm
3.5 kL = __3500__ L
12.9 mL = __0.0129__ L
670 L = __0.67__ kL
92 kg = __92 000__ g
3000 mg = __3__ g
0.4 g = __400__ mg

Solve each problem.

15. Jenny's pot of chili had a mass of 3.2 kg. Meghan's pot of chili had a mass of 2800 g. Whose pot of chili had a greater mass?
__Jenny's__ pot of chili had a mass of __0.4__ kg more.

16. Jaxon has to walk 3.5 km to get home. He has already walked 1900 m. How many more kilometres does he need to walk to get home? How many metres is that?
Jaxon needs to walk __1.6__ more kilometres.
That is __1600__ metres.

271

CHAPTER 8 TEST

Possible Score: 28
Time Frame: 10-15 minutes

Complete the following.

a
1. 4 m = __400__ cm
2. 150 min = __$2\frac{1}{2}$__ h
3. 7 kg = __7000__ g
4. 144 h = __6__ days
5. 4 L = __4000__ mL

b
2 m = __2000__ mm
5 kg = __5000__ g
3 h 14 min = __194__ min
3 km = __3000__ m
8 L = __8000__ mL

Add, subtract, or multiply.

a
6. 4 m
   −2 cm
   ──────
   398 cm

7. 7 g
   −3 mg
   ──────
   6997 mg

b
5 L
+3 mL
──────
5003 mL

4 kg
×5
──────
20 kg

c
3 min 21 s
×2
──────
6 min 42 s

8 h 45 min
+6 h 37 min
──────
15 h 22 min

Round as indicated.

	a nearest ten	b nearest hundred	c nearest thousand
8. 6851	6850	6900	7000
9. 74 883	74 880	74 900	75 000

Write an estimate for each exercise. Then find the answer.

10. 6254   6000   8159   8000   485   500
    +3867 +4000   −6901 −7000   ×64   ×60
    ───── ─────   ───── ─────   ───── ─────
    10 121 10 000   1258   1000   31 040 30 000

272

CHAPTER TESTS
Answers

279

CHAPTER 9 TEST

Possible Score: 15
Time Frame: 5–10 minutes

Use the figures below to answer the questions that follow.

1. Name the diameter of circle B. $\overline{AC}$ or $\overline{CA}$
   Name a radius of circle B. $\overline{BA}$ or $\overline{BC}$

2. Which figures are isosceles triangles? △JKL or △MNO
   Which figure is an acute triangle? △MNO

3. Which figure is a right triangle? △FGH
   Which figure is a scalene triangle? △FGH

Use a protractor to measure each angle. Then describe each angle by writing *acute, obtuse,* or *right.*

4. a: 145, obtuse   b: 90, right   c: 30, acute

Tell whether each pair of lines is *parallel* or *perpendicular.*

5. perpendicular   parallel   perpendicular

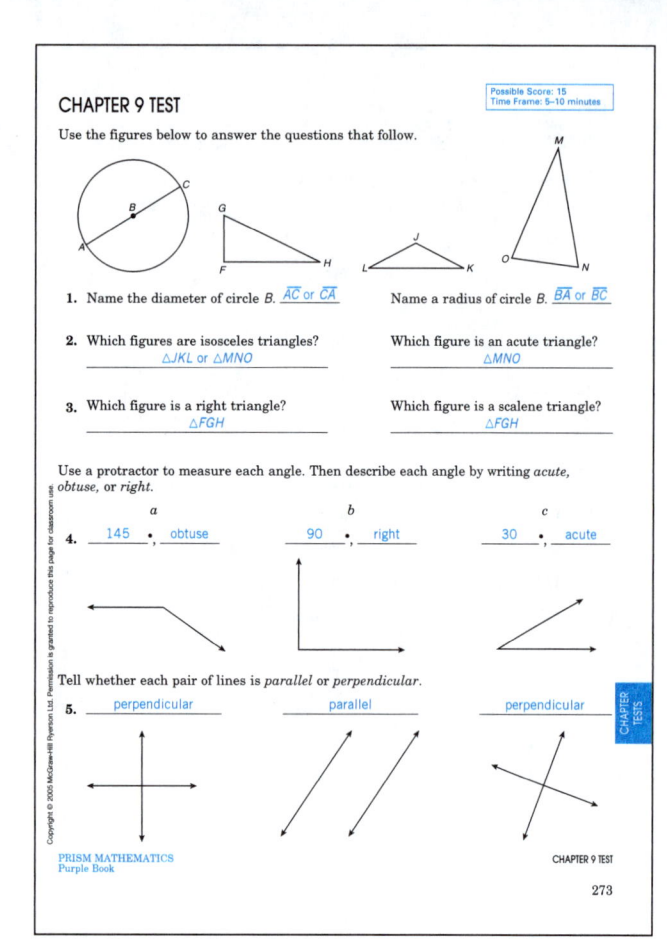

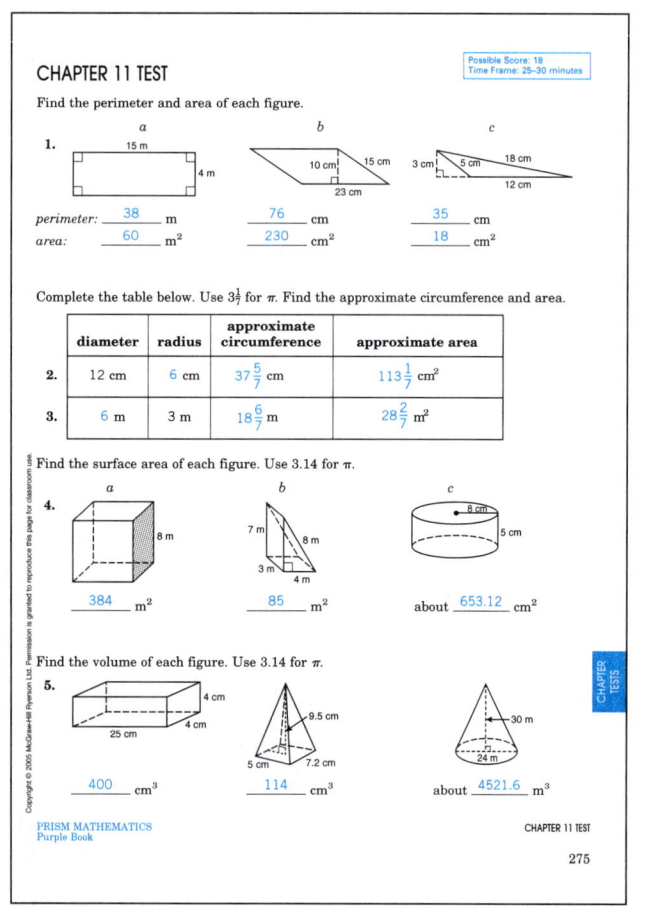

CHAPTER 10 TEST

Possible Score: 10
Time Frame: 25–30 minutes

Use the triangles below to help you complete the following.

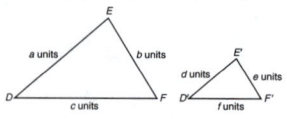

$$\frac{a}{d} = \frac{b}{e} = \frac{c}{f}$$

1. If $a = 5$, $d = 10$, $c = 7$, then $f = $ 14
2. If $b = 45$, $e = 15$, $a = 21$, then $d = $ 7
3. If $c = 2$, $f = 3$, $d = 9$, then $a = $ 6
4. If $d = 15$, $a = 12$, $f = 10$, then $c = $ 8

Use the triangle below and the table on page 149 to help you complete the following.

5. If $a = 3$ and $b = 4$, then $c = $ 5
6. If $a = 9$ and $c = 15$, then $b = $ 12
7. If $b = 10$ and $c = 15$, then $a \approx $ 11.18
8. If $a = 5$ and $c = 9$, then $b \approx $ 7.48

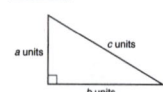

$$c^2 = a^2 + b^2 \text{ or } c^2 = b^2 + a^2$$

Solve each of the following.

9. A building and a person cast shadows as shown below. What is the height of the building? 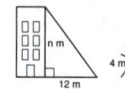
   The height of the building is 16 m.

10. A rectangular mirror is 2 m by 3 m. What is the distance between opposite corners of the mirror?
    The distance is about 3.6 m.

CHAPTER 11 TEST

Possible Score: 18
Time Frame: 25–30 minutes

Find the perimeter and area of each figure.

1. a: perimeter: 38 m, area: 60 m²
   b: 76 cm, 230 cm²
   c: 35 cm, 18 cm²

Complete the table below. Use $3\frac{1}{7}$ for π. Find the approximate circumference and area.

	diameter	radius	approximate circumference	approximate area
2.	12 cm	6 cm	$37\frac{5}{7}$ cm	$113\frac{1}{7}$ cm²
3.	6 m	3 m	$18\frac{6}{7}$ m	$28\frac{2}{7}$ m²

Find the surface area of each figure. Use 3.14 for π.

4. a: 384 m²   b: 85 m²   c: about 653.12 cm²

Find the volume of each figure. Use 3.14 for π.

5. a: 400 cm³   b: 114 cm³   c: about 4521.6 m³

CHAPTER 12 TEST

Possible Score: 15
Time Frame: 15–20 minutes

Use the line graph to answer each question.

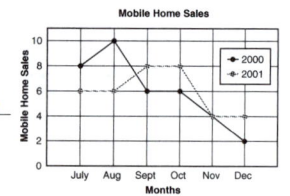

1. In 2000, how many months had an increase in sales? 1
2. In 2001, between which two months was the largest decline in sales? Oct and Nov
3. In what month were the sales in 2000 equal to the sales in 2001? November

Use the bar graph to determine how this graph is misleading.

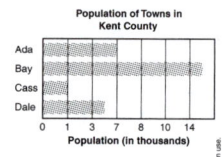

4. By looking at the bars, how many times greater do you expect the population of Ada to be than the population of Cass?
   It would appear to be 3 times greater
5. What is incorrect that brings you to reach the conclusion from question 4?
   The horizontal scale is non standard.

6. Use the chart to determine the number of degrees for each meal time.
   breakfast = 108°
   mid-morning snack = 36°
   lunch = 72°
   dinner = 126°
   late-night snack = 18°

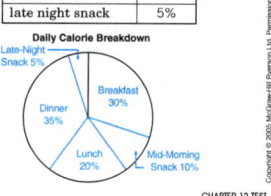

Daily Calorie Breakdown	
	Percent
breakfast	30%
mid-morning snack	10%
lunch	20%
dinner	35%
late night snack	5%

7. Create a circle graph using the information from question 6.

CHAPTER 13 TEST

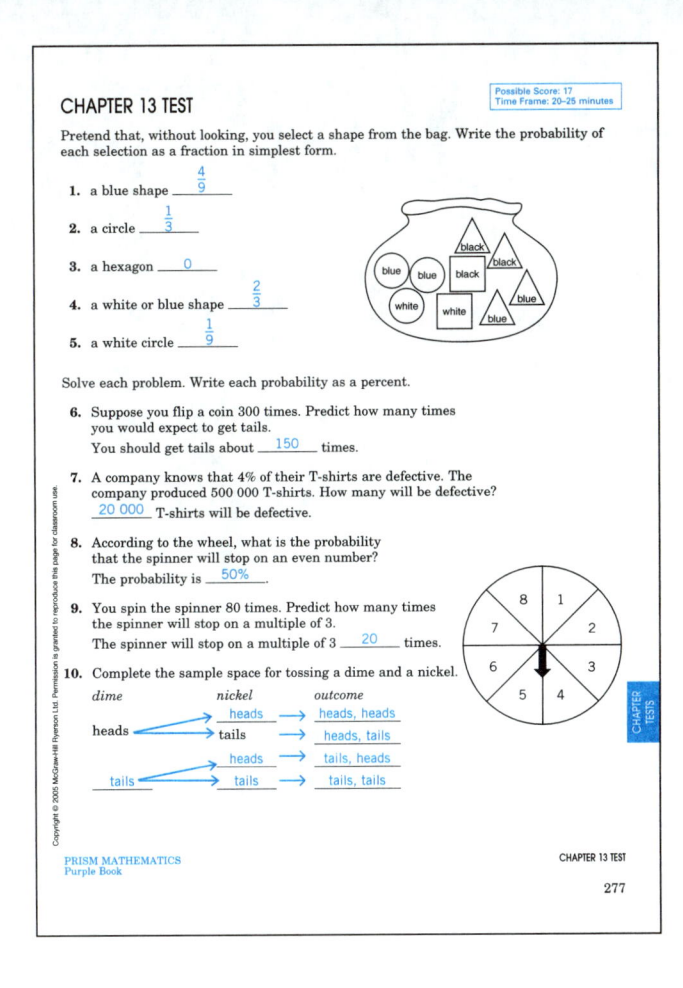

Possible Score: 17
Time Frame: 20–25 minutes

Pretend that, without looking, you select a shape from the bag. Write the probability of each selection as a fraction in simplest form.

1. a blue shape $\frac{4}{9}$
2. a circle $\frac{1}{3}$
3. a hexagon 0
4. a white or blue shape $\frac{2}{3}$
5. a white circle $\frac{1}{9}$

Solve each problem. Write each probability as a percent.

6. Suppose you flip a coin 300 times. Predict how many times you would expect to get tails.
   You should get tails about __150__ times.

7. A company knows that 4% of their T-shirts are defective. The company produced 500 000 T-shirts. How many will be defective?
   __20 000__ T-shirts will be defective.

8. According to the wheel, what is the probability that the spinner will stop on an even number?
   The probability is __50%__.

9. You spin the spinner 80 times. Predict how many times the spinner will stop on a multiple of 3.
   The spinner will stop on a multiple of 3 __20__ times.

10. Complete the sample space for tossing a dime and a nickel.

dime	nickel	outcome
heads	heads	heads, heads
heads	tails	heads, tails
tails	heads	tails, heads
tails	tails	tails, tails